Molecular Structure
The Physical Approach

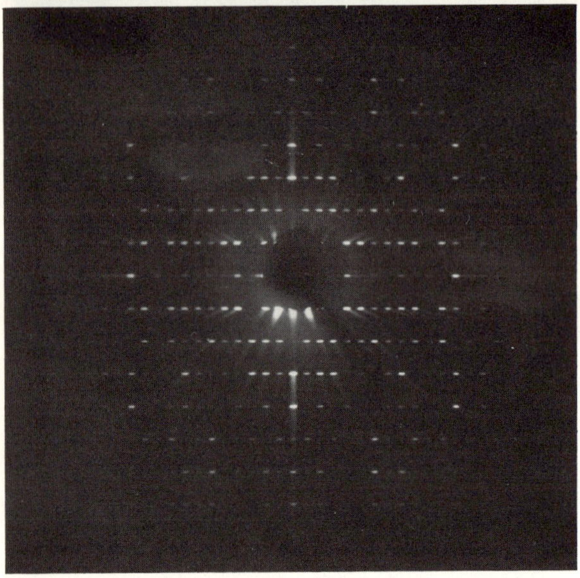

Plate IA. Single-crystal photograph from lead thiocyanate set to rotate about its *c*-axis (Wm. Cook). (The primitive translation, *c*, may be determined from measurements of the vertical distances between corresponding upper and lower layer-lines: see problem No. 23.)

Plate IB. Precession photograph showing the 0*kl* reciprocal-lattice net of a crystal of $[MoOCl_2(PMe_2Ph)_3]$ (Lj. Manojlović–Muir).
(From such photographs the geometry of the reciprocal lattice may be determined, and the relative intensities of the various reflexions estimated for structure analysis. The central, horizontal row of spots in this plate shows systematic absences characteristic of a twofold screw-axis: 0*k*0 absent when *k* is odd.)

Molecular Structure
The Physical Approach

by
J C D Brand and J C Speakman

Second Edition revised by

J. C. Speakman
Formerly Reader in Physical Chemistry, University of Glasgow

and

J. K. Tyler
Senior Lecturer in Physical Chemistry, University of Glasgow

Edward Arnold

First published 1960
Reprinted 1961
Reprinted with corrections and additional material 1964

Second edition 1975

ISBN: 0 7131 2483 0 Boards Edition
0 7131 2491 1 Paper Edition

To the memory of
George Macdonald Bennett, F.R.S.

Typeset by E. W. C. Wilkins Ltd., London & Northampton
and printed in Great Britain by Unwin Brothers Ltd, Woking

Preface to the first edition

The elucidation of molecular structure is one of the two principal tasks of chemical theory. During the past quarter-century the application of physical methods to this task has been developing rapidly and now provides a variety of weapons for the chemist's armoury. Whether or not he uses these methods himself, every chemist has occasion to use the results they yield, and he ought to have some understanding of them, and of their qualities and limitations. The time is therefore ripe for a general account of these methods, written at a level suited to the honours student, or to the research worker in some other branch of Chemistry. With such readers in mind, we have concentrated on principles rather than on practice; and we have dealt in greater detail with those parts of the subject which inherently imply a deeper insight into the nature of the molecule.

We take the word 'structure', of a molecule, to connote more than just the relative positions of its atoms. On the other hand, to do justice to all the ideas that might be associated with the notion of atoms-united-to-form-a-molecule would have inflated our book unreasonably. (Valency theory is in any case thoroughly treated in several existing monographs). Therefore matters connected explicitly with the properties of electrons in molecules have been excluded, though topics such as electronic spectra or electron magnetic resonance might well have come within the scope of our title. The planner of any book must recognise its judicious limits.

* * *

Before his death in February, 1959, George Macdonald Bennett had acceded to a request that we might dedicate our book to him. At different periods of his career, and in different universities, we were each his pupil. We wish affectionately to record our appreciation of a brilliant teacher and a most distinguished chemist.

October, 1959, J.C.D.B.
GLASGOW. J.C.S

Preface to the second edition

The first edition was explicitly restricted to the molecule in its electronic ground state. There was therefore no discussion of molecular electronic spectra – an omission which must have appeared strange to those familiar with J.C.D. Brand's distinguished work in this area. Persisting with this omission in a second edition, which has another spectroscopist-author, may seem even more preposterous; but we felt this to be inevitable unless the book was to be expanded to double its original length.

However we have thought it advisable to introduce a new chapter (4) which gives an elementary survey of molecular spectroscopy as a whole, and in which we have tried to put the more important branches of this very large subject into perspective. The specific areas of molecular spectroscopy covered in subsequent chapters are as before, but there are some changes in detail and in presentation. We have also included in the main text much of the material previously relegated to appendices. An elementary account of nuclear-spin statistical weights is added to chapter 5, on the rotation of molecules. A more formal development of group theory has been added to chapter 2.

The chapters on diffraction methods were, we considered, less in need of change; but chapter 9, on crystal structure analysis, has been amplified in the interests of clarity, as we hope, and brought up-to-date in some details. We are indebted to Dr. B. Beagley for helpful advice on chapter 10, dealing with electron diffraction.

The text has increased by roughly 10%, and the whole book has been re-set. We have added a selection of numerical problems.

Drs R. V. Emanuel, and Ljubica Manojlović-Muir helped with proof correction; and Ms Hilda Thomas typed much of the new material.

GLASGOW
1974

J. K. Tyler
J. C. Speakman

Contents

Chapter 1

Historical and general introduction

1.1 Development of the Molecular Concept

At the beginning of this century it was possible for a good chemist to doubt the existence of atoms and molecules. To be sure, the structural theory in organic chemistry was maturing into one of the most powerful and productive of all scientific theories. But the structures it postulated, though they were extremely detailed in some particulars, were based on indirect evidence—essentially on the analytical compositions of quantities of matter containing enormous numbers of individual molecules. So to suppose, for instance, that benzene was an aggregation of molecules, each consisting of a regular hexagon of carbon atoms with hydrogens attached, was an elegant and convenient way of accounting for its composition, the numbers of its substitution products, and a few other simple chemical facts. But it might well have seemed an improper question to ask whether the atoms were 'really' so arranged; for no one had ever seen a molecule, and it would have seemed extremely unlikely that anyone ever would.

This was the chemist's point of view. That of the physicist appeared very different. Certain properties of matter—the P-V-T relationships of gases, for example—could be rather satisfactorily discussed in terms of 'hard, elastic, spherical' molecules. These entities had little direct correspondence with the molecules of the chemist.

The same incertitude clouded the atomic theory itself. So great a scientist as Faraday had made hardly any use of it. And at that time Ostwald was developing a comprehensive treatment of Chemistry without introducing atoms at all and without mentioning the atomic or molecular theories, except in such derogatory terms as are exemplified in the following quotations. 'The elements and their properties disappeared when the compound was formed, and it is therefore impossible that an element should persist in its compounds. The idea that the elements have disappeared, but are nevertheless present as such, is an indefinite one, too indefinite for scientific use.' (There is) 'the so-called structural theory, in which a distinct difference is assumed between elements that are directly bound to each other and those that are indirectly bound . . . Where substances have been investigated with great minuteness it is found that the constitution becomes less and less certain.'

'Various considerations, all of which contain a larger or smaller number of arbitrary assumptions, or other uncertainties, have been applied with the hope of reaching conclusions about molecular size' (1907).*

About 1895 came the discoveries of radioactivity and of the electron; they led to the rise of atomic physics, and it soon became practicable to detect individual atomic events, and hence to specify more and more of the properties of individual atoms. With molecules the corresponding developments were less dramatic, but they began to seem significant some twentyfive years later. The consequent accessibility of *molecular data* has rendered the concept of the molecule more plausible; and as the concept has become firmer, the physicist's molecule has moved into closer correspondence with the chemist's. Whatever may be thought about the reality behind scientific hypotheses, it is inevitable that the benzene molecule should seem more real—to the scientist, at any rate—when its mass, dimensions, and other properties can be specified: for instance, that the mass is $1{\cdot}29 \times 10^{-22}$ g, that it is based on a regular flat hexagon of carbon atoms 1·397 Å apart, each one linked radially to a hydrogen atom at a distance of $1{\cdot}08_4$ Å, and that the whole system is vibrating with known frequencies and mean amplitudes.

The physical methods that have made such information accessible are the subject of this book.

Parallel with this increasing facility for obtaining precise numerical data on molecules, the application of quantum principles has led to a theory of molecular structure; and, as always, work in each of these complementary fields has stimulated work in the other. Though the theory of intramolecular forces has no part in the plan of this book, some background of wave mechanics is essential for a proper appreciation of the physical methods of measurement, and particularly the spectroscopic methods. In Chapter 3 this background is sketched in. It has to be admitted that, whilst the accumulation of detailed molecular data has made the molecule seem more real, the seeming reality is a little different from anything that even the most prescient chemist is likely to have conceived in 1900.

1.2 The Concept of the Molecule

For the purposes of this book a molecule is normally considered as an isolated entity. It is in the gaseous state, or at least theoretically capable of being so. The molecule is thus to be regarded as an assemblage of a limited number of atoms held together by strong forces—essentially covalencies—and independent of other molecules. In the solid state the molecule is sometimes less clearly defined. But with many substances, and with nearly all organic compounds, it appears to be little affected by its crystalline environment. Crystalline forces may of course be important; in particular, hydroxylic compounds almost always link their molecules together in the solid state by means of hydrogen bonds; there is a marked tendency

* It is fair to add that Ostwald modified these extreme views before the end of his life.

for the structure to be such as to allow full play to this bonding. Where the much stronger electrovalent forces hold sway no discrete molecules exist in the ordinary sense. The methods of crystal structure analysis can still be applied, and its results are then the more valuable in that ionic compounds cannot be so readily studied in any other way.

1.3 The Measurable Properties of Molecules

Given a recognisable molecule, we may enquire which of its properties can be measured by the physical methods with which we are concerned. The most obvious objective is to determine the relative positions of the atoms, or of their nuclei. This is indeed what is colloquially implied by the word 'structure'. The methods all give information of this kind, and it can be expressed as bond lengths, bond angles, etc. Since the various atoms in a molecule are always in a state of mutual vibration, our measurements of a bond length, for instance, are brought into question, especially when they can claim a high precision. What we derive from our measurements may be an *equilibrium* bond length, appropriate to the hypothetical state where all vibrations have ceased; or it may be some sort of *mean* bond length. In general these dimensions will differ slightly; and in highly accurate work the kind of bond length being used must be carefully defined.

Some of the methods give us more or less detailed information about these molecular vibrations, and this can be important in assessing the strengths of the bonds involved. In general it is possible to measure the restoring force caused by given small displacements of the atomic nuclei. The ratio of force to displacement–the force constant–is the best criterion of the bond strength as the bond obtains in the molecule. Another less satisfactory criterion is the energy needed to dissociate the bond, and this too can sometimes be deduced from spectroscopic data.

At an early stage in a spectroscopic or crystallographic study the symmetry of the molecule may become apparent. It is obviously useful to be able to say something about the shape of a molecule, even if we do not know its size. Thus the preliminary study of the spectrum of water vapour indicated that the molecule, H_2O, was non-linear, a fact of profound importance in Chemistry, and indeed throughout the physical world.

When two different kinds of atoms unite to form a diatomic molecule, the product is usually polar; the atoms acquire partial positive and negative charges, and the molecule acquires an electric moment. Similar moments appear to be associated with each bond in any polyatomic molecule, so that, unless it happens to be so highly symmetrical that the individual bond moments cancel one another, the molecule as a whole will possess an electric dipole moment. This property can be derived from various physical techniques.

Atoms may possess magnetic moments, and these may have structural implications. When the molecule embodies unpaired electrons, paramagnetism results. We shall not deal with this topic because electronic paramagnetism is not common amongst the stable, covalent molecules that primarily concern us. However, the occurrence of measurable paramagnetism, and hence of an assessable number of unpaired electron spins, is very significant, and its study is particularly important in structural inorganic chemistry. When there are no unpaired spins, the material is diamagnetic. The study of diamagnetism is occasionally of structural value; notably, when the effect is markedly anisotropic in a crystalline aromatic compound, it enables the orientation of the benzenoid rings to be found. Certain species of atomic nuclei possess magnetic moments, and they are said to show nuclear spin. This property is a subtle one, not easily detected because the moment is minute and well shielded by the orbital electrons. Effects due to nuclear spin can now be measured with high accuracy as a nuclear magnetic resonance ('N.M.R.'). Similarly remote from traditional observation, and now readily studied, is the electric quadrupole moment possessed by some nuclear species. Both these effects are of value in structural work. They yield information, about the environment of a given nucleus, that could not be obtained by other means.

The electrons in a molecule can be raised to higher energy levels, when the molecule becomes excited. Such transitions give rise to spectra in the visible and ultraviolet regions. Though the study of these electronic spectra is of the greatest importance in chemistry, their interpretation is generally partial and qualitative; only for some very simple molecules can the complexities of the spectra be subjected to a detailed and quantitative interpretation; even then the treatment of the rotational or vibrational details of the spectrum—once they have been elucidated—resembles that of the same features as they might appear, uncomplicated, in the infrared spectrum. For these reasons we have restricted ourselves in this book, apart from chapter 4, to *the molecule in its electronic ground state.*

1.4 Survey of the Physical Methods Available for Studying Molecular Structure

Since even large molecules are much too small to be seen by the eye, or by any other conceivable optical device, the problem of making detailed measurements on them may well seem a forbidding one, and at the beginning of the century many thought it insoluble. Nevertheless, since that time a great many indirect methods have been devised for obtaining information of one kind or another. Only a few of them have proved capable of yielding results of high numerical accuracy, and it is with these few that this book is concerned.

Many of the other, less accurate, methods have been important in providing a body of independent, confirmatory evidence on the general sizes and shapes of molecules. From time to time in the past they have supplied useful information, and they may well answer particular questions in the future. For example, substances, such as the long-chain fatty acids, whose molecules are large and mainly of a hydrophobic character but carry a hydrophilic (polar) group at one end, can be spread on a water surface as a monolayer, or unimolecular film. The study of matter in this 'two-dimensional' state by Langmuir and Adam in the period 1917–1927 led to some early estimates of over-all molecular dimensions. A little later such studies provided a first indication—the value of which was not at once recognised—that the structure originally proposed for the steroid skeleton needed revision.

The fundamental reason why they cannot be examined by any ordinary optical system is that molecules—and atoms, *a fortiori*—are much smaller than the wavelengths of light. This limitation may be overborne if alternative radiations of appropriately shorter wavelengths can be used. This is what is done in the X-ray, electron and neutron diffraction methods. In a more direct way it is also done in the electron microscope. Electrons of moderately high energy possess de Broglie wavelengths considerably less than the interatomic distances in molecules. If, therefore, we can contrive some sort of 'lens' for electrons—and this can be done with suitable electromagnetic coils—it should be possible to construct a microscope that uses a beam of electrons instead of a beam of light; and with such an instrument it should be possible to explore the details of molecular structure. In fact, the performance of existing electron microscopes falls short of the theoretical ideal by at least a power of ten. At present the very large molecules, regularly stacked in a crystalline protein, can be individually recognised; and in the future it may become possible to decipher the grosser details of their structures. Though they are large in the context of this book, molecules such as those of platinum phthalocyanine are considerably smaller than those of proteins; the presence of the heavy platinum atom facilitates observation, however, and such molecules are now being individually detected in the electron microscope.

The physical methods that have proved most valuable for winning accurate and detailed information about molecular structure all depend upon interactions between matter and radiations of one kind or another. The matter may scatter the radiation with, or without, change of energy; energy may, or may not, pass between the radiation and the molecules concerned. When there is a transfer of energy, there will in general be a change in the wavelength of the radiation, and usually absorption or emission at specific wavelengths. In the broad sense of the term, these effects are studied by *spectroscopic* techniques. Since spectroscopic measurements of frequency can be made with high accuracy, related molecular parameters can also be determined with high accuracy. As we have already seen, a limitation is imposed by the fact that it is impossible to interpret in detail the spectra of any but simple molecules. Spectroscopic methods give results of high accuracy for simple systems.

When there is no transfer of energy between radiation and the molecule, the scattering occurs without change of wavelength; there is elastic scattering. The rays scattered from different molecules, or from different parts of the same molecule, will reinforce, or interfere with, one another in appropriate directions; *diffraction effects* will occur. This principle can be applied even when the molecules are randomly orientated in the vapour state; and *electron diffraction* as ordinarily practised enables the dimensions of simple molecules to be determined with considerable accuracy. The diffraction method comes to full fruition when the molecules are presented to the radiation in the regular array characteristic of the crystalline state. It is most frequently applied in *crystal structure analysis by X-rays*, though other types of radiation are also used. Structure analysis by crystal diffraction is less accurate than by spectroscopic methods, but it is applicable to a much wider range of materials.

1.5 The Avogadro Number

The fundamental scale-factor between the gross, or macroscopic, or *molar* properties of matter and its fine, or microscopic, or *molecular* properties is the *Avogadro Number*. It is often represented by the symbol N_A, and it is equal to the number of actual molecules in the gram-molecule, or mole, of any substance. Our present system of atomic weights is based on ^{12}C. Thus the Avogadro number is the number of atoms in 12·0000g of this isotopic species. At the time of writing (1973) the officially accepted value is $6{\cdot}02252 \times 10^{23}$ molecules per mole. All the methods for determining molecular dimensions depend, at one point or another, on a knowledge of N_A.

Avogadro's hypothesis (1807) implied that such a constant must exist, but its value was not even roughly known till half a century later. The first reliable estimate was made by Loschmidt in 1865. At that time interest centred rather on molecular size, and Loschmidt obtained a value for the (mean) diameter of the molecules in air, though he clearly recognised the wider implications of his calculations. Transformed into a more modern idiom, his calculations run as follows:

The kinetic theory of gases had led to the formula,

$$l = 3/(4\pi s^2 N_L), \qquad (1)$$

for the mean free path, l, of a molecule of effective diameter, s. N_L is the number of molecules per cm^3, and it is often appropriately called the *Loschmidt Number*. From the viscosity of air at ordinary temperatures, the value of l had been assessed at $1{\cdot}4 \times 10^{-5}$cm, so that equation (1) can be rearranged and evaluated thus:

$$\pi N_L s^2 = 3/(4l) = 5{\cdot}4 \times 10^4 \text{ cm}^{-1}. \qquad (2)$$

The density of air was known to be $1{\cdot}3 \times 10^{-3}$g cm^{-3} under ordinary conditions; and this must be equal to $N_L m$, where m is the mass in grams of the 'average' molecule in air. Next we evaluate the ratio between this effective density of air and the density of the molecules themselves. The volume of a spherical molecule would be $\pi s^3/6$, and, the mass being m, its density would be $6m/(\pi s^3)$; so that the ratio is $(\pi N_L s^3)/6{:}1$. The density of liquid air was not known to Loschmidt; from Kopp's systematised data for the molar volumes of various liquids at their boiling points, he estimated it at 1·24g cm^{-3}. After correcting this to room temperature and after making allowance for the fact that packed spheres do not fill space completely, he found an amended value of 1·50g cm^{-3}. Hence

$$\frac{\text{Density of air}}{\text{Density of molecules}} = \frac{\pi N_L s^3}{6} = \frac{1{\cdot}3 \times 10^{-3}}{1{\cdot}50};$$

so that

$$\pi N_L s^3 = 5{\cdot}2 \times 10^{-3}. \qquad (3)$$

Finally from (2) and (3),

$$s = (5{\cdot}2 \times 10^{-3})/(5{\cdot}4 \times 10^4) = 9{\cdot}7 \times 10^{-8}\,\text{cm},$$

and

$$N_L = (5{\cdot}4 \times 10^4)/\pi(9{\cdot}7 \times 10^{-8})^2 = 1{\cdot}8 \times 10^{18}\ \text{molecules cm}^{-3}.$$

Since a mole of gas occupies about 23 700 cm^3 under room conditions, this value for N_L corresponds to $4{\cdot}3 \times 10^{22}$ for N_A. Loschmidt's value for s was too high by a factor of about three; and when squared, this error makes his value for N_A about ten times too small. But for a first estimate, his result was not unsatisfactory.

Subsequently a great variety of methods have been applied to the direct or indirect determination of this number. Most of them are of low, or moderate, accuracy; but the general consistency of results obtained in such diverse ways is persuasive evidence for the validity of the molecular concept.

The accepted value of N_A was based until recently upon two methods which had been developed to give results of high precision. The first depends on the determination of the unit of electric charge—the charge, e, carried by the electron, which is related to N_A *via* the faraday, $\mathfrak{F}$: $\mathfrak{F} = N_A \times e$. The second depends on the absolute determination of the wavelength of X-rays of a particular type; when these rays are then used to produce diffraction effects from a suitable crystal of known structure and accurately known density, and in a manner which will be explained in Chapter 9, a value of N_A can be deduced. The former method was used by Millikan in a series of classical experiments that led to a value of $6{\cdot}06 \times 10^{23}$, which was generally accepted for a number of years until about 1940. (It should be remembered that all the earlier atomic and molecular data are based on this value of N_A.) The latter method was found to yield a lower value, close to that now adopted. The discrepancy, which was beyond the supposed limits of error of either method, was traced to Millikan's having used the then accepted, but erroneous, value for the viscosity of air; when this was amended, the two methods came into excellent agreement.

Since about 1950 the development of radio-frequency techniques has made it possible to determine with high precision the magnitudes of certain quantities which are combinations of the fundamental physical constants. The best current values for these constants are based on the rather complex interrelations between these measured quantities. The value for N_A thus derived differs only slightly from that obtained by the more direct routes outlined above.

1.6 Units and Some Other Numerical Constants

Besides the Avogadro number, several other important constants will be needed, and values for these—taken from the 1966 edition of Kaye and Laby's Tables— are listed at the end of this chapter.

Certain units and conventions will be used, and some explanation of them should be given. The velocity (c), frequency (ν), and wavelength (λ) of radiation are related by the familiar equation,

$$c = \nu \times \lambda.$$

In the X-ray and visible regions of the spectrum, wavelengths are usually expressed as *Ångstrom units* or *ångstroms*; strictly this unit is now defined by reference to the wavelength of a particular line in the spectrum of krypton, but its magnitude is very close indeed to 10^{-8} cm, the difference being wholly negligible for our purposes. In the infrared region the *micron*, or the *millimicron*, are sometimes used: $1\,\mu = 10^{-4}$ cm; $1\,\text{m}\mu = 10^{-7}$ cm.

Spectroscopists often express their frequency results as reciprocal wavelengths or wavenumbers: $\sigma = 1/\lambda(\text{cm}^{-1})$. Any uncertainty in the wavenumber depends only on the uncertainty of the wavelength measurement, and not on the precision with which the velocity of light may be known. Like the wavelength however, the wavenumber of a particular radiation does depend upon the medium in which it is measured. If we wish to convert σ to true frequency, ν, by multiplying by the velocity of light in vacuum ($\nu = \sigma c$), then σ must be the *vacuum wavenumber*. Wavelengths in air have to be corrected to vacuum by adding $(n - 1)\lambda_{\text{air}}$, where n is the refractive index of air. The reciprocal of the wavelength thus corrected is the vacuum wavenumber.

Radiation in the visible, infrared, and microwave regions with wavelengths of 4000 Å, 10μ and 1 cm, respectively, would have wavenumbers of $2{\cdot}5 \times 10^4$, 1000 and 1 cm^{-1}.

By the basic equation of quantum theory the passage of energy between matter and radiation occurs in quanta of magnitude ΔE, where $\Delta E = h\nu$. When ΔE is expressed in terms of the erg and ν as reciprocal seconds, Planck's constant, h, has the value $6{\cdot}6256 \times 10^{-27}$ erg sec. With frequency in wavenumbers, $\Delta E = hc\sigma$; and spectroscopists often express their energy terms as wavenumbers by the latter equation, according to which 1 cm^{-1} is equivalent to $hc = 1{\cdot}9863 \times 10^{-16}$ erg.

Another practical energy unit is the *electron-volt* (eV), which is the amount of energy acquired by an electron with charge $4{\cdot}803 \times 10^{-10}$ electrostatic unit, falling through a potential difference of 1 volt (= 1/299·79 electrostatic unit of potential). Hence 1 eV is equivalent to $4{\cdot}803 \times 10^{-10}/299{\cdot}79 = 1{\cdot}602 \times 10^{-12}$ erg. Further, 1 eV is equivalent to 8066 cm^{-1}. Chemists are accustomed to expressing their energy quantities as calories or kilocalories per mole; and on this basis, 1 eV is equivalent to

$$\frac{1{\cdot}602 \times 10^{-12} \times 6{\cdot}0225 \times 10^{23}}{4{\cdot}184 \times 10^7 \times 10^3} = 23{\cdot}06 \text{ kcal mole}^{-1}.$$

RT is also a certain quantity of energy per mole; and correspondingly kT is the energy per *molecule*, where k is here the *Boltzmann constant* with the value $1{\cdot}3805 \times 10^{-16}$ erg molecule^{-1} degree^{-1}. Hence it is sometimes convenient to relate the quantum of energy to a *characteristic temperature*: for example, 1 eV is equivalent to

$$\frac{1{\cdot}602 \times 10^{-12}}{1{\cdot}3804 \times 10^{-16}} = 11\,604\ \text{K}.$$

Similarly the three types of radiation specified above would, respectively, correspond to temperatures of $3{\cdot}597 \times 10^4$, 1439 and 1·439 K.

The distance between the nuclei of two bonded atoms A and B is known as the *bond length*; and it will sometimes be written as r_{AB}, and sometimes simply as A—B. The angle subtended at nucleus B by the nuclei of two other atoms A and C which are bonded to B may similarly be represented as A—B—C. The distance between atoms that are not directly bonded is often shown as A . . . B.

1.7 SI Units

The *Systéme Internationale* has now received the approval of the International Union of Pure and Applied Chemistry and will, presumably, gradually displace c.g.s. units. However *SI* has so far had an unenthusiastic reception from chemists active in research, who tend–conservatively–to regard it (despite its greater tidiness and self-consistency, especially in electromagnetic theory) as an unnecessary reform. At the time of writing (1973) a large majority of research papers continue to report their results in c.g.s. Furthermore, nearly all the major monographs in Physical Chemistry and Chemical Physics are in c.g.s. So are most current compilations of data, and so is all the scientific literature of the past century. Now, and for the foreseeable future, therefore the advanced student of Chemistry has to be bilingual in this matter; if he comes from school knowing only *SI*, he has to learn c.g.s. What is more, some of the basic equations of chemical theory undergo a strange metamorphosis when *SI* is adopted–the familiar expression first derived by Bohr for the Rydberg constant, for instance.

For all these reasons, this second edition retains the style of the first in respect of units. But the table of universal constants which follows carries *SI* values, as well as c.g.s, and we have appended a short list of equivalents.

Table 1.7.1. Some Universal Constants*

	Value	c.g.s.	SI
Avogadro Number, N_A	6·022 52	10^{23} mole^{-1}	10^{23} mol^{-1}
Mass of Unit Atomic Weight ($1/N_A$)	1·660 435	10^{-24} g	10^{-27} kg
Electronic Charge, e	4·802 98	10^{-10} e.s.u.	—
	1·602 10	—	10^{-19} C
Planck's Constant, h	6·625 59	10^{-27} erg sec	10^{-34} J s
Velocity of Light, c	2·997 93	10^{10} cm sec^{-1}	10^{8} m s^{-1}
Nuclear Magneton, n	5·050 50	10^{-24} erg gauss^{-1}	10^{-27} J T^{-1}
Ice-point on Absolute Scale	273·15	K	K
Gas Constant, R	8·314 34	10^{7} erg deg^{-1} mole^{-1}	10^{0} J K^{-1} mol^{-1}
Boltzmann Constant, k	1·380 54	10^{-16} erg deg^{-1} molecule^{-1}	10^{-23} J K^{-1} molecule^{-1}
Permittivity ('Dielectric Constant') of a Vacuum	—	1 (e.s. system)	$8{\cdot}8542 \times 10^{-12}$ F m^{-1}
Permeability of a Vacuum	—	1 (e.m. system)	$1{\cdot}256\,64 \times 10^{-6}$ H m^{-1}

* According to Taylor, Parker and Langenberg (*Rev. Mod. Phys.*, 1969, **41**, 375), who have re-determined the ratio e/h, some of the accepted values given in the above Table may need surprisingly large emendations—by several times their estimated standard deviations in some cases. In particular, N_A would become $6{\cdot}022\,17 \times 10^{23}$. The change, of 58 parts per million, is considerably larger than the change needed when the basis of 'chemical' atomic weights was moved from ordinary oxygen to ^{12}C. Planck's constant would need correction by 91 parts per million.

Table 1.7.2. Interconverting Energy Quantities Expressed in Different Units

erg molecule^{-1}	*eV*	*cm*$^{-1}$	*K*	*kcal mole*$^{-1}$
1	$6{\cdot}242 \times 10^{11}$	$5{\cdot}035 \times 10^{15}$	$7{\cdot}244 \times 10^{15}$	$1{\cdot}439 \times 10^{13}$
$1{\cdot}602 \times 10^{-12}$	1	$8{\cdot}066 \times 10^{3}$	$1{\cdot}1604 \times 10^{4}$	$2{\cdot}306 \times 10^{1}$
$1{\cdot}986 \times 10^{-16}$	$1{\cdot}240 \times 10^{-4}$	1	$1{\cdot}439 \times 10^{0}$	$2{\cdot}859 \times 10^{-3}$
$1{\cdot}380 \times 10^{-16}$	$8{\cdot}616 \times 10^{-5}$	$6{\cdot}951 \times 10^{-1}$	1	$1{\cdot}986 \times 10^{-3}$
$6{\cdot}947 \times 10^{-14}$	$4{\cdot}336 \times 10^{-2}$	$3{\cdot}498 \times 10^{2}$	$5{\cdot}033 \times 10^{2}$	1

Table 1.7.3. Some c.g.s./SI Equivalents

1 Å	100 pm, 0·1 nm
1 D (Debye Unit)	$3{\cdot}336 \times 10^{-30}$ C m
1 kcal	4·184 kJ
1 cal	4·184 J
1 dyne	10^{-5} N

Chapter 2

Symmetry

2.1 Introduction

Considerations of symmetry occupy an essential place in the study of molecular structure. When a molecule is being studied by any method, to determine its symmetry is obviously a desirable first objective; but in interpreting its spectrum, this symmetry is of more fundamental importance. In a crystal there is a repetitive pattern in space, and the symmetry of this pattern overlies that of the molecule itself.

The treatment of symmetry by the spectroscopist differs from that of the crystallographer: superficially because they generally use different symbols, but more significantly because they each emphasise different aspects of the subject. Nevertheless, the principles are the same, and in this chapter they will be outlined first in such a way as to bring out their unity. Afterwards there will be some account of those special developments of symmetry theory required by the respective needs of the spectroscopist and the crystallographer. A treatment of symmetry from the spectroscopic angle is included in Herzberg's book (1945), and a more detailed account from the crystallographic angle has been given by Buerger (1956). An elementary, though masterly, survey of the subject in its most general context is embodied in Weyl's *Symmetry* (1952). Details of an elegant *Stereoscopic Guide*, by Bernal, Hamilton and Ricci (1972), are also included in the Bibliography.

2.2 Point Symmetry in Two Dimensions

The symmetry properties of two-dimensional objects are readily apprehended, and they are simple to represent in diagrams. The plan will therefore be to open with an account of symmetry in this simplified form, and afterwards to indicate how the conclusions reached need elaboration in three dimensions.

First we consider simple flat objects such as those drawn in Fig. 2.2.1. Each of them, except (*a*), will be recognised to possess some symmetry in the informal sense. (Since they are supposed to be strictly two-dimensional, the question of possible mirror-symmetry across the plane of the paper does not arise.) Our purpose is to define what is meant by symmetry, and to elucidate the possible types of it that can occur in flat objects.

When we say that an object has symmetry, we mean that by movements of a certain kind the object assumes an aspect indistinguishable from that presented in its original position. For instance, an imaginary line may be considered to bisect the acute angle of the isosceles triangle in Fig. 2.2.1.(*b*); if, then, every point on the left-hand side of the triangle were translated normally across the line to a new position equidistant from it on the right, and *vice versa,* the new figure could not be distinguished from the old. A movement of this kind is an example of a *symmetry operation;* this particular one is the operation of *reflexion* across the line. It is evident that the line chosen is a special one; for, were we to draw any arbitrary line through the object, and then perform the operation of reflexion about that line, the object would no longer present the same appearance. In fact the isosceles triangle possesses only the one line across which the symmetry operation of

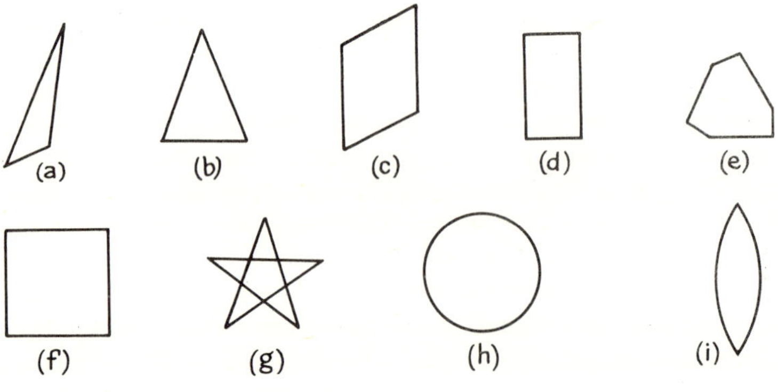

Fig. 2.2.1. Objects showing various types of two-dimensional symmetry.

reflexion may be carried out. This line is an example of an *element of symmetry;* it is known as a *line* (or mirror-line) *of symmetry.* Objects (*a*), (*c*), and (*e*) possess no such lines; (*d*) has two, mutually perpendicular, and so has (*i*); (*f*) has four; (*g*) five; and (*h*) an infinite number. Quite generally, a symmetry operation sends an object into a configuration *equivalent* to the original one, so that an observer who has not actually witnessed the operation is unable to tell whether the object has been moved or not. (For some purposes—as we shall see later—we suppose that certain points on the object are ear-marked with imaginary labels; when this has been done, then in a formal sense we may be able to differentiate between original and final configurations of the object.) A symmetry operation is to be carefully distinguished from the corresponding symmetry element about which the operation is performed.

The second type of symmetry operation in two dimensions comprises rotations. If the objects (*c*), (*d*), and (*i*) in Fig. 2.2.1 are rotated through 180° about axes perpendicular to the paper and passing through their centres (or, more strictly in two dimensions, are rotated about a point), they will assume indistinguishable positions. These axes are called *two-fold axes of symmetry,* because equivalent positions of the object are related by rotation through 360°/2. Similarly object (*e*) possesses a *three-fold axis*; rotation through 360°/3 produces an equivalent configuration. Objects (*f*) and (*g*) possess *four-* and *five-fold* axes respectively; whilst (*h*) possesses an axis of *infinite order*, since rotation about its centre produces an infinite number of indistinguishable aspects. In all these examples the axis is a symmetry element of the object: the corresponding symmetry operation consists in rotating the object through 360°/*n*, where *n* is the *order* of the axis, or the number of equivalent configurations ocurring during a complete revolution. Axes of order two, three, four, and six are often referred to as digonal (or diad), trigonal (or triad), tetragonal (or tetrad), and hexagonal (or hexad), respectively.

Every object, no matter how unsymmetrical, possesses a one-fold (monad) axis, since rotation through 360°/1 necessarily leads to self-coincidence. (To be more exact, it possesses an infinite number of such axes–one passing through each point in the object; with a three-dimensional object, an infinite number of monad axes passes through each point.) In two dimensions, but not in three, a two-fold axis and a centre of symmetry always occur together, and are effectively identical. In the corresponding operation of inversion at a centre, each point of the object is projected through the centre of symmetry to an equal distance beyond. The objects in Fig. 2.2.1, (*c*), (*d*), (*f*), (*h*), and (*i*) each possesses a centre of symmetry.

Any object with a four-fold axis must also have a coincident two-fold axis. Similarly the presence of two, and only two, mirror-lines of symmetry in (*d*) implies that these lines must be at right angles to one another, and that a two-fold axis must run through their point of intersection. We see therefore that certain elements of symmetry require the presence of others of a different kind. The reverse is also true: certain elements preclude the possibility of certain others. For instance, a two-dimensional object such as (*g*), where the highest axis is of order five, cannot have a three-fold axis, nor can it have a two-fold axis or a centre of symmetry. The corresponding operations would not send the object into a position indistinguishable from its original one, and therefore they are not symmetry operations. Clearly the symmetry operations that can be performed on any object must form a self-consistent set; and, as we shall show later, each such set then constitutes a *group* in the mathematical sense. With the exception of (*d*) and (*i*) which belong to the same group (despite their difference in shape), the objects in Fig. 2.2.1 all belong to different symmetry groups; they each possess a different set of possible symmetry operations and elements. The symmetry elements of any object must intersect at a point or, in other words, the symmetry operations must leave one point unchanged. Were this not so, a repetitive pattern would result. For this reason the symmetry displayed by single objects is known as *point-symmetry,* and the symmetry groups are known as *point groups.*

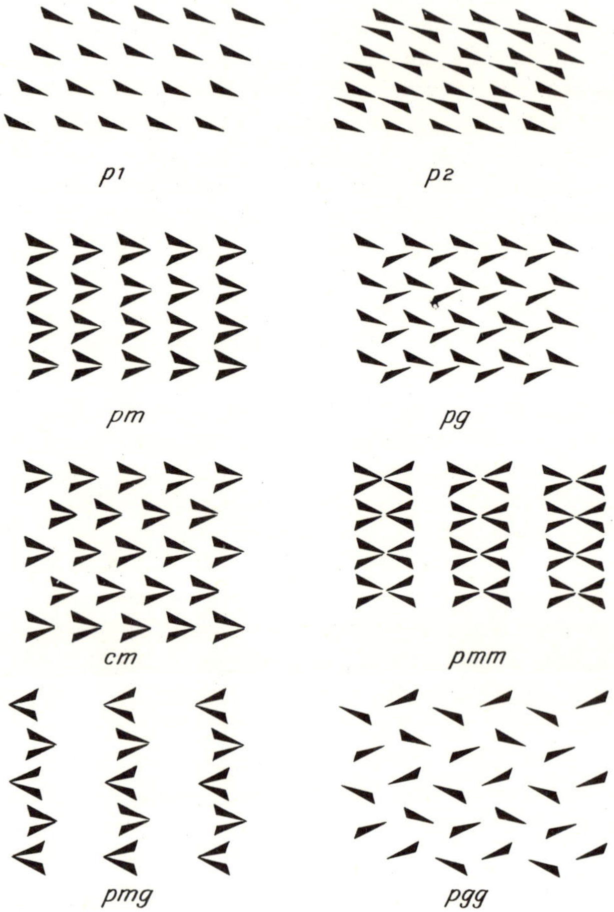
p1
p2
pm
pg
cm
pmm
pmg
pgg

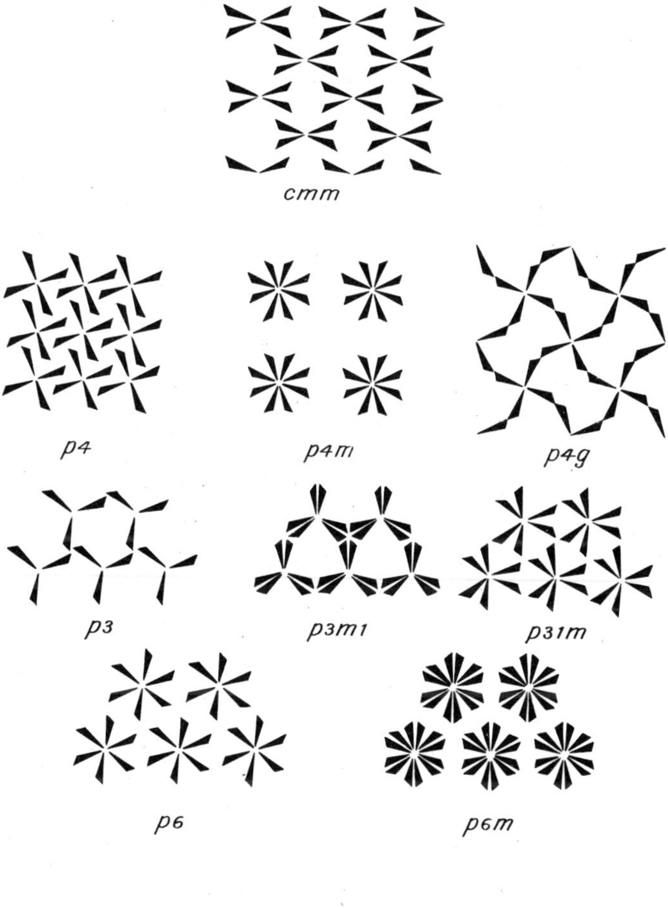

Fig. 2.3.1. Patterns exemplifying the 17 plane groups—the only possible types of pattern repetitive in two dimensions. (Each of these designs must be supposed to extend indefinitely.)

[The symbols given here correspond to the Hermann–Mauguin system for the three-dimensional space groups as described on pp. 33–39, with the addition of *g* for the glide and *c* for a centred cell.]

For a reason that will be explained shortly, axes of orders 1, 2, 3, 4, and 6 only can occur in crystals. It is an interesting exercise to determine what are the possible point groups in two dimensions, when the orders of the axes are limited in this way. There are in fact just *ten planar, crystallographic point groups,* and it is easy to demonstrate how they arise: there are five types of axes, and each of them may be with, or without, mirror-lines of symmetry. Six of these point groups are illustrated amongst the objects in Fig. 2.2.1; and the reader should construct objects exemplifying the other four. Objects (*g*) and (*h*) belong to "non-crystallographic" groups, since they embody axes of orders five and infinity.

2.3 Repetitive Symmetry in Two Dimensions

We have next to consider how these principles are affected when we have to do with repetitive patterns, such as may occur in a wallpaper design, for instance. Patterns of this kind are shown in Fig. 2.3.1, where it is to be supposed that each design is continued indefinitely in all directions in the plane of the paper. The essence of the repetition is that any point is repeated exactly, including its environment, at regularly recurring distances. This quality can be associated with the operation of *translation;* and it can be represented in isolation by an array of points. One possible array is shown in Fig. 2.3.2, which covers the repetitive operations in the first two patterns of Fig. 2.3.1. Any point whatsoever in the repeat-unit may be chosen—the sharpest corner of a particular triangle, for instance; then all other

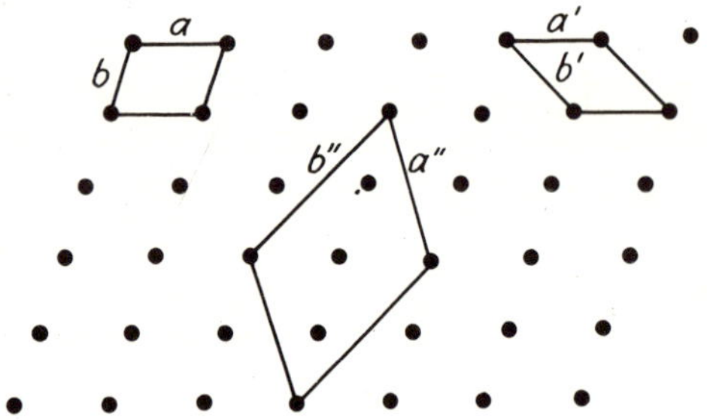

Fig. 2.3.2. A plane net, with some alternative choices of unit cell.

exactly equivalent points in the pattern are plotted so as to give the array. This array is known generally as a *lattice,* or, more properly in two dimensions, as a *plane net.* Just as the points of a single object can be regarded as generated by the symmetry operations of its point group, so the points of the net, or lattice, are generated by the translational operations.

The vector from any lattice point to the next, in a given direction, is known as a *primitive translation.* Some of these are marked in Fig. 2.3.2–e.g. *a* and *b*. In a planar net, two independent primitive translations define a *unit cell,* such as that marked out by *a* and *b*. The whole net can be regarded as constructed from these unit cells, just as a pavement might be constructed from similar tiles. Other pairs of

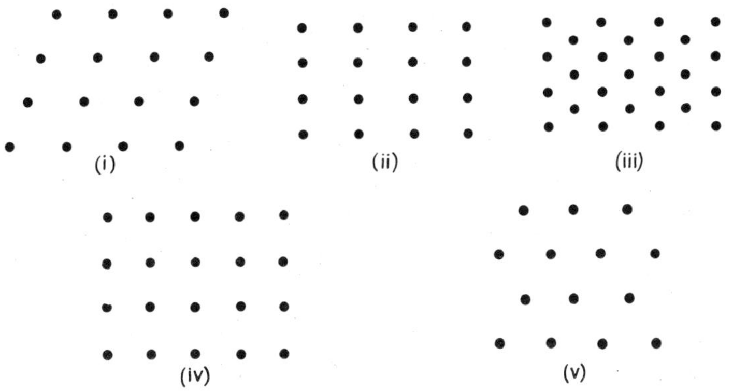

Fig. 2.3.3. The five types of plane nets.

primitive translations can be chosen to define alternative unit cells: for instance, a' and b', or a'' and b''. It is usual to choose a cell of minimum area, which is then called a *primitive cell,* and that particular one which most nearly approaches a rectangle. The translations *a* and *b* define a primitive cell of this kind.

There are only five types of two-dimensional nets. They are shown in Fig. 2.3.3, and all of them are exemplified amongst the patterns of Fig. 2.3.1; *viz.* the unit cell must be either (*i*) a general parallelogram, (*ii*) a rectangle, (*iii*) a centred rectangle, (*iv*) a square, or (*v*) a special parallelogram derived from two equilateral triangles. Case (*iii*) can formally be dispensed with, as this net could be described in terms of a primitive parallelogram of half the area of the centred rectangle. However the quality of orthogonality–*i.e.* 90° angles– would then be lost, or at least hidden; and, to avoid this inconvenience, the centred, non-primitive cell is usually adopted.

Axes of orders 2, 3, 4, and 6 occur in these nets; and in fact only these axes– along with that of order unity, when a asymmetric object is placed at each lattice point–can be built into such repetitive patterns. Suppose, for example, we try to graft a five-fold axis on to a plane net. In Fig. 2.3.4, A and B denote neighbouring net points. Place a five-fold axis at A. Then a similar axis must also be placed at B, since A–B is a primitive translation, at which interval all details of the pattern must be repeated. The operation of the axis at A on B will produce another net point at C, complete with a five-fold axis, the angle BAC being $360/5 = 72°$; and similarly the axis at B operates on A to produce a fourth net-point at D, with $ABD = 72°$. The two new points, C and D, do not coincide, nor are they separated by a primitive translation. This inconsistency, which would destroy the net or lattice, can be avoided only by restricting axes to the orders 1, 2, 3, 4, and 6.

(All this does not mean that an object of five-fold inherent symmetry cannot occur in a crystal lattice. It would be perfectly possible to make a pattern in which

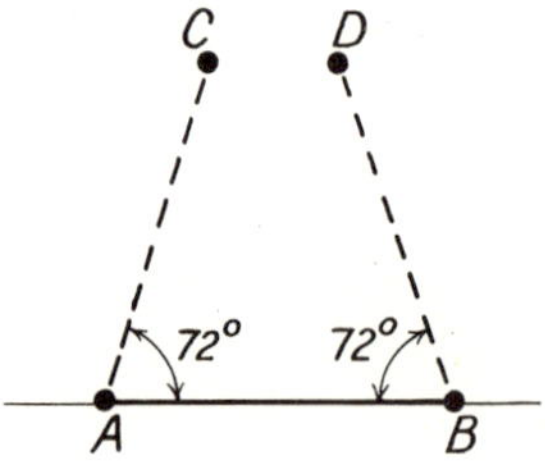

Fig. 2.3.4. Impossibility of building a five-fold axis into a plane net.

the five-pointed star of Fig. 2.2.1. (*g*) was sited at each point of the lattice in Fig. 2.3.2. But the five-fold symmetry of the star would bear no relation to the symmetry of the whole pattern. Each star would be placed on a point of initially two-fold symmetry, thereby debasing that symmetry to one-fold; and this environmental symmetry would bear no relationship to that of the star. With a molecule in a crystal, the lower symmetry of the environment would–formally at least–cause slight distortion of the molecule, whose five-fold symmetry would thus be degraded by that of the lattice.)

Another way of expressing this limitation is to state that identical regular polygons can be packed together, so as to cover a plane surface completely, only when they are either parallelograms, equilateral triangles, squares, or hexagons. These polygons have respectively 2-, 3-, 4-, and 6-fold symmetry. The *law of rational indices* in crystallography is a consequence of this same principle. Indeed the discovery of this law, by Haüy in 1784, was strong evidence for a lattice structure in crystals, and hence for a molecular theory of matter.

This limitation on the permissible orders of axes is offset by the availability of a new type of symmetry operation in repetitive patterns. This is the *glide*, which is exemplified in the fourth pattern of Fig. 2.3.1, for example, and which is explained in more detail in Fig. 2.3.5. The triangle I is converted into II by a translation of half the primitive (to III) followed by reflexion across the dotted glide-line. When the glide is added to the six kinds of operation already noted (*i.e.* five kinds of axes and the mirror-line) it turns out that there are 17 possible sets of self-consistent operations. These are known as the 17 *plane groups;* and Fig. 2.3.1, which is based on one due to Buerger, shows patterns corresponding to each of these these groups. Every two-dimensional pattern that is regularly repeated must belong to one of these groups. It is said that all 17 have been recognised amongst designs surviving from antiquity.

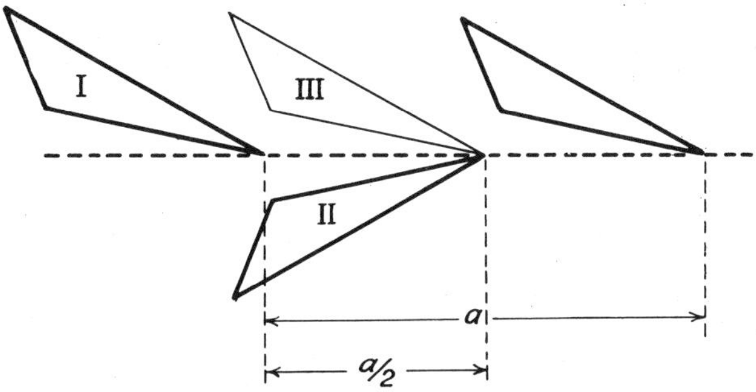

Fig. 2.3.5. The glide operation.

2.4 The Band Groups

If the repetitive elements in a planar pattern are restricted to one dimension only, the permissible symmetry operations, besides translation in the one direction, are one- and two-fold axes, mirror-lines either parallel or perpendicular to the translation, and the glide parallel to the translation. The various combination of these operations give rise to the *seven band groups.* All possible types of frieze-patterns, for example, must belong to one of these seven groups. As Pabst has suggested, the student should find it a suitable exercise to derive these groups for himself, and to draw examples of them–or to recognise examples in his everyday environment.

2.5 Symmetry in Three Dimensions

In three dimensions the range of available symmetry operations is considerably wider, and there is a great deal more variety in their possible combinations. The symmetry of a single object such as a molecule–or a crystal considered externally, and without reference to its internal structure–is usually described in terms of symmetry elements of three types: (*i*) axes of symmetry, (*ii*) planes of symmetry, and (*iii*) the centre of symmetry.

Axes can now be *of two sorts.* The first is the simple, or *proper, axis* of rotation. Such an axis is said to have order *n* if *n* equivalent configurations appear during a complete rotation through 360°. The molecules of ethane (Fig. 2.5.1), ferrocene (2.5.2) and benzene respectively possess proper axes of orders 3, 5, and 6. For reasons already given, the orders of crystallographic axes are limited to $n = 1, 2, 3, 4$, and 6, but with molecules there is no such restriction.

The second sort of axis is the *improper axis,* which may be considered in either of two ways: as an *alternating axis,* or as an *inversion axis.* The definitions of these axes are best given in relation to examples. Fig. 2.5.1 represents the ethane molecule, C_2H_6, in its equilibrium, *staggered* conformation. If the molecule is rotated

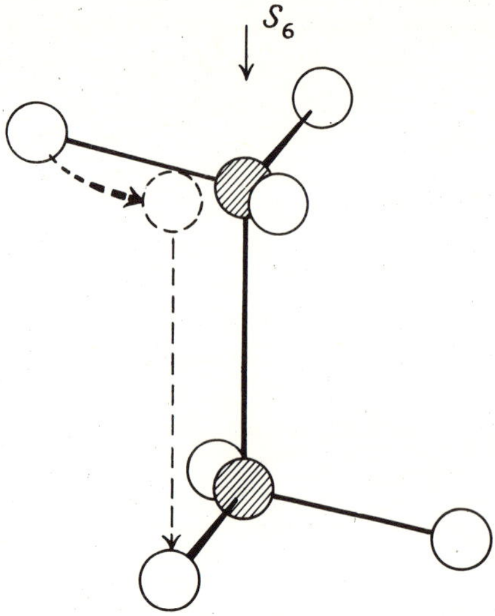

Fig. 2.5.1. The ethane molecule in its 'staggered' conformation. (The shaded circles represent carbon atoms, the others hydrogens.)

counterclockwise through $360°/6 = 60°$ about an axis coincident with the C–C direction, and each point is then reflected in a plane perpendicular to the axis of rotation and midway between the two carbon atoms (as is suggested by the broken lines for one hydrogen atom), it attains a configuration equivalent to the original one. The symmetry element corresponding to this two-stage operation is the *alternating axis*–in this example a six-fold alternating axis. The reader should verify for himself, preferably with the aid of models, that allene, $CH_2{=}C{=}CH_2$, and ferrocene, $(C_5H_5)Fe(C_5H_5)$ (Fig. 2.5.2), possess a four-fold, and a ten-fold, alternating axis respectively. Each coincides with a proper axis of half that order, and this is generally true for improper axes of even order. It should be emphasised that the reflexion, which constitutes the second stage of the total operation, is carried out across a plane which is not itself necessarily a symmetry element of the molecule. The ethane molecule in Fig. 2.5.1 does not possess an equatorial plane of symmetry, for instance.

When there is rotation through $360°/n$ followed by inversion through a centre on the axis, the total operation is known as that of an *inversion axis* of order *n*. Thus the ethane molecule in Fig. 2.5.1 possesses a three-fold inversion axis, since rotation through 120° with inversion through the centre leads to an indistinguishable aspect. Wherever there is an alternating axis there is always an inversion axis, though–as in this example–their orders are not necessarily the same. In allene however, both alternating and inversion axes are of the same order–*viz*. four-fold. Alternating and inversion axes are simply different descriptions of the same symmetry element, the improper axis, and which is used is a matter of choice. Discussion in group theory is always based on alternating axes, whilst crystallographers generally use inversion axes.

The three types of symmetry elements listed at the beginning of this section can be reduced to two when it is realised that a plane of symmetry and a centre can equally well be described in terms of improper axes. As the reader should confirm for himself, the operation of reflexion in a plane, of a one-fold alternating axis, and of a two-fold inversion axis, are all identical. Thus (*i*), (*ii*), and (*iii*) can now be reclassified as (*a*) simple, or proper, rotations, and (*b*) as improper rotations; or simply as operations of the *first* and *second sorts.*

Seen through a mirror, a right hand looks like a left hand seen directly.

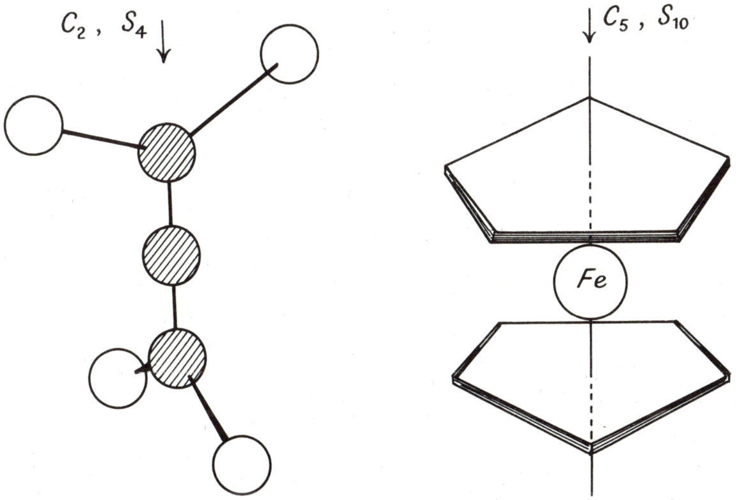

Fig. 2.5.2. Molecules of allene, $H_2C{=}C{=}CH_2$, and ferrocene. (The shaded circles in the former represent carbon atoms; the pentagons in the latter rings of five CH-groups.)

Any operation of the second sort transforms a right-handed object into its enantiomorph, whilst an operation of the first sort leaves it unchanged in this respect. It follows that an object and in particular a molecule, possessing only symmetry elements of the first sort is not superposable upon its mirror image. Such a molecule is optically active–or at least potentially so. Following Pasteur, chemists often describe a molecule of this kind as *dissymmetric,* though it will not necessarily lack all symmetry as the word literally implies. The most general definition of a dissymmetric molecule is therefore that it lacks any symmetry element of the second sort. This includes the familiar, but incomplete, rule that the molecule must possess neither a plane nor a centre of symmetry. A compound has been described by McCasland and Proskow (1955), whose molecules lack these two symmetry elements, but are not optically active because they have a four-fold alternating axis.

2.6 Three-Dimensional Point Groups and their Notation

As in two dimensions, the set of operations that can be performed on a single three-dimensional object constitutes a point group. When axes are restricted to the orders 1, 2, 3, 4, and 6, there are 32 possibilities, corresponding to the 10 crystallographic point groups in two dimensions. As will be explained in more detail in Section 2.10; the macroscopic properties of crystals must always conform to the symmetry of one of these *32 cystallographic point groups.* With objects in general, including molecules, there is no restriction upon the orders of the axes, though axes of order 5 are rare amongst molecules and orders higher than 6 extremely rare: consequently a majority of non-linear molecules do in fact belong to one of these 32 point groups.

Two systems of notation are used to represent symmetry elements and point groups. The older one, due to Schoenflies (1888), is generally applied to the symmetry of molecules and is always used by spectroscopists; but it has been replaced in crystallography by a system due to Hermann and Mauguin (1931). Because the symmetry of a single molecule is readily apprehended, we shall describe the Schoenflies system first and use its symbols to discuss molecular symmetry from a group-theoretical standpoint. We shall return to the Hermann-Mauguin system later, in connexion with the extension of symmetry theory to three-dimensional repeat-patterns. In Table 2.14.1 at the end of this chapter, the 32 crystallographic point groups are listed with both Schoenflies and Hermann-Mauguin symbols; and, where examples can be found, molecules having these symmetries are indicated.

2.7 The Schoenflies Notation

The possible symmetry elements present in a molecule are given the following symbols: C_n for an n-fold proper axis; σ for a plane of symmetry; i for a centre of symmetry; and S_n for an n-fold alternating axis. The same symbols are also used to represent the corresponding symmetry operations. Thus i may represent either the element of a centre of symmetry, or the operation of inversion at the centre. Although the double meaning attaching to the symbols introduces the possibility of confusion, it will always be clear from the text which meaning obtains. (It may have mnemonic value to point out that C stands for "cyclic", the Greek σ (= s) for "Spiegel" (= mirror), and i for "inversion".)

At this point it is useful to introduce an operation which at first sight appears trivial. This is the *identity operation,* symbolised by I which consists in leaving all atoms of the molecule unmoved. The corresponding symmetry element is possessed by all molecules, no matter how unsymmetrical. The operation I can be regarded either as that of "doing nothing", or as C_1^k–rotation k times through 360°; both leave the atoms in a configuration *identical* with–and not merely *equivalent* to–the original. The latter viewpoint will be adopted as it is sometimes a convenient formality to relate the operation I with the trivial element C_1.

The ideas in this section should become clearer when we consider examples.

The molecule H_2O_2 adopts a *gauche* form: when viewed along the O—O direction, the two O—H bonds make a dihedral angle of about 100° with one another. The only non-trivial symmetry element possessed by this molecule is the two-fold axis, C_2; and so the symmetry operations that can be performed are only $I(\equiv C_1^k)$ and C_2. These two operations comprise the point group C_2. Another representative molecule in this group is 1:2-dichloroethane in its *gauche* conformation. The point group C_2 is one member of a class of point groups C_n, in which the possible operations are C_n^k, with $k = 1, 2 \ldots n$–*i.e.* rotation k times through $360°/n$. (This set of operations includes the identity operation, since $C_n^n \equiv C_1 \equiv I$.) For reasons explained above, all molecules belonging to these C_n groups are dissymmetric; they exist as distinct, optically active forms. But in all the examples mentioned, interconversion of *d*- and *l*-forms would take place far too rapidly to permit isolation of optical enantiomorphs. When there is no non-trivial symmetry element, as with the lactic acid molecule, the only possible operation is $I(\equiv C_1)$, and the point group is C_1.

Several axes may be combined in the same point group, though in a strictly limited number of ways. The proper axis of highest order is known as the *principal axis.* The *dihedral* class of point groups has two-fold axes perpendicular to the principle axis; and they are symbolised as D_n, n being the order of the principal axis. For the axes to operate on each other in a self-consistent way, the two-fold axes must be perpendicular to the principal axis; otherwise they would operate to generate more than one principal axis. Furthermore the number of two-fold axes must equal the order n; as the reader should confirm for himself, a C_2-axis perpendicular to the C_3-axis in D_3 automatically requires the presence of two more axes, C_2' and C_2''. If the upper half of the ethane molecule were rotated through some arbitrary angle (say 20°) from the staggered conformation shown in Fig. 2.5.1, the resulting conformation would be an example of the point group D_3; and similarly a partly-rotated molecule of ferrocene (Fig. 2.5.2) would be a representative of D_5.

Some dihedral groups are not so readily exemplified in molecules (though examples are known amongst crystals), because axes are often combined with planes of symmetry. When a molecule possesses both axes and planes, the reader should visualise it as set with its principal axis vertical–*i.e.* coincident with the z-axis of a frame of cartesian coordinates. Three kinds of planes are recognised: (*i*) perpendicular to the principal axis, designated by σ_h (h standing for "horizontal"); (*ii*) containing the principal axis (and–in a dihedral group–one of the equatorial C_2-axes also), designated σ_v (v standing for "vertical"); and (*iii*) in a dihedral group, containing the principal axis and bisecting the angles between pairs of equatorial C_2-axes, designated σ_d (d standing for "diagonal"). These possibilities give rise to four important classes of point groups; C_{nh}, C_{nv}, D_{nh}, and D_{nd}. Examples of molecules belonging to these groups will be found in Table 2.14.1, and the possible symmetry operations are indicated in the collected table in Appendix I. Since these groups occur so frequently, it is worth while considering two cases in some detail.

In Fig. 2.7.1 (*a*) (compare Fig. 2.5.1) we show the *staggered* conformation of ethane viewed along its principal C_3 axis; the further elements of symmetry are S_6

(coincident with C_3), three C_2-axes (perpendicular to C_3 written C_2, C_2' and C_2''), i, and three σ_d. The C_3, C_2 and σ_d imply the other elements so that all are covered by the point-group symbol D_{3d}. In the *eclipsed* conformation of the molecule the

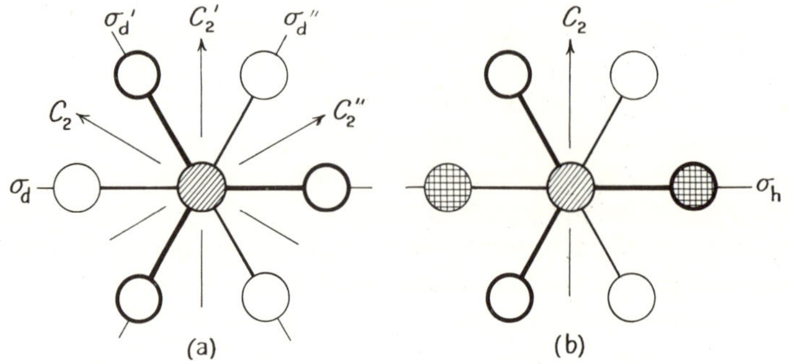

Fig. 2.7.1. (*a*) The ethane molecule, in its "staggered" conformation, projected along its C—C-axis. (*b*) The molecule of 1, 2-dichloroethane, $ClH_2C—CH_2Cl$, similarly projected. (The shaded circle represents two superposed carbon atoms, the hatched circles, chlorine atoms.)

hydrogen atoms of one methyl group lie vertically above those of the other; as the reader should verify, the symmetry elements then are one C_3, three C_2 axes, one σ_h, and three σ_v, comprised in the group D_{3h}.

Now consider that two antipodal hydrogen nuclei are replaced by chlorine, yielding the *trans* conformation of 1:2-dichloroethane: the molecule then presents the aspect shown in Fig. 2.7.1 (*b*). Substitution has destroyed the C_3 axis present in ethane, and other elements of symmetry disappear also: those elements remaining are C_2, σ_h, and i. Only two of these elements are independent, in the sense that any pair requires the presence of the third also, and hence the point-group symbol is C_{2h}. Since every operation embodied in the group C_{2h} is contained also in the group D_{3d}–though not in D_{3h}–C_{2h} is described as a sub-group of D_{3d}.

Two more complex combinations of axes are those characteristic of the tetrahedron and octahedron, the class of groups being designated T and O. In both, four three-fold axes are directed towards the corners of a cube; T-groups have only certain two-fold axes in addition, whilst O have four-fold axes also. Representative molecules in these groups usually possess planes of symmetry in addition to the axes. The methane and carbon tetrachloride molecules, for example, belong to the point group T_d, the possible operations in which can be seen in the appropriate table in Appendix 1.

Linear molecules fall into two point groups. If they lack a centre of symmetry, like HCN, the elements are the axis of infinite order, C_∞, and an infinite number of vertical symmetry planes, σ_v: the group is symbolised as $C_{\infty v}$. If they possess a centre of symmetry, like acetylene, H—C≡C—H, they possess the additional elements σ_h, and i, and an infinite number of two-fold axes perpendicular to C_∞, as well as S_∞ coincident with C_∞. The point group is therefore $D_{\infty h}$.

2.8 The Algebra of Symmetry Operations

In the next two sections we will develop our ideas of symmetry operations in a more quantitative way. The purpose of this somewhat formal exercise may not be obvious at this stage, but when we come to consider the details of molecular spectroscopy in later chapters, the power of simple group theoretical methods will become apparent. This will be especially so when we discuss the *selection rules* which govern the absorption and emission of radiation by molecular systems.

We have seen that the application of a single symmetry operation delivers a molecule into a new configuration equivalent to its original one; if the same operation is performed twice, or if two different operations are performed successively, the initial and final configurations are still equivalent.

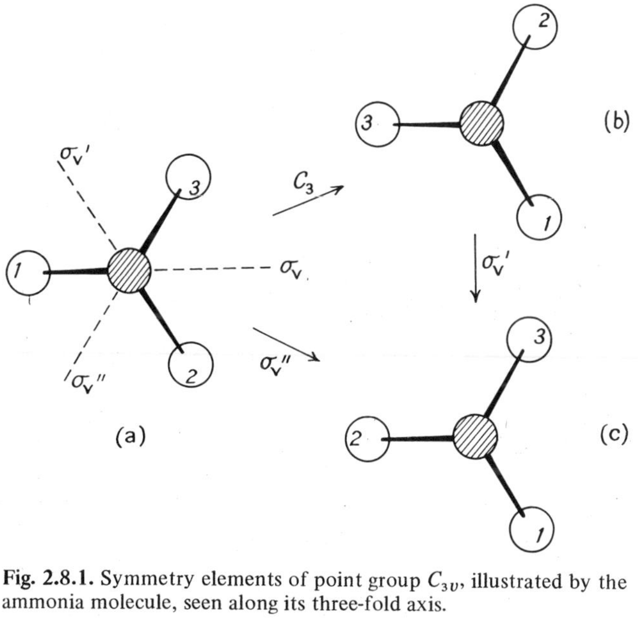

Fig. 2.8.1. Symmetry elements of point group C_{3v}, illustrated by the ammonia molecule, seen along its three-fold axis.

The symmetry elements present in the ammonia molecule, NH_3, are the three-fold axis C_3, and three (vertical) planes of symmetry σ_v, σ_v', and σ_v''. Fig. 2.8.1 represents a view of this pyramidal molecule along its C_3-axis. When we apply the operation C_3 *counterclockwise* (conventionally the positive direction of rotation), we pass from configuration (*a*) to (*b*) by the 120° rotation.

Before we can write down the result of a second operation we need a convention as to whether the symmetry elements remain fixed in space or move with the molecule during an operation. Our convention will be that the elements remain *fixed in space*; they retain the orientation shown at (*a*) regardless of how many operations are subsequently performed on the molecule itself.

If we follow the operation C_3 by σ_v', the molecule is carried into configuration (*c*), which might have been reached by a simple application of σ_v''. The operation σ_v'' is therefore called the *product* of C_3 and σ_v'. It is important to realise that the order of the operations is significant; for had we taken σ_v' first and followed it with C_3, the consequent configuration would have been that produced by a single application of σ_v. *Symmetry operations*, like algebraic, *do not necessarily commute*. Symbolically we write,

$$\sigma_v' \times C_3 = \sigma_v'' \qquad (1)$$

and

$$C_3 \times \sigma_v' = \sigma_v, \qquad (2)$$

so that $\sigma_v' \times C_3 \neq C_3 \times \sigma_v'$. The convention is the same as that for algebraic operations generally: the operation applied first is written to the right. The factors in such a product are not of course limited to two. In a sequence of three operations the associative law of multiplication holds; *e.g.*

$$(C_3 \times \sigma_v') \times C_3 = C_3 \times (\sigma_v' \times C_3), \qquad (3)$$

as the reader should confirm.

It may happen that a sequence of operations carries the molecule into a configuration identical with—not merely equivalent to—the original. For example, this occurs when the operation of reflexion about any one symmetry plane is performed twice. The result is that of the identity operation, *I*; and therefore

$$\sigma_v \times \sigma_v = \sigma_v' \times \sigma_v' = \sigma_v'' \times \sigma_v'' = I. \qquad (4)$$

After a single application of the operation C_3, however, two further applications are needed to restore the original configuration. Hence

$$C_3 \times C_3 \times C_3 = C_3^2 \times C_3 = C_3 \times C_3^2 = I. \qquad (5)$$

It can thus be established that, for every operation that can be performed on the molecule, there is some other which returns the molecule to its original configuration. This other operation is called the *inverse* of the first. The inverse of C_3 is thus C_3^2; that of σ_v is σ_v itself.

The set of operations that can be performed on a molecule, or on any single object, constitutes a *group* in the mathematical sense. In mathematics a set of operations, which need not be symmetry operations, comprise a group under the following conditions: (*i*) the product of two or more operations is equivalent to a single operation which also belongs to the set; (*ii*) the operations obey the associative law of multiplication; (*iii*) the set includes the identity operation; and (*iv*) the inverse of every operation is also a member of the set. Evidently the set of operations, I, C_3, $C_3^{\,2}$, σ_v, σ_v', and σ_v'', possible with ammonia satisfy these conditions. It comprises the group C_{3v}.

A group is said to be *commutative* if all pairs of operations, A, B, commute:

$$\text{i.e. } A \times B = B \times A. \tag{6}$$

The group C_{2v}, to which the water molecule belongs, is an example of this type.

Operations A and B are *conjugate* and belong to the same *class* if

$$X \times A = B \times X \tag{7}$$

for all operations X of the group. The identity, I, and the inversion, i, are always in classes by themselves since they commute with all other symmetry operations,

$$\text{e.g. } X \times I = I \times X;\ X \times i = i \times X. \tag{8}$$

Generally speaking the individual classes contain geometrically similar operations.

2.9 The Representation of a Group

A group is defined by its multiplication table. Let us consider the group C_{2v}, to which many simple molecules belong. The symmetry elements are I, C_2, σ_v and σ_v'. The products of all pairs of operations can be ascertained by the method illustrated for ammonia in Fig. 2.8.1; they are set out in Table 2.9.1. The symbols in the body of the table represent the product of the operation in the appropriate part of the left-hand column with that in the top row. In this simple group every operation is its own inverse, with the result that the diagonal elements are all I; and the reader should have no difficulty in verifying the other entries. Whenever, as here, the order of the principal axis does not exceed 2, the operations commute, and the array of products is symmetrical about its diagonal.

Table 2.9.1 Multiplication Table for the Group C_{2v}

	I	C_2	σ_v	σ_v'
I	I	C_2	σ_v	σ_v'
C_2	C_2	I	σ_v'	σ_v
σ_v	σ_v	σ_v'	I	C_2
σ_v'	σ_v'	σ_v	C_2	I

When a symmetry operation is applied to a molecule the coordinates (x, y, z) of a given point in the molecule with respect to a space-fixed set of axes will be altered to new values (x', y', z'). We may write this change,

$$\begin{pmatrix} x' \\ y' \\ z' \end{pmatrix} = A \times \begin{pmatrix} x \\ y \\ z \end{pmatrix}. \tag{1}$$

Here A will be a 3×3 matrix the form of which depends on the particular operation it *represents*. We may thus represent the identity operation, I, by

$$I = \begin{pmatrix} 1 & 0 & 0 \\ 0 & 1 & 0 \\ 0 & 0 & 1 \end{pmatrix}. \tag{2}$$

A general rotation about the z axis by an angle α in a positive sense $C(\alpha)$ takes the form,

$$C(\alpha) = \begin{pmatrix} \cos\alpha & -\sin\alpha & 0 \\ \sin\alpha & \cos\alpha & 0 \\ 0 & 0 & 1 \end{pmatrix}. \tag{3}$$

Other examples are

$$\sigma_h = \begin{pmatrix} 1 & 0 & 0 \\ 0 & 1 & 0 \\ 0 & 0 & -1 \end{pmatrix}; \tag{4}$$

$$S(\alpha) = \begin{pmatrix} \cos\alpha & -\sin\alpha & 0 \\ \sin\alpha & \cos\alpha & 0 \\ 0 & 0 & -1 \end{pmatrix}; \tag{5}$$

$$i = \begin{pmatrix} -1 & 0 & 0 \\ 0 & -1 & 0 \\ 0 & 0 & -1 \end{pmatrix}. \tag{6}$$

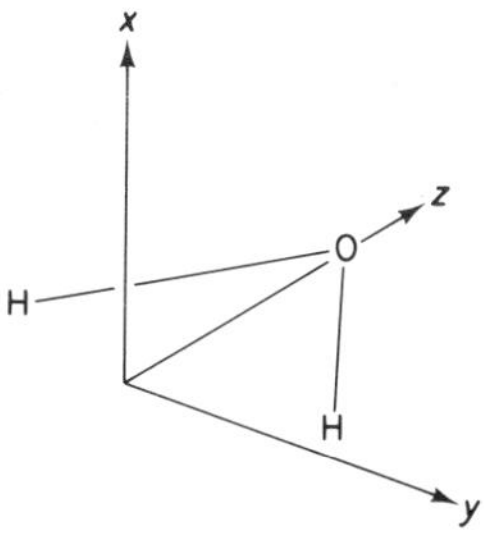

Fig. 2.9.1. Cartesian axis system for the water molecule.

Fig. 2.9.1 shows a water molecule (point group C_{2v}) and a space-fixed set of axes. Let us represent the symmetry operations for this case in the manner just outlined. We obtain (Table 2.9.2) a *representation*, Γ, of the group C_{2v} which describes how the coordinates of a point in the water molecule are changed by the symmetry operations. It is neither a unique nor the simplest representation of C_{2v}.

Table 2.9.2

C_{2v}	I	C_2	σ_v	σ_v'
Γ	$\begin{pmatrix} 1 & 0 & 0 \\ 0 & 1 & 0 \\ 0 & 0 & 1 \end{pmatrix}$	$\begin{pmatrix} -1 & 0 & 0 \\ 0 & -1 & 0 \\ 0 & 0 & 1 \end{pmatrix}$	$\begin{pmatrix} 1 & 0 & 0 \\ 0 & -1 & 0 \\ 0 & 0 & 1 \end{pmatrix}$	$\begin{pmatrix} -1 & 0 & 0 \\ 0 & 1 & 0 \\ 0 & 0 & 1 \end{pmatrix}$

In this case the matrices are all diagonal and we can represent the behaviour of x, y, z independently as in Table 2.9.3.

Table 2.9.3

C_{2v}	I	C_2	σ_v	σ_v'	
Γ_1	1	−1	1	−1	x
Γ_2	1	−1	−1	1	y
Γ_3	1	1	1	1	z
Γ_4	1	1	−1	−1	

This table has an important property. If we multiply the entry under a particular operation by the entry under the corresponding operation in another representation, and repeat this process for all operations, then the result must also be a representation of the group. A product of two representations of this type is called a *direct product* which we will write $\Gamma_i \times \Gamma_j$. When we take the direct product of Γ_1 with Γ_2 in Table 2.9.3 we immediately generate a new representation of C_{2v}, Γ_4. All other direct products lead to one of the representations, Γ_1, Γ_2, Γ_3, Γ_4, but no new ones, and we have thus found a *complete set* of representations for C_{2v}. The representation, Γ, of Table 2.9.2 is said to be *reducible* in that it can be separated into the simpler representations Γ_1, Γ_2 and Γ_3. These latter representations, together with Γ_4, cannot be further simplified and they are termed the *irreducible representations* or *symmetry species* of C_{2v}.

Not all point groups have such simple irreducible representations. Let us consider the group C_{3v}, to which the ammonia molecule belongs (Fig. 2.9.2).

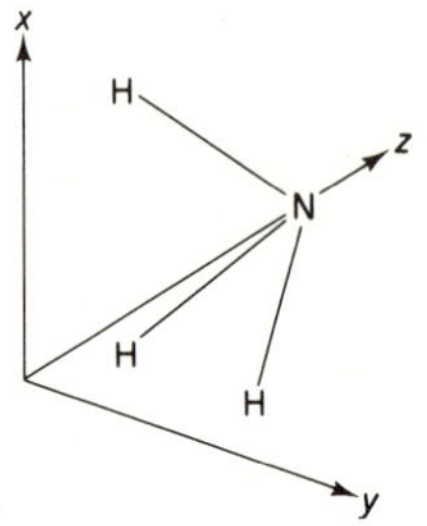

Fig. 2.9.2. Cartesian axis system for the ammonia molecule.

The symmetry operations fall into three classes. The first contains I the second C_3 and $C_3{}^2$, and the third σ_v, σ_v' and σ_v''. For our present purpose we need only consider a single member of each class and we shall take I, C_3 and σ_v. In σ_v the plane of reflection is taken as xz. Let us now write down representations for these operations as we did for the C_{2v} case.

Table 2.9.4

C_{3v}	I	$2C_3$	$3\sigma_v$
Γ	$\begin{pmatrix} 1 & 0 & 0 \\ 0 & 1 & 0 \\ 0 & 0 & 1 \end{pmatrix}$	$\begin{pmatrix} -\frac{1}{2} & -\frac{\sqrt{3}}{2} & 0 \\ \frac{\sqrt{3}}{2} & -\frac{1}{2} & 0 \\ 0 & 0 & 1 \end{pmatrix}$	$\begin{pmatrix} 1 & 0 & 0 \\ 0 & -1 & 0 \\ 0 & 0 & 1 \end{pmatrix}$

For this case, with the axes chosen as in Fig. 2.9.2 we note that only two of our 3×3 matrices are diagonal. If we were to take any other choice of axes we would obtain different matrices representing the operations, but they would never be all diagonal for any one choice. However the trace or *character* (the sum of the diagonal elements) of the various matrices representing a particular operation will be invariant to the choice of axes. We use this character to represent our operations. Γ can be broken down into a set of 2×2 matrices and a set of 1×1 matrices as indicated by the broken lines in Table 2.9.4. We will tabulate the characters of these to give us two irreducible representations Γ_1 and Γ_2.

Table 2.9.5

C_{3v}	I	$2C_3$	$3\sigma_v$	
Γ_1	2	−1	0	x,y
Γ_2	1	1	1	z
Γ_3	1	1	−1	
$\Gamma_1 \times \Gamma_1$	4	1	0	

Taking the direct product $\Gamma_1 \times \Gamma_1$ we obtain the result at the foot of Table 2.9.5. This must be a representation of C_{3v} though it is not necessarily an irreducible one. Indeed, if we introduce a new irreducible representation Γ_3 we can express $\Gamma_1 \times \Gamma_1$ as $\Gamma_1 + \Gamma_2 + \Gamma_3$. Γ_1, Γ_2 and Γ_3 form the complete set of irreducible representations for the group C_{3v}. We note that Γ_1 contains the characters of 2×2 matrices and this representation is said to be *doubly degenerate* or of *dimension* two. Γ_2 and Γ_3, like all the irreducible representations of C_{2v}, contain the characters of 1×1 matrices and are said to be *non-degenerate* or of dimension one. Tables such as 2.9.3 and 2.9.5 are called *character tables*.

We are now able to make some general statements about these character tables.

(i) If there is an axis of symmetry higher than 2-fold then one or more of the irreducible representations will be doubly degenerate.

(ii) If there are more than one 3-fold, or higher, non-coincident symmetry axes there will be a triply degenerate irreducible representation.

(iii) There are as many irreducible representations as there are classes of symmetry operations.

(iv) The sum of the squares of the dimensions of the irreducible representations is equal to the number of operations in the group (*e.g.* C_{3v} must contain two one-dimensional and one two-dimensional irreducible representations).

(v) The character is the same for each member of a class; hence we only enter each class once.

Designation of Symmetry Species. Rather than use the non-informative labels Γ_1, Γ_2, etc., it is helpful to designate the symmetry species (irreducible representations) in a more meaningful way. The general rules are:

(i) A indicates symmetry with respect to the axis of highest symmetry (z).

(ii) B indicates antisymmetry with respect to z.

(iii) E indicates a doubly degenerate species.

(iv) F (or T) indicates a triply degenerate species.

(v) Subscripts g and u indicate symmetry and antisymmetry with respect to a centre of symmetry.

(vi) Subscripts *1* and *2* denote symmetry and antisymmetry with respect to a rotation (or rotation-reflexion) axis other than z, or, in some point groups where there is no second axis, with respect to a plane of symmetry.

(vii) Single and double prime superscripts ($'$ and $''$) denote symmetry and anti-symmetry with respect to a plane of symmetry.

The character tables for C_{2v} and C_{3v} using this more descriptive labelling of the symmetry species are reproduced in Table 2.9.6.

Table 2.9.6

C_{2v}	I	C_2	σ_v	σ_v'	
A_1	1	1	1	1	z
A_2	1	1	−1	−1	
B_1	1	−1	1	−1	x
B_2	1	−1	−1	1	y

C_{3v}	I	$2C_3$	$3\sigma_v$	
A_1	1	1	1	z
A_2	1	1	−1	
E	2	−1	0	x,y

The positions of the coordinates x, y, z on the right-hand side of these tables indicate how these properties *transform*. In C_{2v} we say that z transforms as species A_1; in C_{3v} we say x and y together transform as species E. The character tables for several common point groups are set out in Appendix I. The transformation properties of several quantities useful in spectroscopic applications are included in these tables.

It is possible to summarise the calculation of the direct product of symmetry species by certain simple rules:

(i) $A \times A = A$ (ii) $1 \times 1 = 1$

$B \times B = A$ $2 \times 2 = 1$

$A \times B = B$ $1 \times 2 = 2$

$A \times E = E$

$B \times E = E$

(iii) $g \times g = g$ (iv) $' \times ' = '$

$u \times u = g$ $'' \times '' = '$

$g \times u = u$ $' \times '' = ''$

We also note that the dimension of a direct product is the product of the dimensions of the two symmetry species involved.

Using these rules it is a simple matter to write down the result of a sequence of direct products. For example, suppose we need to evaluate the product $A_1 \times A_2 \times B_1 \times B_2$. Multiplying from the right we have: $B_1 \times B_2 = A_2$; $A_2 \times A_2 = A_1$; $A_1 \times A_1 = A_1$ which is the required product. For certain point groups there are exceptions and extensions to the general notation and direct product rules outlined here; these are given in Appendix I.

2.10 Crystallographic Symmetry

After this account of the specialised applications of symmetry theory needed in molecular spectroscopy, we return to three-dimensional symmetry in crystals. Crystallographers now always use Hermann–Mauguin symbols, though a corresponding Schoenflies symbol is often added in parentheses.

In the Hermann–Mauguin system an axis is represented by a digit, 1, 2, 3, 4, or 6, for its order. An inversion axis carries a bar: $\bar{1}$, $\bar{2}$. $\bar{3}$, $\bar{4}$, or $\bar{6}$. A centre of symmetry is represented by its equivalent, $\bar{1}$; but a plane is usually symbolised by m ("mirror"), though $\bar{2}$ would be more logical. When necessary, the same symbol may be used for symmetry operation or element, and for the group. Thus the group with only a single two-fold axis (2, or C_2), and which is represented by C_2 in Schoenflies symbols, appears simple as 2. Most point groups embody several non-trivial operations; and the group symbol is then constructed by combining symbols for the operations. Usually a shortened symbol is used; in accordance with certain rules, only a minimum number of the operations is explicitly shown, the rest being implied. When there are axes, a digit denoting the order of the principal axis comes first; digits for other axes may follow. When there is a plane of symmetry parallel with and including the axis, m follows the axial symbol; whilst a plane perpendicular to the axis is denoted by $/m$ following the axial symbol.

Many monoclinic crystals belong to the point group C_{2h} of the Schoenflies system. In this group there is a two-fold axis (2) perpendicular to a plane of symmetry ($/m$), and a centre of inversion ($\bar{1}$) where axis and plane intersect. As we have already stressed, any two of these elements imply the presence of the third, so that the group can be adequately described by two. It is in fact represented by $2/m$, which is to be read "two upon em".

(Occasionally the symbol is written $\dfrac{2}{m}$).

Combinations of symmetry operations can be represented algebraically as we described earlier (2.8). For the symmetry operations in the point group $2/m$, we would write:

$$2 \times \bar{1} = m,$$

which means that successive applications of the two operations on the left-hand side are equivalent to a mirror-plane. Similarly for point groups,

$$2 \times \bar{1} = 2/m$$

means that the addition of the centre of symmetry to the point group 2 ($= C_2$) yields the group $2/m$ ($= C_{2h}$).

The point group C_{2v}, with a plane parallel to and containing the two-fold axis, necessarily includes a second plane at right-angles to the first. The Hermann–Mauguin symbol is therefore $2m$, or its equivalent mm.

There are only ten possible point-symmetry operations in crystallography: 1, 2, 3, 4, 6, $\bar{1}$, $\bar{2}$, $\bar{3}$, $\bar{4}$, and $\bar{6}$. There are 32 ways of combining these, constituting the *32 crystallographic point-groups*, corresponding– as we have already stated–to the 10 possibilities in two dimensions. What may be described (rather roughly) as the external properties of any crystal must have the symmetry of one of these groups. The most obvious example is the geometry of the external faces–a feature of crystals that has attracted attention since remote times. Actual crystals are of course almost always imperfectly developed in their outward shapes, because growth of the various faces has been more or less hampered by the environment during crystallisation. The angles between corresponding pairs of faces are constant, however, and from measurements of these angles the ideally developed crystal can be reconstructed. It is these ideal forms that are drawn in textbooks of crystallography. The crystal symmetry reveals itself in many other properties besides form: for example, in mechanical properties, in electrical behaviour, in thermal expansion or conductivity, in refractive indices, and in the response to etching. In Table 2.14.1 at the end of this chapter, Hermann–Mauguin symbols for these 32 point groups are correlated with the Schoenflies symbols.

2.11 Repetitive Symmetry in Three Dimensions

We have dealt with the crystallographic point groups at once, because it was convenient to do so as soon as possible after our consideration of molecular point symmetry. But the special features of the former result from the repetitive elements of symmetry in the internal make-up of crystals. A study of repeat-patterns in three dimensions is thus fundamental to an understanding of crystallography.

An infinite, planar array of points, all of them identical in their environment, is known as a plane net, and we saw that five types of such nets are possible (see Fig. 2.3.3). In three dimensions the problem is to combine an infinite number of plane nets together, whilst maintaining the condition that no point can be distinguished from any other in its environment. Obviously the nets must be parallel to one another, and they must also be equidistantly spaced. In 1848 Bravais proved that there are 14 possibilities. They are the *space lattices*. These fourteen are represented in Fig. 2.11.1 in a rudimentary form; and one of them is shown in a more extended form in Fig. 2.11.2. In both figures, it should be emphasised, the pattern is supposed to be repeated indefinitely in all directions. (In an actual crystal, big enough to be visible even under the microscope, the repetition, though not literally infinite, occurs many thousands of times along each line. Points at the surfaces of a real crystal no longer satisfy the requirement of identity in environment; but the properties of these surfaces are well known to be anomalous; and in any case the proportion of points in surface sites is insignificant). It may also be stressed that the array of points constitutes the lattice; the lines connecting the points in Figs. 2.11.1 and 2.11.2 are meant to make the drawing more intelligible; they are not an essential–or even a strictly relevant–part of the lattice.

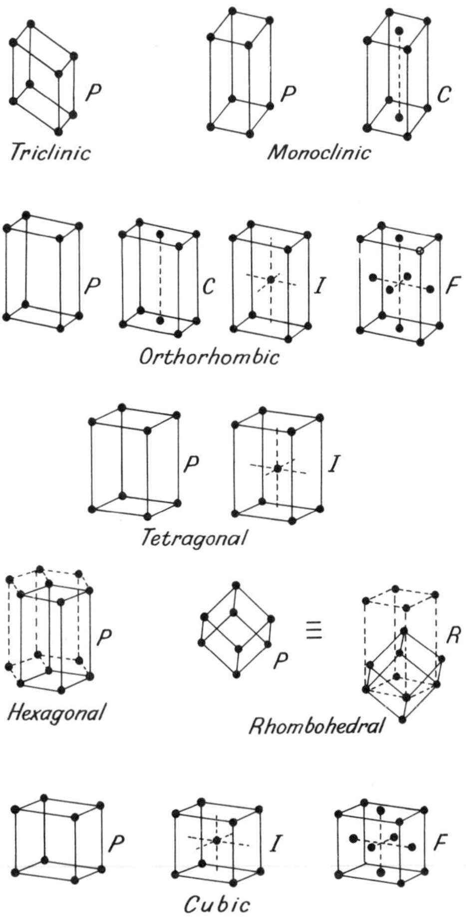

Fig. 2.11.1. The 14 Bravais space lattices.

As in a plane net, the distance from any lattice point to the next is known as a *primitive translation*. At double this distance, in the same direction, a third point must lie; and so on. When the lengths and directions of three independent primitive translations are known, the lattice is defined. Three such translations, all starting from the same point, mark out a *unit cell*. The whole lattice may be regarded as built with such identical unit cells. In Fig. 2.11.2 one possible cell, defined by the translations *a, b,* and *c,* is indicated by the heavier lines. An infinite number of alternative unit cells can be chosen, as was suggested for two-dimensional nets in Fig. 2.3.2. Almost always a particular cell is chosen with two ends in view: (*a*) so that it shall have the minimum volume; and (*b*) so that it shall embody the maximum symmetry of the lattice. Taking the latter aspect first, we can appreciate that, when–in two particular directions–rows of lattice points are precisely, and identically, at right angles to one another, it would be undesirable to select a cell which took no advantage of this orthogonality.

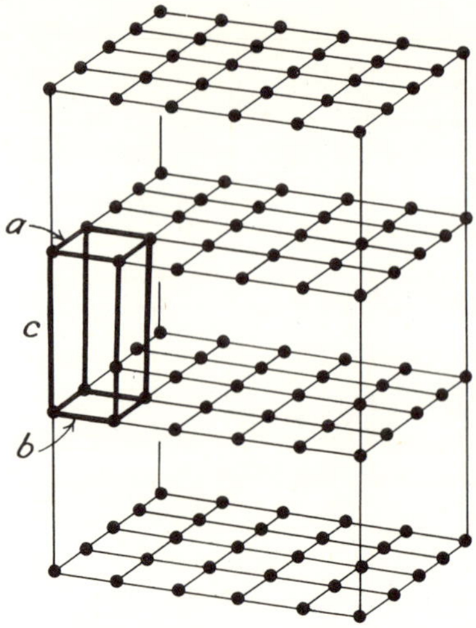

Fig. 2.11.2. More extended representation of a lattice. (The actual points constitute the lattice; the lines drawn to connect some of the points have been added to make the perspective of the diagram clearer, but they are not a necessary part of the lattice.)

As for the former object, the cell demarcated by *a, b,* and *c* is one of the infinite number possessing the minimum volume. Such a cell is called *primitive*, and represented by *P*. Overall a primitive cell comprises one lattice point only. It is true that eight lattice points lie at the corners of the cell as shown; but each corner also appertains to seven other neighbouring cells, so that only one-eighth of each point can be allocated to any one cell; and $8 \times \frac{1}{8} = 1$. (Alternatively the origin of the cell need not be placed at a lattice point; and, when the origin is displaced, one lattice point only will lie wholly within the cell.) As was explained in Section 2.3, for the plane net, orthogonality can sometimes be preserved by selecting a larger, non-primitive cell. This has been done in eight of the lattices represented in Fig. 2.11.1. Four types of non-primitive cells occur: (*i*) with a lattice point at the centres of only one pair of opposite faces; if these be the *A*-faces (100), the cell is said to be *A-centred* and designated by *A*, and correspondingly for *B*- or *C-centring*; (*ii*) with a lattice point at the cell-centre; this is known as *body-centring*, and designated by *I* ("inner"); (*iii*) with lattice points at the centres of all faces; the cell is then said to be *face-centred*, and is designated by *F*; (*iv*) with two lattice points, one-third and two-thirds of the way along one diagonal of an hexagonal cell; such a lattice can alternatively be marked out in terms of a *rhombohedral* cell, as is indicated in the eleventh and twelfth diagrams in Fig. 2.11.1, and is then designated by *R*. Cells of types *A* (*B*, or *C*) and *I* are doubly primitive; they comprise *two* lattice points. Those of type *R*, when referred to hexagonal axes, are triply primitive, and those of type *F* quadruply so.

2.12 Crystal Systems

Crystals can be ordered in *six systems* as indicated in Table 2.12.1. In the earlier literature seven systems were recognised, and some authors still use seven; but it is now general practice to put rhombohedral crystals into the hexagonal system. The systems are commonly defined by reference to the shapes of their unit cells. In all crystals of the cubic (or regular) system, for instance, it is possible to choose a cell of cubic shape, and this is invariably done; its edges are therefore all equal ($a = b = c$), and the angles (α between b and c, β between c and a, and γ between a and b) all right-angles. In the hexagonal system, space is divided into hexagonal blocks of height c, with six sides of equal length, making angles of 120° with one another; the primitive cell is one-third of this block, and consists of a parallelepiped of height c, on a diamond-shaped base having sides $a = b$ at an angle γ of 120°. In the simple hexagonal lattice, successive hexagonal plane nets are arranged with the points of each vertically above those of its neighbour. In the rhombohedral lattice, successive nets are displaced laterally by one-third of the *ab*-diagonal, so that lattice points of only every third net lie vertically above one another.

Table 2.12.1. The Six (or Seven) Crystal Systems

System	*Unit-cell Parameters*	*Lattice Types*	*Minimum Symmetry*	*Numbers of Point Groups*	*Numbers of Space Groups*
Triclinic (Anorthic)	$a \neq b \neq c$; $\alpha \neq \beta \neq \gamma \neq \pi/2$	*P*	1	2	2
Monoclinic	$a \neq b \neq c$; $\alpha = \gamma = \pi/2$; $\beta \neq \pi/2$	*P*, *C* (or *A*)	one 2 (or $\bar{2}$)	3	13
Orthohombic	$a \neq b \neq c$; $\alpha = \beta = \gamma = \pi/2$	*P*, *C* (or *A* or *B*), *I*, *F*	three 2 (or $\bar{2}$)	3	59
Tetragonal	$a = b \neq c$; $\alpha = \beta = \gamma = \pi/2$	*P*, *I*	one 4 (or $\bar{4}$)	7	68
Hexagonal	$a = b \neq c$; $\alpha = \beta = \pi/2$ $\gamma = 2\pi/3$	*P*	one 3 (or $\bar{3}$), or one 6 (or $\bar{6}$)	12	52
Rhombohedral	$a' = b' = c'$; $\alpha' = \beta' = \gamma' \neq \pi/2$	*R*			
Cubic (Isometric)	$a = b = c$; $\alpha = \beta = \gamma = \pi/2$	*P*; *I*, *F*	four 3	5	36
Totals		14		32	230

As stated above, rhombohedral crystals are usually put into the hexagonal system, giving a triply-primitive cell. They may alternatively be referred to a primitive, rhombohedral cell, which has $a' = b' = c'$, and $\alpha' = \beta' = \gamma'$ but $\neq 90°$, and which can thus be regarded as a deformed cube. Whilst a rhombohedral crystal can always be alternatively described in terms of a hexagonal lattice, the converse is not generally true. The other four systems are similarly set out in the table.

In the Table we have represented a right-angle by $\pi/2$ rather than by 90°, and similarly for some other angles. This is to stress the necessity, in a monoclinic crystal for example, for α and γ to be *identically* right-angles, whilst β need not be. A monoclinic crystal might indeed have a β angle whose value was experimentally not distinguishable from 90°; it would not thereby become orthorhombic. For that to be true, the structure would have to possess such internal symmetry as to force the angle to be identically $\pi/2$. Similarly one angle which came out as 90.00° would not make a triclinic crystal monoclinic.

A more fundamental description of the crystal systems is based on their minimum symmetries. Thus a monoclinic crystal always has at least one two-fold axis; this may be either a simple rotation axis (2), or an inversion axis ($\bar{2}$) which is equivalent to a plane (m). This minimum requirement may be exceeded; those monoclinic crystals belonging to the point group $2/m$ have both 2 and m. Likewise an orthorhombic crystal must possess at least three mutually perpendicular two-fold axes, simple (2) or inversion ($\bar{2} = m$). The highest possible symmetry occurs in the most symmetric ("holosymmetric") class of cubic crystals; and it is exemplified in the unit cell itself, as represented by the last-but-two of the lattices shown in Fig. 2.11.1. As the reader should verify for himself, a cube has three four-fold axes, four three-fold, and six two-fold; it also has a centre, and nine planes of symmetry, amounting in all to 23 symmetry elements, and constituting the point group $m3m$. A crystal of rock-salt possesses this symmetry, which is reflected in all its properties. But it by no means follows that every crystal belonging to the cubic system has this full symmetry. Crystals of alum belong to the point group $m3$ for example, and those of hexamethylene tetramine to $\bar{4}3m$. The irreducible minimum, which all cubic crystals must possess, is four three-fold axes, directed along the four diagonals of a cube. These minimum requirements for all the systems are listed in the fourth column of Table 2.12.1. The distribution of the fourteen Bravais lattices is shown in the third column, and that of the 32 point groups in the fifth.

2.13 Space Groups

The classical approach to crystallography depended chiefly on measuring the angles between pairs of faces (goniometry). This was sometimes supplemented by observations of other macroscopic properties, such as those involving the behaviour of light whilst passing through the crystal. By such means it is always possible to assign a given crystal to its proper system, and in principle to its proper point group. (The results of classical crystallography are collected in the five monumental volumes of

Groth's *Chemische Krystallographie*, issued between 1906 and 1919.) However, in repeat-patterns, there are various translational-symmetry operations, besides the ten types of point-symmetry operations. These additional operations increase the number of possibilities from the 32 point groups to the *230 space groups*.

Apart from the operation of simple translation, which gives rise to the lattices, the translational operations in three dimensions are *glide-planes* and *screw-axes*. (In two dimensions both degenerate to the glide described in 2.3.) A glide-plane involves the combined operation of reflexion across a plane followed by translation parallel to the plane and through a distance equal to a simple fraction of a primitive translation–nearly always one-half. Like all translational operations, this is repetitive: given a single object and the glide-plane, we must generate an infinite array of similar objects. The glide is symbolised by *a, b,* or *c* according as the fractional translation is parallel to the respective axes, and by *n* (or in certain circumstances by *d*) when the fractional translation is along a diagonal of the unit cell. A screw-axis involves the combined operation of rotation about an axis of order 2, 3, 4 or 6, and translation, in the direction of the axis, through fractions (which must be multiples of $\frac{1}{2}$, $\frac{1}{3}$, $\frac{1}{4}$, or $\frac{1}{6}$ respectively) of the primitive translation. The operation is symbolised as 2_1, for example, where the principal figure gives the axial order, and the subscript the fractional translation (here $= \frac{1}{2}$). The symbol is read "two-one", or "two-sub-one". We may note that the operations 3_1 and 3_2 differ only in that they correspond to right- and left-handed screws. (The operation of a screw-axis is sketched in Fig. 9.5.1.)

These two types of translational symmetry elements may be present in the internal structure of a crystal, but they produce no effects in the external properties, and hence could not be detected by classical means. As will be explained in Chapter 9, it was only with the development of X-ray diffraction after 1913 that the consideration of glide-planes and screw-axes moved from the realm of the purely academic. Nevertheless, by 1894–a date even before the discovery of X-rays themselves–the possibilities of translational symmetry had been fully worked out by three independent workers, Federov, Schoenflies, and Barlow. There can be few better examples of the unforeseen applicability of work apparently of purely theoretical interest.

When the translational symmetry operations are added to the 10 crystallographic point-symmetry operations, the number of self-consistent groups rises to 230–the *space groups*. Just as every repetitive pattern in two dimensions must possess the symmetry of one of the 17 plane groups represented in Fig. 2.3.1, so every repetitive three-dimensional pattern must possess the symmetry of one of these space groups. In particular this is true of any crystal structure. The assignment of a crystal to its proper space group is an essential, early step in the study of its structure. As such the problem will be discussed in Chapter 9. The numbers of space groups within each system are recorded in the last column of Table 2.12.1.

Space groups were originally symbolised by the Schoenflies symbol for the corresponding point group, with a numerical superscript–*e.g.* C_{2h}^5 for one of the monoclinic space groups. Crystallographers, following a recommendation of their International Union, now use symbols based on the Hermann–Mauguin system, though the Schoenflies symbol is sometimes added as well. In a book intended for the non-specialist, it is unnecessary to detail the rules for representing space groups. They are developed in the introduction to the *International Tables for X-Ray Crystallography* (1952), in Buerger's *Elementary Crystallography*, and in less detail in various other works dealing with X-ray crystallography. It will suffice to discuss one representative space group as an illustration.

We shall select $P2_1/a$ (C_{2h}^5). The P in the Hermann–Mauguin symbol indicates a

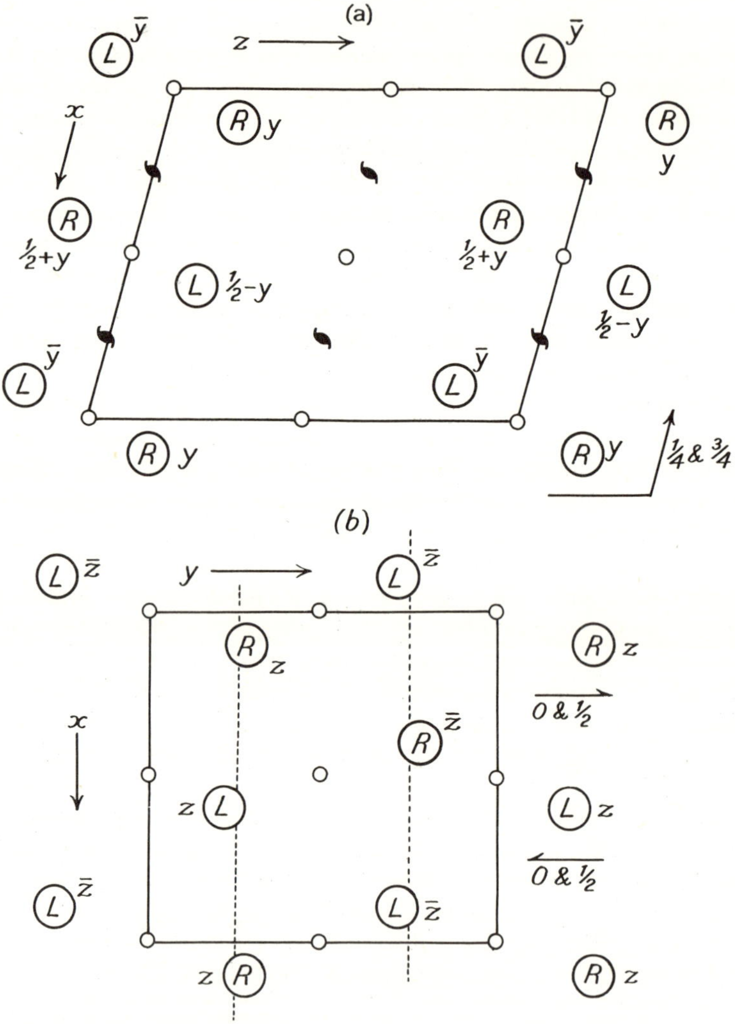

Fig. 2.13.1. The space group $P2_1/a$ shown (*a*) in projection along the *b*-axis, and (*b*) in projection along the *c*-axis, this latter being the projection usually adopted in *International Tables*.

primitive lattice; the presence of the symbol for only one two-fold axis implies the monoclinic system; the subscript indicates a two-fold screw-axis; and the rest of the symbol indicates that there is an *a*-glide-plane perpendicular to the screw-axis. In the convention adopted by nearly all practising crystallographers and mineralogists, the unique two-fold axis in the monoclinic system is taken to be parallel to the *b*-axis. (This is admittedly illogical, since the unique axis in the tetragonal and hexagonal systems is *c*.) The space group $P2_1/a$ includes other symmetry elements besides those shown explicitly, but they are all implicit in the symbol. They are represented in Fig. 2.13.1, which shows one unit cell plus some of its immediate environment, and in two alternative projections: (*a*) as seen along the *b*-axis, and (*b*) as seen along the *c*-axis—the usual aspect shown in the *International Tables*.

A number of centres of symmetry occur, and their projected positions are indicated by small circles. When there are centres, it is always convenient to choose one of them as the origin of coordinates, at $x = y = z = 0$. Another will then be located at $x = \frac{1}{2}a, y = 0, z = 0$ (which may be shortened to $\frac{1}{2}00$); and the remainder up to the limits of the cell shown, at $00\frac{1}{2}$, $\frac{1}{2}0\frac{1}{2}$, $0\frac{1}{2}0$, $\frac{1}{2}\frac{1}{2}0$, $0\frac{1}{2}\frac{1}{2}$, and $\frac{1}{2}\frac{1}{2}\frac{1}{2}$. These centres always occur in pairs, superposed in any axial projection; apart from this superposition, they are all shown in Fig. 2.13.1. The screw-axes are represented by the conventional sign ⬬ when they are normal to the plane of projection, and by the half-barbed arrow when parallel to it; they lie along the *y*-direction at $\frac{1}{4}y0$, $\frac{1}{4}y\frac{1}{2}$, $\frac{3}{4}y0$, and $\frac{3}{4}y\frac{1}{2}$. The glide-planes are at $y = \frac{1}{4}b$ and $\frac{3}{4}b$, the amount of the glide being $a/2$; when they lie parallel to the plane of projection, as in (*a*), they are indicated by the bent arrow at the lower, right-hand corner of the diagram, and, when they are perpendicular to the plane of projection, by the dotted lines appearing in (*b*). This collection of symmetry operations is repeated throughout space by the primitive translations; for instance, the extra centres shown at 100, 010, and 001 arise in this way.

If now any object, which in a crystal would be an atom or a group of atoms, and which for generality may be supposed asymmetric and represented by Ⓡ, is placed in this environment, the object—or its enantiomorph Ⓛ—is reproduced by the symmetry operations as is shown in the figure. For instance, the centre at 000 converts Ⓡ at any position xyz into Ⓛ at $-x, -y, -z$ (usually written $\bar{x}\bar{y}\bar{z}$), which appears at the upper left-hand corner of the diagram, outside the unit cell frame. (In Fig. 2.13.1, x has in fact been put at 0·10, y at 0·28, and z at 0·16.) By the primitive translations, Ⓛ also appears within the frame at $(1 - x)\bar{y}(1 - z)$, etc. The screw-axis at $\frac{1}{4}y\frac{1}{2}$ reproduces the original Ⓡ—again as Ⓡ—at $(\frac{1}{2} - x)(\frac{1}{2} + y)(1 - z)$, whilst the axis at $\frac{1}{4}y0$ gives Ⓡ at $(\frac{1}{2} - x)(\frac{1}{2} + y)\bar{z}$. These two Ⓡ's are of course directly related by the *c*-translation. The glide-plane at $y = \frac{1}{4}$, in a double operation, converts the Ⓡ at xyz, *via* Ⓛ at $x(\frac{1}{2} - y)z$, into Ⓛ shown at $(\frac{1}{2} + x)(\frac{1}{2} - y)z$.

This (L) is also related to (R) at $(\frac{1}{2}-x)(\frac{1}{2}+y)(1-z)$ by the centre at $\frac{1}{2}0\frac{1}{2}$. In this way equal numbers of (R)'s and (L)'s are reproduced through space. The scheme is summed up by the four sets of related coordinates: xyz, $\bar{x}\bar{y}\bar{z}$, $(\frac{1}{2}-x)(\frac{1}{2}+y)\bar{z}$, $(\frac{1}{2}+x)(\frac{1}{2}-y)z$. (R), or (L), may be called the *asymmetric unit*; and within the limits of any unit cell, four such units occur, two (R) and two (L). In this sense the cell is of fourth order. If the cell contains only two atoms of a particular kind, these must lie in *special positions*, on a symmetry element; in fact they must occupy centres of symmetry, 000 and $\frac{1}{2}\frac{1}{2}0$, or the alternative pair, $00\frac{1}{2}$ and $\frac{1}{2}\frac{1}{2}\frac{1}{2}$.

This space group is very common amongst simple organic compounds, and naphthalene, $C_{10}H_8$, is a well known example. A general view of the structure is shown (without its hydrogen atoms) in Fig. 2.13.2. Since there are only *two* $C_{10}H_8$ molecules in the unit cell, the asymmetric unit must be the half-molecule, C_5H_4. When these nine atoms are suitably placed in the unit cell, the whole structure springs into being by the symmetry operations. In particular a centre of symmetry doubles the C_5H_4 unit to yield the complete molecule, which must therefore be centro-symmetric.

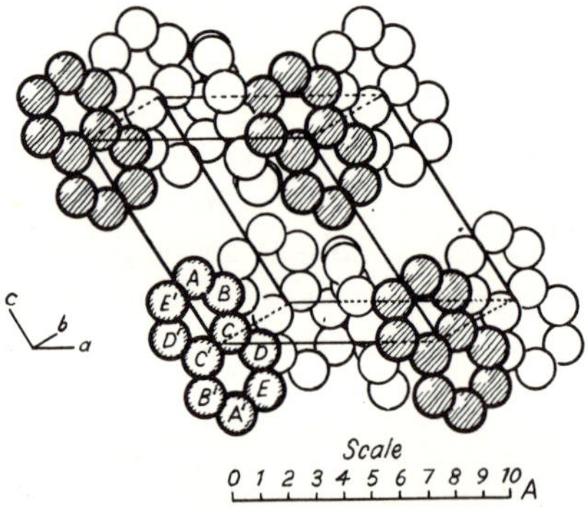

Fig. 2.13.2. Packing and arrangement of molecules of naphthalene, $C_{10}H_8$, in the crystal, illustrating the space group $P2_1/a$.

In the monoclinic system the *b*-axis is unique. The others can be chosen in arbitrary ways. In particular *a* and *c* may be interchanged so that the glide is parallel to *c*; and, when this is done, the space group symbol $P2_1/a$ becomes $P2_1/c$; these are alternative orientations of the same space group. The equivalent point-positions become xyz, $\bar{x}\bar{y}\bar{z}$, $\bar{x}(\frac{1}{2}+y)(\frac{1}{2}-z)$, and $x(\frac{1}{2}-y)(\frac{1}{2}+z)$. When the axes are chosen so that the glide occurs, neither along *a* nor along *c*, but along the *ac*-diagonal, we get the third possible orientation, with symbol $P2_1/n$.

This space group has been chosen as a representative example. Corresponding information about the other 229 groups is contained in the *International Tables*.

2.14 Correlation of Schoenflies and Hermann–Mauguin Symbols

In the following table corresponding symbols are given for the 32 crystallographic point groups. Certain of the H.–M. symbols often appear in equivalent alternative, or fuller, forms; for instance, *mm* can be 2*m* or *mm*2, and 42 can be 422. Where familiar examples can be found, molecules possessing these symmetries in their equilibrium conformations are listed. But with many of the groups no corresponding molecule has yet been discovered. (By assuming that certain molecules adopt particular idealised conformations, some of the gaps may be filled. For example, the molecule of neopentane, $C(CH_3)_4$, with its methyl groups suitably orientated, would be a representative of group 23.)

Table 2.14.1

Schoenflies	*Hermann–Mauguin*		*Schoenflies*	*Hermann–Mauguin*	
C_1	1	*d*-glucose	C_3	3	
C_i	$\bar{1}$	*meso*-BrClHC.CHClBr	$C_{3i}(S_6)$	$\bar{3}$	
			D_3	32	
C_2	2	H_2O_2, *gauche*-$CH_2Cl.CH_2Cl$	C_{3v}	3*m*	NH_3, CH_3Cl
C_s	*m*	HOD, H.COOH	D_{3d}	$\bar{3}m$	staggered-C_2H_6
C_{2h}	2/*m*	*trans*-CHCl=CHCl, *trans*-$CH_2Cl.CH_2Cl$	C_6	6	
D_2	222		C_{3h}	$\bar{6}$	
C_{2v}	*mm*	H_2O, CH_2Cl_2, *cis*-CHCl=CHCl	C_{6h}	6/*m*	
			D_6	62	
D_{2h}	*mmm*	$H_2C{=}CH_2$, naphthalene	C_{6v}	6*m*	
C_4	4		D_{3h}	$\bar{6}m$	BF_3, eclipsed-C_2H_6
S_4	$\bar{4}$		D_{6h}	6/*mmm*	benzene
C_{4h}	4/*m*		*T*	23	
D_4	42		T_h	*m*3	
C_{4v}	4*m*	IF_5	*O*	43	
D_{2d}	$\bar{4}2m$	allene	T_d	$\bar{4}3m$	CH_4, P_4
D_{4h}	4/*mmm*	square complexes, *e.g.* $[NiX_4]^=$	O_h	*m*3*m*	SF_6, octahedral complexes

BIBLIOGRAPHY

Buerger, *Elementary Crystallography*, Wiley & Chapman-Hall, New York and London, 1956.
Herzberg, *Infrared and Raman Spectra*, Van Nostrand, New York, 1945.
International Tables for X-ray Crystallography, Kynoch Press, Birmingham, 1952.
Weyl, *Symmetry*, Princeton Univer. Press, 1952.
Bernal, Hamilton and Ricci, *Symmetry, a Stereoscopic Guide for Chemists*, Freeman, San Francisco, 1972.

Chapter 3

Elementary wave mechanics

Several later chapters of this book are concerned with the derivation of molecular shape and size from observations on spectra. This is a field where interpretation of the experimental results would be almost impossible without the guiding principles of quantum mechanics. Indeed the new quantum mechanics, or *wave mechanics*, was developed to explain the details of atomic spectra; and the position today is still that in its application to spectroscopy the theory is deployed in its most elegant form and enjoys its fullest measure of success. The advance that flowed from the solution of the central problems of spectra left no one in doubt that the correct formula for quantisation had been discovered.

To keep the account of the principles of wave mechanics within bounds, we shall here confine our attention deliberately to one-dimensional systems. This has the disadvantage that we are cut off from consideration of most physically real systems; but, while the mathematical techniques are the same, the manipulation of the algebra is enormously simplified when we deal with only one coordinate instead of three.

3.1 Mathematical Preliminaries

As the concept of an *operator* plays an important part in the mathematics of wave-mechanics, it is convenient to introduce an account of wave-mechanical principles with a summary of some of the properties of operators. We shall consider functions of the variable x and will designate them by $u(x)$ or, more simply, by u: for the present it will be sufficient to visualise $u(x)$ as an elementary function, such as $u(x) = \sin x$ or $u(x) = e^{-\frac{1}{2}ax^2}$, though this is not of course a necessary restriction. The expression

$$\frac{\mathrm{d}}{\mathrm{d}x}u(x) \tag{1}$$

can be thought of as having two parts, the operator $\mathrm{d}/\mathrm{d}x$ and the *operand* $u(x)$. Likewise in the expressions $xu(x)$ and $au(x)$, x and a may be regarded as operators governing the operand $u(x)$. The process of operating with the operators $\mathrm{d}/\mathrm{d}x$, x, and a is, respectively, that of differentiation with respect to x, multiplication by x, and multiplication by a constant a: thus, if $u(x) = e^{-\frac{1}{2}ax^2}$, the result of operating with $\mathrm{d}/\mathrm{d}x$ is $-axe^{-\frac{1}{2}ax^2}$, and so on. The subtleties of the operator concept does not appear fully until we employ operators rather less simple in form than these used for the present illustration.

Operators can be combined into sums and products. The term $(-\mathrm{d}^2/\mathrm{d}x^2 + x^2)$ in the expression $(-\mathrm{d}^2/\mathrm{d}x^2 + x^2)u(x)$ is an operator representing the *sum* of the operators $-\mathrm{d}^2/\mathrm{d}x^2$ and x^2. As we shall see, $(-\mathrm{d}^2/\mathrm{d}x^2 + x^2)$ is an operator with properties of its own, distinct from those of $-\mathrm{d}^2/\mathrm{d}x^2$ and x^2. Similarly, the *product* of the operators x and $\mathrm{d}/\mathrm{d}x$ is a new operator $x(\mathrm{d}/\mathrm{d}x)$, and the product of the operators x and a is a new operator xa. It is important to realise that when a differential operator, say $\mathrm{d}/\mathrm{d}x$, is compounded into a product the order of the factors is significant; for instance $x(\mathrm{d}/\mathrm{d}x)$ is an operator with properties different from $(\mathrm{d}/\mathrm{d}x)x$. If, for example, $u(x) = e^{-\frac{1}{2}ax^2}$ the result of operating with $x(\mathrm{d}/\mathrm{d}x)$ is $-ax^2e^{-\frac{1}{2}ax^2}$, whereas the product of operating with $(\mathrm{d}/\mathrm{d}x)x$ is $(1 - ax^2)e^{-\frac{1}{2}ax^2}$. Note that operation by a product, such as $x(\mathrm{d}/\mathrm{d}x)$, entails consecutive operations; first one operates on $u(x)$ with the operator $\mathrm{d}/\mathrm{d}x$, and then one operates on the result of this operation with the operator x.

Formally, two operators α and β are defined to be equal when, for every operand $u(x)$ on which both can operate,

$$\alpha u(x) = \beta u(x). \tag{2}$$

It will be clear from the example above that the operators $x(\mathrm{d}/\mathrm{d}x)$ and $(\mathrm{d}/\mathrm{d}x)x$ are not equal,

$$x(\mathrm{d}/\mathrm{d}x) \neq (\mathrm{d}/\mathrm{d}x)x. \tag{3}$$

Differential operators do not obey the commutative law of multiplication.

The operator $(\alpha\beta - \beta\alpha)$ is known as the *commutator* of the operators α and β. The commutator of $(\mathrm{d}/\mathrm{d}x)x$ and $x(\mathrm{d}/\mathrm{d}x)$ is the operator

$$\frac{\mathrm{d}}{\mathrm{d}x}x - x\frac{\mathrm{d}}{\mathrm{d}x}. \tag{4}$$

The usual rule for differentiation of a product tells us that

$$\frac{\mathrm{d}}{\mathrm{d}x}x = 1 + x\frac{\mathrm{d}}{\mathrm{d}x}, \tag{5}$$

and hence that

$$\frac{\mathrm{d}}{\mathrm{d}x}x - x\frac{\mathrm{d}}{\mathrm{d}x} = 1. \tag{6}$$

Eqn. (6) states that the commutator of the operators $(\mathrm{d}/\mathrm{d}x)$ and x is the operator $+1$. To regard a numeral, or a constant, as an operator is of course largely a mathematical formality: the convention is adopted so that expressions like (5) and (6) can be described as *operator equations*.

Certain classes of operators and operands are of special interest. When an operator α and an operand $u(x)$ are such that operating with α regenerates the function $u(x)$ multiplied by a constant a,

$$\alpha u(x) = au(x), \tag{7}$$

we say that $u(x)$ is an *eigenfunction* of the operator α belonging to the *eigenvalue* a. In wave mechanics supplementary conditions are imposed upon $u(x)$: it must be finite, continuous, and single-valued throughout the whole range of possible values of x, from $x = +\infty$ to $x = -\infty$. Only those solutions $u(x)$ of (7) which conform to these *boundary conditions* are acceptable eigenfunctions of the operator α.

To develop the concept of an acceptable eigenfunction, let us suppose that α is the operator $-\mathrm{d}^2/\mathrm{d}x^2$. Although the operand $u(x) = e^x$ satisfies eqn. (7),

$$-\frac{\mathrm{d}^2}{\mathrm{d}x^2}e^x = -1.e^x, \tag{8}$$

the function e^x is not an acceptable eigenfunction because it violates the boundary conditions by becoming infinite as $x \to \infty$. On the other hand $u(x) = \sin 3x$ is an acceptable eigenfunction of $-\mathrm{d}^2/\mathrm{d}x^2$, since

$$-\frac{\mathrm{d}^2}{\mathrm{d}x^2}\sin 3x = 9\sin 3x \tag{9}$$

and $\sin 3x$ is a function that satisfies the boundary conditions. The eigenvalue bellonging to the eigenfunction $\sin 3x$ is the number 9. Likewise $\sin 4x$, $\sin 5x$, . . . are also eigenfunctions of $-\mathrm{d}^2/\mathrm{d}x^2$, belonging respectively to the eigenvalues 16, 25, . . . We see that in fact the operator $-\mathrm{d}^2/\mathrm{d}x^2$ possesses a whole spectrum of eigenfunctions and eigenvalues.

It frequently happens that a symbol must be used in two senses, to denote an operator and to denote a variable. For instance, the letter x may represent the operator x or the coordinate x. To comprehend the equations it will be necessary to read into them the correct significance of the symbol, as operator or variable, though it will usually be obvious from the text which is meant.

Compound operators, like $(-\mathrm{d}^2/\mathrm{d}x^2 + x^2)$, will generally be written in parentheses to distinguish them clearly from the operand.

3.2 The Operator $(-\mathrm{d}^2/\mathrm{d}x^2 + x^2)$

To illustrate the methods of operator algebra, let us determine the eigenfunctions and eigenvalues of the operator

$$-\frac{d^2}{dx^2}+x^2, \tag{1}$$

which is of central importance in the wave-mechanical treatment of the linear oscillator. The problem is to discover the functions $u = u(x)$–which must satisfy the standard boundary conditions–and the possible values of the constant a for which the differential equation

$$\left(-\frac{d^2}{dx^2}+x^2\right)u = au, \tag{2}$$

has solutions. We enquire first what form u adopts for large values of x. In (2), x^2 and a are both multipliers of u; consequently a can be neglected in comparison with x^2 in any region where x is large, and eqn. (2) then simplifies to

$$-\frac{d^2u}{dx^2}+x^2u = 0. \tag{3}$$

An approximate solution of (3) is $u = e^{\pm\frac{1}{2}x^2}$; for

$$(d^2/dx^2)e^{\pm\frac{1}{2}x^2} = (x^2 \pm 1)e^{\pm\frac{1}{2}x^2},$$

and the term ± 1 can be neglected when x is large. The solution $e^{\frac{1}{2}x^2}$ violates the boundary conditions by becoming infinite as $x \to \infty$ and is of no further interest. The solution $u = e^{-\frac{1}{2}x^2}$ does satisfy the boundary conditions (see Fig. 3.2.1.), but as yet is only known to be a solution of (2) when x is large. However, by substituting $u = e^{-\frac{1}{2}x^2}$ into (2) we obtain

$$\left(-\frac{d^2}{dx^2}+x^2\right)e^{-\frac{1}{2}x^2} = 1.\,e^{-\frac{1}{2}x^2} \tag{4}$$

so that the function $e^{-\frac{1}{2}x^2}$ is one possible eigenfunction of the operator (1), belonging to the eigenvalue + 1. Likewise if we try the substitution $u = xe^{-\frac{1}{2}x^2}$ we have

$$\left(-\frac{d^2}{dx^2}+x^2\right)xe^{-\frac{1}{2}x^2} = 3.\,xe^{-\frac{1}{2}x^2} \tag{5}$$

and thus $xe^{-\frac{1}{2}x^2}$ is also an eigenfunction of (1), belonging to the eigenvalue + 3. These trial substitutions suggest that the general solution of (2) may have the form

$$u = ve^{-\frac{1}{2}x^2}, \tag{6}$$

where $v = v(x)$ is normally a function of x though it may reduce to a constant—as in (4)—in a special case.

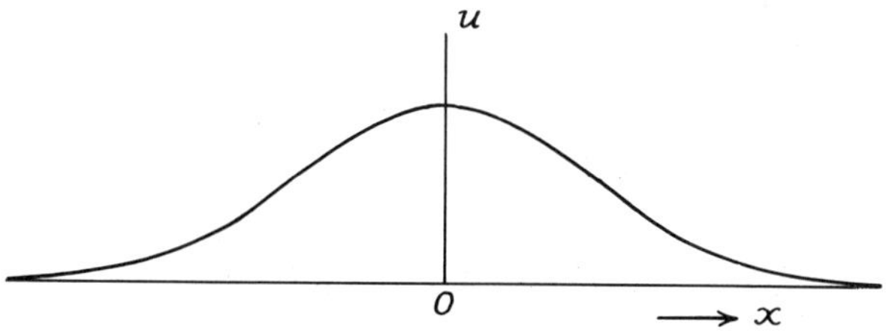

Fig. 3.2.1. Graph of the function $u = e^{-\frac{1}{2}x^2}$. The function is continuous, finite, and single-valued throughout the range $\infty > x > -\infty$.

To discover when (6) is a solution of (2), we substitute the expression (6) into the differential equation (2). The condition for compatibility that emerges is

$$\frac{d^2 v}{dx^2} - 2x\frac{dv}{dx} + (a-1)v = 0. \tag{7}$$

Since this is a general relation it must embrace the two solutions discussed earlier. Thus when $v = x$, eqn. (7) tells us that $a = 3$, that is, $u = xe^{-\frac{1}{2}x^2}$ is an eigenfunction belonging to the eigenvalue + 3 as we have previously established. The equation (7) is similar in form to a classical differential equation known as Hermite's equation,

$$\frac{d^2 v}{dx^2} - 2x\frac{dv}{dx} + 2vv = 0, \tag{8}$$

the solutions of which are shown in Appendix II to be

$$v = (-1)^v e^{x^2}\frac{d^v}{dx^v}e^{-x^2} \tag{9}$$

for all positive integers v, including $v = 0$. The expression on the right-hand side of (9) is usually denoted by the symbol $H_v(x)$ and is described as a Hermite polynomial of degree v. The first six Hermite polynomials, for $v = 0$ to 5, are:

$$\left.\begin{aligned} H_0(x) &= 1 & H_1(x) &= 2x \\ H_2(x) &= 4x^2 - 2 & H_3(x) &= 8x^3 - 12x \\ H_4(x) &= 16x^4 - 48x^2 + 12 & H_5(x) &= 32x^5 - 160x^3 + 120x \end{aligned}\right\}. \tag{10}$$

By substituting $v = H_0(x)$ and $v = H_1(x)$ into (6), we see that (apart from a numerical constant) the first two Hermite polynomials are the factors of $e^{-\frac{1}{2}x^2}$ in the trial solutions (4) and (5) of the differential equation (2). The most general solution of (2), that is, the manifold of acceptable eigenfunctions of the operator $-\mathrm{d}^2/\mathrm{d}x^2 + x^2$, is

$$u_v = N_v H_v(x) e^{-\frac{1}{2}x^2} \tag{11}$$

with N_v a numerical constant.* Graphs of the function u_v with

$$N_v = (2^v v! \sqrt{\pi})^{-\frac{1}{2}}$$

for $v = 0$ to 3 are given in Fig. 3.2.2. Evidently u_v satisfies the standard

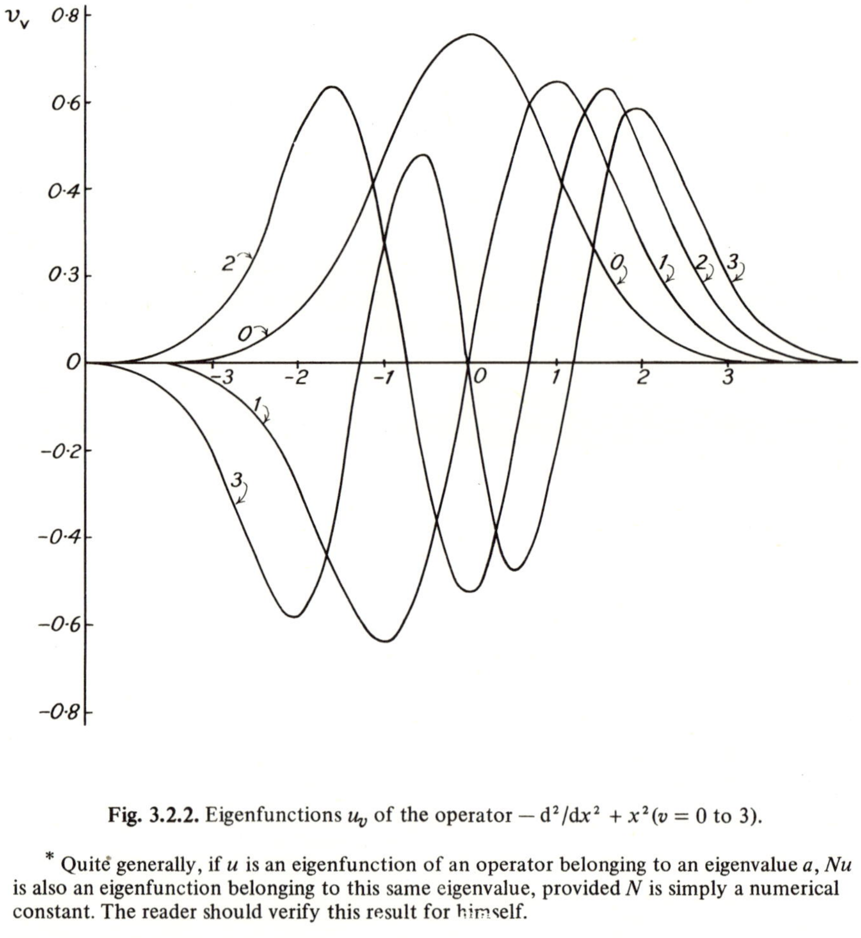

Fig. 3.2.2. Eigenfunctions u_v of the operator $-\mathrm{d}^2/\mathrm{d}x^2 + x^2 (v = 0$ to 3).

* Quite generally, if u is an eigenfunction of an operator belonging to an eigenvalue a, Nu is also an eigenfunction belonging to this same eigenvalue, provided N is simply a numerical constant. The reader should verify this result for himself.

boundary conditions by being everywhere finite, continuous, and single-valued.

The eigenvalues of $-\mathrm{d}^2/\mathrm{d}x^2 + x^2$ are obtained from the condition that (11) is a solution of the differential equation (2) for all positive integral values of v including the value zero. Comparing eqns. (8) and (7), we see that v is related to the eigenvalues a by the expression

$$a = 2v + 1; \qquad v = 0, 1, 2, \ldots \tag{12}$$

The eigenvalues of the operator are therefore the odd positive integers, $a = 1, 3, 5 \ldots$

3.3 Some Remarks on Classical Mechanics

In Newtonian mechanics, a particle of mass m in motion along a coordinate x and subject to a force f obeys the equation

$$f = m\ddot{x}, \tag{1}$$

where $\ddot{x}$ stands for $\mathrm{d}^2x/\mathrm{d}t^2$. For systems of interest to us, the force f to which the particle is subject is the negative value of the gradient of the potential energy $V(x)$,

$$f = -\mathrm{d}V(x)/\mathrm{d}x. \tag{2}$$

Elimination of f between (1) and (2) yields

$$\frac{\mathrm{d}V(x)}{\mathrm{d}x} + m\ddot{x} = 0. \tag{3}$$

Eqn. (3) can be integrated directly by conversion to the variable $\dot{x} = \mathrm{d}x/\mathrm{d}t$,

$$\mathrm{d}V(x) + m\dot{x}\,\mathrm{d}\dot{x} = 0, \tag{4}$$

which integrates to

$$V(x) + \tfrac{1}{2}m\dot{x}^2 = E. \tag{5}$$

E in (5) appears as the constant of integration; it is the sum of the kinetic energy $T = \frac{1}{2}m\dot{x}^2$ and the potential energy $V(x)$, and is known as the *total* or *Hamiltonian* energy of the system.

For a linear harmonic oscillator (Fig. 3.3.1) in classical mechanics,

$$V(x) = \tfrac{1}{2}Fx^2, \tag{6}$$

in which F is known as the *force-constant* of the oscillator, and

$$T = \tfrac{1}{2}m\dot{x}^2, \tag{7}$$

and so the Hamiltonian energy E is given by

$$E = T + V = \tfrac{1}{2}m\dot{x}^2 + \tfrac{1}{2}Fx^2. \tag{8}$$

Once the oscillator is set in motion it continues to vibrate with the total energy E, provided there is no exchange of energy with the surroundings. The condition that there shall be no loss or gain of energy is, of course, implicit in (5) and (8) which show that E is a *constant of motion*. The *frequency* of the motion can be obtained directly from the equation of motion in the differential form (3). Since, for a simple oscillator $\mathrm{d}V(x)/\mathrm{d}x = Fx$, (3) becomes

$$m\ddot{x} + Fx = 0, \tag{9}$$

a possible solution of which is

$$x = A\cos(\sqrt{\lambda}t + \zeta), \tag{10}$$

whence

$$\ddot{x} = -\lambda x. \tag{11}$$

λ, the square of the angular frequency of vibration, is found by substituting (11) into (9),

$$(F - m\lambda)x = 0 \tag{12}$$

or

$$\lambda = 4\pi^2\nu^2 = F/m. \tag{13}$$

ν is called the *classical frequency* of the oscillator.

The coordinate x, the velocity $\dot{x} = \mathrm{d}x/\mathrm{d}t$, the momentum $p = m\dot{x}$, the kinetic energy $T = \tfrac{1}{2}m\dot{x}^2$, *etc*., of any dynamical system are described as *dynamic variables*. It is important to realise that the Hamiltonian energy, besides being a constant of motion, is also a dynamical variable; for although a dynamical system possessing the Hamiltonian energy E maintains this energy indefinitely, there is an infinite range of values to which E may be adjusted initially. When we refer to the Hamiltonian energy as a dynamical variable we shall use the symbol H, to distinguish the meaning clearly from the numerical magnitude of the energy which we shall continue to designate by E. The expression for the Hamiltonian energy as a dynamical variable is that given by (5) with H substituted for E,

$$H = \tfrac{1}{2}m\dot{x}^2 + V(x). \tag{14}$$

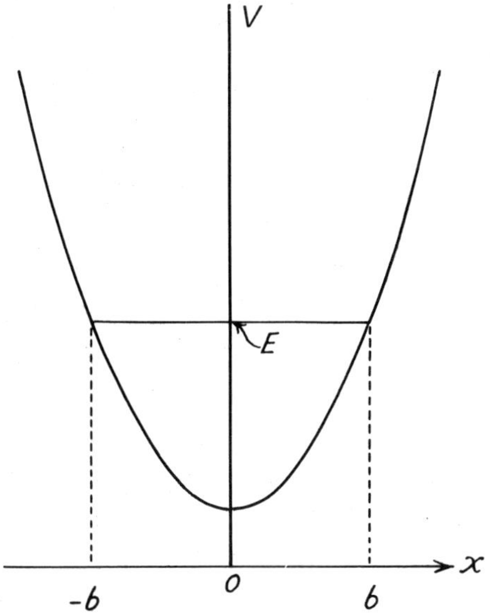

Fig. 3.3.1. Energy diagram for the harmonic oscillator according to classical mechanics. The division of the total energy between kinetic and potential energies varies with the phase of the motion . At the mid-point ($x = 0$) the energy is wholly kinetic; at the turning-points ($x = \pm\, b$) it is wholly potential energy.

Further, we shall often write $\frac{1}{2}m\dot{x}^2 = p^2/2m$, where the *momentum* $p = m\dot{x}$, whence

$$H = p^2/2m + V(x). \tag{15}$$

Differentiation of (15) with respect to p yields

$$\frac{\partial H}{\partial p} = \frac{p}{m} = \dot{x}, \tag{16}$$

and with respect to x,

$$\frac{\partial H}{\partial x} = \frac{\mathrm{d}V(x)}{\mathrm{d}x} = -f. \tag{17}$$

But, according to (1), $f = m\ddot{x} = \dot{p}$, and so

$$\frac{\partial H}{\partial x} = -\dot{p}. \tag{18}$$

Eqns. (16) and (18) are *Hamilton's canonical equations*. In general, two dynamical variables q and s are said to be *canonically conjugate* if

$$\frac{\partial H}{\partial q} = \dot{s}\,; \quad \text{and} \quad \frac{\partial H}{\partial s} = -\dot{q}. \tag{19}$$

3.4 Schroedinger Operators

We shall introduce the system of mechanics known as *Schroedinger mechanics*, or *wave mechanics*, through a series of postulates which, like Newton's laws of motion, are not proved; the reader is asked to take them for granted. The postulates as stated here are adequate for an elementary treatment.

Postulate I. To every dynamical variable q there must be assigned an operator q. The physical properties of a dynamical variable are deducible from the mathematical properties of the operator assigned to the variable. In particular, the possible results of an exact experimental measurement of a dynamical variable q are the eigenvalues of the operator q, and conversely.

The choice of operator to be associated with a dynamical variable requires a second postulate which introduces Planck's constant, and is known as the *quantum condition*.

Postulate II. The operators q and r associated with two canonically conjugate dynamical variables q and r must satisfy the equation

$$qr - rq = ih/2\pi. \tag{1}$$

In (1), h is Planck's constant, and $i^2 = -1$. The left-hand side of the equation represents the commutator of the operators q and r (Section 3.1).

Special interest attaches to the canonically conjugate variables x and p which characterise a one-dimensional dynamical system. If we assign the operator x to the variable x, and the operator $-(ih/2\pi)\mathrm{d}/\mathrm{d}x$ to the variable p, it is evident that we have a solution of (1), for then,

$$\begin{aligned} xp - px &= x\left[-(ih/2\pi)\frac{\mathrm{d}}{\mathrm{d}x}\right] - \left[-(ih/2\pi)\frac{\mathrm{d}}{\mathrm{d}x}\right]x \\ &= (ih/2\pi)\left(\frac{\mathrm{d}}{\mathrm{d}x}x - x\frac{\mathrm{d}}{\mathrm{d}x}\right), \end{aligned} \tag{2}$$

and it has already been shown (eqn. 3.1.6) that the operator $\frac{\mathrm{d}}{\mathrm{d}x}x - x\frac{\mathrm{d}}{\mathrm{d}x}$ equals the operator $+1$. In the Schroedinger method, therefore, the operator x is associated with the variable x, and the operator $-(ih/2\pi)\mathrm{d}/\mathrm{d}x$ with the momentum p (along x).

Schroedinger operators assigned to other dynamical variables of a system are derived from the basic x and p operators by ordinary processes of algebra. Thus the operator assigned to the variable x being x, that assigned to x^2 is the operator x^2; likewise the operator associated with the variable p^2 is $[-(ih/2\pi)\mathrm{d}/\mathrm{d}x]^2 = -(h^2/4$ $-(h^2/4\pi^2)\mathrm{d}^2/\mathrm{d}x^2$, and so forth. Some of our conclusions up to this point are summarised in Table 3.4.1.

Table 3.4.1. Schroedinger Operators for a One-dimensional System

Dynamical variable	Operator
x	x
x^2	x^2
$p = m\dot{x}$	$-(ih/2\pi)\mathrm{d}/\mathrm{d}x$
p^2	$-(h^2/4\pi^2)\mathrm{d}^2/\mathrm{d}x^2$
H	$-(h^2/8\pi^2 m)\mathrm{d}^2/\mathrm{d}x^2 + V(x)$
ϕ	ϕ
P	$-(ih/2\pi)\mathrm{d}/\mathrm{d}\phi$

The operator H assigned to the Hamiltonian variable H is of central importance in the Schroedinger scheme. The classical expression for the Hamiltonian energy is that given by eqn. (3.3.15). In Schroedinger mechanics, this equation,

$$H = p^2/2m + V(x), \tag{3}$$

is an operator equation which may be expanded by substituting in the right-hand side the operators p and $V(x)$ expressed in terms of the operator x. The operator $V(x)$ changes from one dynamical system to another and for the moment will be left unspecified. Introducing $p^2 = -(h^2/4\pi^2)\mathrm{d}^2/\mathrm{d}x^2$, however, (3) becomes

$$H = -(h^2/8\pi^2 m)\mathrm{d}^2/\mathrm{d}x^2 + V(x). \tag{4}$$

Eqn. (4) is the general form of the expression for the Hamiltonian operator of a system possessing one degree of freedom.

According to postulate I the results of an exact experimental measurement of the energy of a system whose classical Hamiltonian energy is given by (3.3.15) are the eigenvalues of the operator H in (4); hence we may write symbolically

$$H\psi = E\psi. \tag{5}$$

This is *Schroedinger's first equation*, sometimes called *Schroedinger's amplitude equation*. The operands ψ are eigenfunctions of the operator H and are subject to the boundary conditions outlined in Section 3.1; that is, they must be finite, continuous, and single-valued. In future we shall refer to eigenfunctions of the Hamiltonian operator as *amplitude wavefunctions*.

3.5 The Linear Harmonic Oscillator

For the linear oscillator the classical potential energy, eqn. (3.3.6), is $V(x) = \frac{1}{2}Fx^2$, and the Schroedinger operator $V(x)$ is therefore the operator $\frac{1}{2}Fx^2$. The Hamiltonian operator for the oscillator is obtained by substituting the operator $\frac{1}{2}Fx^2$ for $V(x)$ in eqn. (3.4.4), whence

$$H = -\frac{h^2}{8\pi^2 m}\frac{\mathrm{d}^2}{\mathrm{d}x^2} + \frac{F}{2}x^2. \tag{1}$$

The amplitude wavefunctions ψ and permitted energy levels E of the oscillator are those which satisfy the Schroedinger amplitude equation (3.4.5) with H given by (1); namely, the equation

$$\left(-\frac{h^2}{8\pi^2 m}\frac{\mathrm{d}^2}{\mathrm{d}x^2} + \frac{F}{2}x^2\right)\psi = E\psi. \tag{2}$$

The algebra of (2) is to a large extent already solved in section 3.2. To take advantage of the results obtained there, we make the substitutions

$$\eta = x/r \tag{3}$$

and

$$E = \frac{ah}{4\pi}\sqrt{\frac{F}{m}} \tag{4}$$

where

$$r = \sqrt[4]{h^2/4\pi^2 Fm}. \tag{5}$$

Here, as in section 3.3, m represents the mass and F the force-constant of the oscillator. Introducing the substitutions (3)–(5) into the Schroedinger equation (2) yields, after some simplification,

$$\left(-\frac{\mathrm{d}^2}{\mathrm{d}\eta^2} + \eta^2\right)\psi = a\psi, \tag{6}$$

which in form is identical with eqn. (3.2.2). Evidently the eigenvalues of (6) are the same as those of (3.2.2), that is, they are the odd positive integers 1, 3, 5 . . . Hence

$$a = (2v + 1); \qquad v = 0, 1, 2 \ldots \tag{7}$$

If we now substitute this result into eqn. (4) we obtain the eigenvalues E_v of the Hamiltonian operator,

$$E_v = (v+\tfrac{1}{2})\frac{h}{2\pi}\sqrt{\frac{F}{m}}, \tag{8}$$

the subscript v being attached to E_v as a reminder that (8) specifies a manifold of energy levels defined by the running number $v = 0, 1, 2 \ldots$ The expression (8) for E_v can be further reduced, for in section 3.3 it was shown that $\lambda^{\frac{1}{2}}$, the angular frequency of the oscillator in classical mechanics, is related to F and m by eqn. (3.3.13), $\lambda = F/m$. Denoting by ν the classical absolute frequency of the oscillator, we have

$$\nu = \frac{1}{2\pi}\lambda^{\frac{1}{2}} = \frac{1}{2\pi}\sqrt{\frac{F}{m}} \tag{9}$$

and

$$E_v = (v+\tfrac{1}{2})h\nu. \tag{10}$$

According to wave mechanics the minimum energy E_0 of the oscillator is not zero for, letting $v = 0$, we find

$$E_0 = \tfrac{1}{2}h\nu. \tag{11}$$

E_0, the energy of the lowest state of the oscillator, is termed the *zero-point energy*.

The amplitude wavefunctions ψ_v of (2) are the eigenfunctions (3.2.11) modified to take account of the change of variable,

$$\psi_v = N_v H_v(\eta)e^{-\frac{1}{2}\eta^2}, \tag{12}$$

that is, they are given by eqn. (3.2.11) with $\eta = x/r$ substituted for x. The functions ψ_v for $v = 0$ to 3 are represented in Fig. 3.5.1.

We shall now proceed to evaluate the constant N_v. This requires a third postulate which we will present in two parts.

Postulate IIIA. One-dimensional Schroedinger wavefunctions ψ have the property that $\psi^*\psi\mathrm{d}x$ is the probability that the variable x lies within the limits x and $x + \mathrm{d}x$. (ψ^* denotes the complex conjugate of ψ, *i.e.* the function obtained by replacing i by $-i$ whenever it occurs. This is not a detail of immediate concern for the wavefunctions (12) of the oscillator are wholly real, and thus $\psi^* \equiv \psi$.)

In a one-dimensional system the variable x must lie somewhere within the totality of points comprising the x-axis, and hence

$$\int_{-\infty}^{\infty} \psi^*\psi\mathrm{d}x = 1 \tag{13}$$

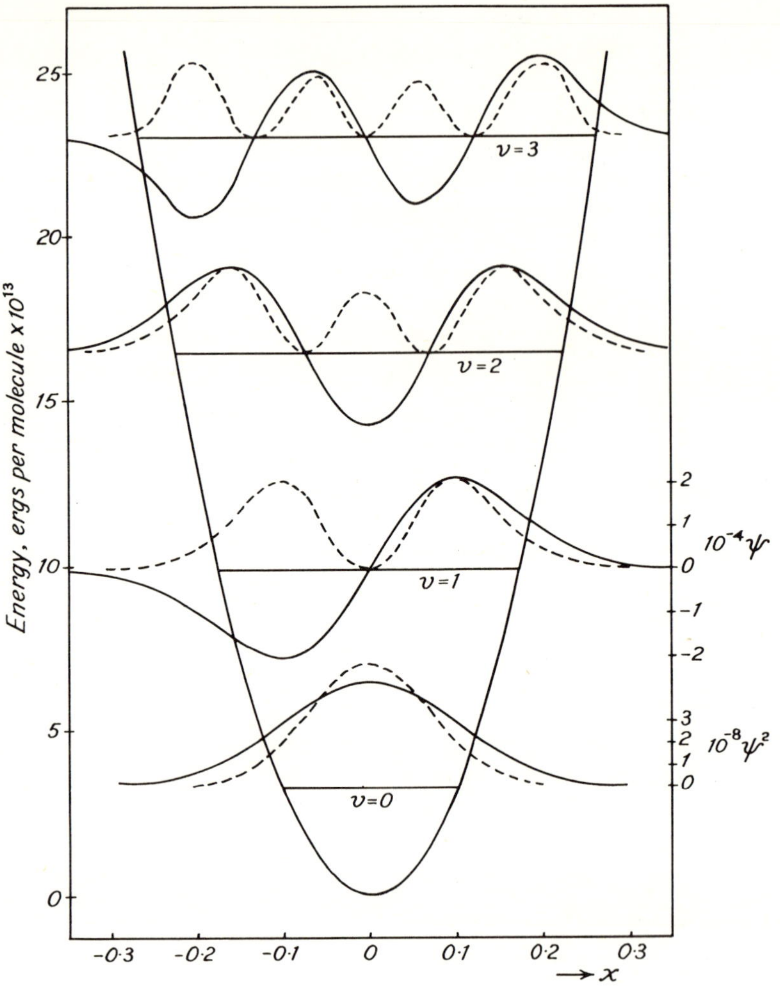

Fig. 3.5.1. Energy levels, wavefunctions (solid curves), and probability distribution (broken curves) for the harmonic oscillator. The wavefunctions, eqn. (3.5.12), are calculated for $m = 1{\cdot}008$ atomic wt. units and $\sigma = 3000\ \text{cm}^{-1}$, *i.e.* they are appropriate to the vibration of a hydrogen nucleus against a much larger mass. Only the scale of the diagram is altered by changes in m or σ.

since the probability that x lies between the limits $x = \pm\infty$ is unity. (13) is known as the *normalising condition.* It is shown in Appendix II. 2 that the amplitude wave-functions (12) obey eqn. (13) if

$$N_v = (r2^v v!\sqrt{\pi})^{-\frac{1}{2}}; \qquad v = 0, 1, 2 \ldots \tag{14}$$

The wavefunction belonging to the energy eigenvalue E_0 of the oscillator in its lowest, zero-point, energy state is, therefore,*

$$\psi_0 = \{(r\sqrt{\pi})^{-\frac{1}{2}}\}e^{-\frac{1}{2}(x/r)^2} \tag{15}$$

A corollary of IIIA is that, if ψ_v is a *normalised* amplitude wavefunction, $\psi_v{}^*\psi_v$ specifies the probability distribution of a one-dimensional dynamical system in the v-th state with respect to the coordinate x. Functions $\psi_v{}^*\psi_v \equiv \psi_v{}^2$ for the harmonic oscillator are represented by the dotted lines in Fig. 3.5.1. In each state of the oscillator there is a finite (though small) probability that the particle will be found by experiment outside the classical limits, that is, outside the area bounded by the parabola $V = \frac{1}{2}Fx^2$. The probability of penetrating to the non-classical region is greater the smaller the value of v.

3.6 The Free Particle: Degenerate Eigenvalues

The classical Hamiltonian function of a particle of mass m in motion along the coordinate x and subject to no force is $H = p^2/2m + V$. As $f = 0$, the potential energy V is independent of x (eqn. 3.3.2), and thus we are at liberty to choose the scale of potential energy so that $V = 0$. The Schroedinger amplitude equation (3.4.5) is then

$$-(h^2/8\pi^2 m)\frac{\mathrm{d}^2}{\mathrm{d}x^2}\psi = E\psi. \tag{1}$$

Since the operator $h^2/8\pi^2 m$ commutes with $\mathrm{d}^2/\mathrm{d}x^2$, (1) rearranges to

$$-\frac{\mathrm{d}^2}{\mathrm{d}x^2}\psi = \frac{8\pi^2 mE}{h^2}\psi. \tag{2}$$

We see that the wavefunctions and energy eigenvalues of (1), apart from a multiplicative constant, are those of the simple operator $-\mathrm{d}^2/\mathrm{d}x^2$.

The operator $-\mathrm{d}^2/\mathrm{d}x^2$ was discussed briefly in section 3.1, where it was shown that $\sin\sqrt{a}x$, with a a constant, is an acceptable eigenfunction (eqn. 3.1.9); equally, $\cos\sqrt{a}x$ is a possible eigenfunction, for sine and cosine functions differ only in phase. A property of eigenfunctions, easily verified by trial, is that if two functions $u_1(x)$ and $u_2(x)$ are eigenfunctions of the same operator with the same eigenvalue, their sum $u_1(x) + u_2(x)$ is also an eigenfunction. Further, recalling that an eigenfunction can be multiplied by any arbitrary constant, the solution of (2) is seen to have the general form

$$N_1 \sin\sqrt{a}x + N_2 \cos\sqrt{a}x \tag{3}$$

* It may help the reader to note that, when $v = 0$, by definition $v! = 1$.

with

$$a = 8\pi^2 mE/h^2, \tag{4}$$

so that we may write

$$\psi = N_1 \sin\left(\frac{2\pi\sqrt{2mE}}{h}\right)x + N_2 \cos\left(\frac{2\pi\sqrt{2mE}}{h}\right)x. \tag{5}$$

The only restriction upon E stems from the requirement that ψ must conform to the standard boundary conditions; in particular, that ψ must remain finite as $x \to \pm\infty$. This condition is satisfied if $2\pi\sqrt{2mE}/h$ is wholly real or, in other words, if E is *any positive number* since both m and h are positive and real. Our first conclusion, therefore, is that the energy eigenvalues of (1) include all positive numbers.

Secondly, if $E = 0$, (5) becomes

$$\psi = N_2. \tag{6}$$

As N_2 is simply a constant, the wavefunction (6) satisfies the boundary condition, and hence we can add $E = 0$ to the list of permitted energy values. If E is negative, however, $E^{\frac{1}{2}}$ is a purely imaginary quantity. To develop the argument in this situation let

$$2\pi\sqrt{2mE}/h = i\gamma, \tag{7}$$

so that γ is wholly real. Now

$$\cos(i\gamma x) = \tfrac{1}{2}(e^{\gamma x} + e^{-\gamma x}). \tag{8}$$

and

$$\sin(i\gamma x) = (i/2)(e^{\gamma x} - e^{-\gamma x}). \tag{9}$$

Substitution of the expanded sine and cosine functions into (5) yields, when $E^{\frac{1}{2}}$ is imaginary,

$$\psi = \tfrac{1}{2}(iN_1 + N_2)e^{\gamma x} - \tfrac{1}{2}(iN_1 - N_2)e^{-\gamma x}. \tag{10}$$

Neither of the terms on the right-hand side of (10) is acceptable as a wavefunction, for the first becomes infinite as $x \to +\infty$ and the second as $x \to -\infty$. Hence the free particle energy cannot be negative (that is, $E \nless V_0$ since, by convention V_0 was set equal to zero), the permitted energy eigenvalues being

$$E \geqslant 0. \tag{11}$$

Wave mechanics places no other restriction on the energy levels, which may be thought of as infinitely fine-grained. More detailed considerations yield the same result for motion in three dimensions; the free translation of molecules is *unquantised.*

Two further aspects of (5) are noteworthy. First, the energy eigenvalue $E = 0$ corresponds to a solution of the cosine part only of the wavefunction (5): all other energies occur twice, once as a solution of the sine term and once as a solution of the cosine term. Because of this property the non-zero energy eigenvalues are described as *doubly degenerate*. Physically, every non-zero eigenvalue occurs twice because the particle may move with the same energy in either the positive or the negative direction of x; but the value $E = 0$ occurs only once, because in this state the particle is motionless. Secondly, both terms on the right-hand side of (5) are periodic with a 'frequency'

$$\nu = \sqrt{(2mE)}/h.$$

We recall that the energy of the free particle is wholly kinetic, whence $E = \frac{1}{2}m\dot{x}^2$, and therefore

$$\sqrt{(2mE)} = m\dot{x} = p. \tag{12}$$

Hence we may associate with the particle a characteristic wavelength $\lambda = 1/\nu$ given by

$$\lambda = h/m\dot{x} = h/p. \tag{13}$$

This is the de Broglie relation between the wavelength of a particle of mass m and its velocity $\dot{x}$. The idea that particles obey the laws of wave motion underlies the whole of wave mechanics and is strikingly verified by observation of the diffraction of electrons (Chapter 10).

3.7 The Fixed-axis Rotator

A rigid body able to rotate about an axis fixed in space is a system with one degree of freedom. The classical kinetic energy $T = \frac{1}{2}I\omega^2$ involves the moment of inertia I and the angular velocity ω, in place of the mass m and the linear velocity $\dot{x}$ of the free particle. As the total energy of the rotator can be taken as wholly kinetic without loss of generality, the classical Hamiltonian function is

$$H = \tfrac{1}{2}I\omega^2 = (I\omega)^2/2I. \tag{1}$$

Let the axis of rotation be denoted by z, then $I\omega = P_z$, is the *angular momentum* of the system. The angular momentum operator associated with the classical dynamical variable P_z can be shown to be

$$P_z = -(ih/2\pi)\mathrm{d}/\mathrm{d}\phi, \tag{2}$$

where ϕ is the angular coordinate defining the position of the rotator. Hence the Schroedinger amplitude equation, $H\psi = E\psi$, for the rotator is

$$\left(-\frac{h^2}{8\pi^2 I}\frac{d^2}{d\phi^2}\right)\psi = E\psi. \tag{3}$$

(3) is an equation precisely like (3.6.1). The solution can be written as a sum of sine and cosine functions, as in eqn. (3.6.5), but it is here more convenient to set it in the alternative form

$$\psi = Ne^{iM\phi} \tag{4}$$

in which

$$M = \pm 2\pi\sqrt{2IE}/h. \tag{5}$$

However, a significant difference between the free particle and the rotator is that, whereas the coordinate x of the particles may lie anywhere between $x = \pm\infty$, ϕ is confined to the limits 0 and 2π. In order that ψ shall be single-valued, each time ϕ becomes a multiple of 2π, ψ must begin to repeat itself, a restriction that has no analogue in the treatment of the free particle. For ψ to be a single-valued function we must have

$$\psi(\phi) = \psi(\phi + 2\pi) \tag{6}$$

or

$$e^{iM\phi} = e^{iM(\phi+2\pi)}. \tag{7}$$

This requires $e^{i2\pi M}$ to be unity, which is only true when M is a positive or negative integer, $\ldots +2, +1, 0, -1, -2 \ldots$

The energy eigenvalues obtained from eqn. (5) are then

$$E = M^2\frac{h^2}{8\pi^2 I}; \qquad M = 0, \pm 1, \pm 2 \ldots \tag{8}$$

The lower part of the energy manifold is shown in Fig. 3.7.1. Owing to the cyclic character of the coordinate the spectrum of energy values is discrete, quite unlike that of the free particle. Yet in other respects there is a close resemblance between the rotator and the particle: thus all energy eigenvalues except the lowest are doubly degenerate—because rotation may be either clockwise or counterclockwise with the same energy—and negative eigenvalues are excluded. The normalisation condition, eqn. (3.5.13), is here

$$\int_0^{2\pi} \psi^*\psi d\phi = N^2\int_0^{2\pi} d\phi = 1 \tag{9}$$

and gives the value $1/(2\pi)^{\frac{1}{2}}$ for N.

The wavefunctions (4), besides being eigenfunctions of the operator H, also happen to be eigenfunctions of the angular momentum operator $P_z = -(ih/2\pi)\mathrm{d}/\mathrm{d}\phi$. The eigenvalues $\boldsymbol{P}_z$ of the operator P_z are, as the reader should check for himself,

$$\boldsymbol{P}_z = Mh/2\pi; \qquad M = 0, \pm 1, \pm 2, \pm 3 \ldots \tag{10}$$

that is, the angular momentum of the rotator is restricted in wave mechanics to values which are integral multiples of $h/2\pi$.

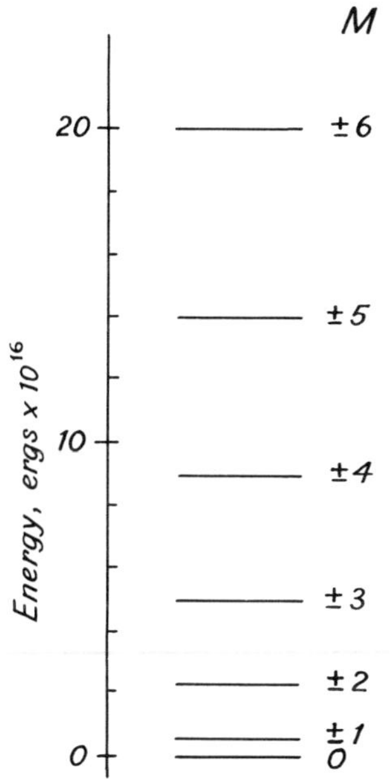

Fig. 3.7.1. Energy levels of a fixed-axis rotator. The energy scale corresponds to $I = 10^{-40}$ gcm^2.

3.8 The Heisenberg Uncertainty Principle

To introduce the Heisenberg Uncertainty Principle it is necessary to complete the third postulate, the first part of which was given in section 3.5.

Postulate IIIB. For a system confined to a linear coordinate x in a state described by a normalised amplitude wavefunction ψ, the average value $\bar{q}$ of a dynamical variable q is

$$\bar{q} = \int_{-\infty}^{\infty} \psi^* q \psi \, dx, \tag{1}$$

where q is the operator associated with the dynamical variable q. The integration in (1) is taken over the whole range of possible values of x, that is, between $x = \pm\infty$.

Consider the linear harmonic oscillator. According to eqn. (1) the average value, $\bar{x}_v$, of the variable x in the v-th quantum state of the oscillator is given by

$$\bar{x}_v = \int_{-\infty}^{\infty} \psi_v^* x \psi_v \, dx. \tag{2}$$

In this instance it happens that $\bar{x}_v$ can be obtained by a simple yet powerful argument. Referring to eqns. (3.5.12) and (3.2.10), or to Fig. 3.5.1, we see that the functions ψ_v fall into two classes. For v even, substitution of $-x$ for x wherever it occurs in the function ψ_v leaves ψ_v unchanged in magnitude and sign; whereas for v odd the substitution leaves ψ_v unaltered in magnitude but changes in sign. The first class of functions is said to be *symmetric-in-x* and the second *antisymmetric-in-x*, the substitution of $-x$ for x being equivalent to the simple operation of reflection at the origin of coordinates (section 3.2). Now the integrand in (2) contains three factors, ψ_v^*, x, and ψ_v; two of them, ψ_v^* and ψ_v, are either both symmetric (v even) or both antisymmetric (v odd), while the third, x, is antisymmetric. Therefore the integrand $\psi_v^* x \psi_v$ in (2) must change sign when the operation of reflection is carried out. But the limits of integration are symmetrical about $x = 0$ and so the magnitude of the integral cannot depend on whether the operation of reflection has been applied. This is only possible if the value of the integral is identically zero. Thus, for the linear oscillator,

$$\bar{x}_v = 0. \tag{3}$$

A similar argument applied to the average value $\bar{p}$ of the momentum p yields

$$\bar{p}_v = \int_{-\infty}^{\infty} \psi_v^* \{-(ih/2\pi) d/dx\} \psi_v \, dx = 0, \tag{4}$$

since the operator $-(ih/2\pi)d/dx$, like the operator x, changes sign under the operation of reflection at the coordinate origin.

Symmetry arguments do not require the average of the dynamical variables x^2 and p^2 of the linear oscillator to be zero, as the reader should check for himself. To determine $\overline{x^2}$ and $\overline{p^2}$, then, it is necessary to evaluate the integrals

$$\overline{x_v^2} = \int_{-\infty}^{\infty} \psi_v^* x^2 \psi_v \, dx \tag{5}$$

and

$$\overline{p_v^2} = \int_{-\infty}^{\infty} \psi_v^* \{-(h^2/4\pi^2) d^2/dx^2\} \psi_v \, dx. \tag{6}$$

The integrals are solved easily for the zeroth ($v = 0$) state of the oscillator, when

$$\psi_0^* = \psi_0 = \{1/(r\pi^{\frac{1}{2}})^{\frac{1}{2}}\}e^{-\frac{1}{2}(x/r)^2}. \qquad (3.5.15)$$

As the operator x^2 commutes with ψ_0 the order of the factors under the integral sign in (5) is immaterial. After some rearrangement we obtain

$$\overline{x_0^2} = \{1/(r\pi^{\frac{1}{2}})\}\int_{-\infty}^{\infty} x^2 e^{-(x/r)^2}\,dx = r^2/2, \qquad (7)$$

the integral $\int_{-\infty}^{\infty} x^2 e^{-(x/r)^2}dx = 2\int_0^{\infty} x^2 e^{-(x/r)^2}\,dx = \frac{1}{2}r^3\pi^{\frac{1}{2}}$ being in a standard form. As to eqn. (6), the presence of the differential operator

$$-(h^2/4\pi^2)d^2/dx^2$$

renders the order of its factors important. But as

$$-(d^2/dx^2)e^{-\frac{1}{2}(x/r)^2} = (1/r^2)\{1-(x/r)^2\}e^{-\frac{1}{2}(x/r)^2}, \qquad (8)$$

and

$$\int_0^{\infty} \{1-(x/r)^2\}e^{-(x/r)^2}\,dx = \tfrac{1}{4}r\pi^{\frac{1}{2}} \qquad (8a)$$

(see tables of standard integrals) the equation can be brought into a form that integrates directly,

$$\overline{p_0^2} = (h^2/4r^3\pi^{5/2})\int_{-\infty}^{\infty}\{1-(x/r)^2\}e^{-(x/r)^2}\,dx = h^2/8\pi^2r^2. \qquad (9)$$

The results obtained in eqns. (3), (4), (7), and (9) are summarised in Table 3.8.1.

The quantities x_v, p_v, x_v^2, and p_v^2, are to be thought of as the average value of a large number of experimental measurements of the variable, made on an oscillator which before each measurement was in the state ψ_v.

Table 3.8.1. Average values and Uncertainties for the Zeroth ($v = 0$) State of the Oscillator

Variable, q	Average value, $\bar{q}$	Uncertainty, Δq
x_0	0	$r/\sqrt{2}$
x_0^2	$r^2/2$	
p_0	0	$h/2\sqrt{2}r\pi$
p_0^2	$h^2/8\pi^2r^2$	

According to statistical theory, when the repeated measurement of some variable

q gives rise to a spread of values the *uncertainty*, Δq, associated with an individual measurement is taken to be

$$\Delta q = \sqrt{\overline{q^2} - (\bar{q})^2}, \tag{10}$$

in which $\overline{q^2}$ is the average value of the variable q^2, and $(\bar{q})^2$ is the square of the average value of q. Clearly both $\overline{q^2}$ and $\bar{q}$ must be available in order to obtain Δq.

Let us now use the results in Table 3.8.1 to determine the uncertainties, Δx_0 and Δp_0, in a possible measurement of the coordinate x and the momentum p of an oscillator in its ground state. For the coordinate, we find

$$\Delta x_0 = r/\sqrt{2} \tag{11}$$

and for the momentum,

$$\Delta p_0 = h/2\sqrt{2}r\pi \tag{12}$$

Δx_0 and Δp_0 measure the spread of repeated experimental determinations of x and p made on an oscillator in its ground state. The product of the uncertainties,

$$\Delta x_0 \cdot \Delta p_0 = h/4\pi, \tag{13}$$

is then a constant independent of the parameters of the system. Eqn. (13) represents the *Heisenberg Uncertainty Principle* applied to the ground state of an harmonic oscillator. Whatever experiment might be devised to measure simultaneously the values of x and p pertaining to a given oscillator (in its ground state), the limit of accuracy is always determined by (13). Thus if measurements of p are relatively precise, so that Δp is small, the uncertainty in x must be proportionately great in order that eqn. (13) is satisfied.

The scope of the Uncertainty Principle is much broader than its application to the harmonic oscillator by which we have chosen to introduce it. The Principle establishes a relation between any pair of canonically conjugate variables belonging to a given dynamical system. These include the variables x and p of the oscillator and the free particle, the coordinate ϕ and the momentum P_z of the fixed-axis rotator, and—as we shall presently see—the energy E and the time t in systems where t is considered explicitly. The form of the general relation between the uncertainties in any two canonically conjugate variables q and s then is

$$\Delta q \cdot \Delta s \approx h/2\pi. \tag{14}$$

The Heisenberg Principle helps to explain why certain dynamical systems are endowed with zero-point energy. Of the three systems we have considered, the harmonic oscillator, the simple rotator, and the free particle, only the first has energy greater than zero in its lowest state. This is in accord with the Principle: for if the oscillator were allowed zero energy it would be located exactly at the position of minimum of potential energy, and then the uncertainty in x and p would simultaneously both be zero, which is contrary to (13). The existence of zero-point energy can therefore be regarded as necessary if the oscillator is to satisfy the Heisenberg Principle. As to the rotator, the wavefunction $\psi_0 = 1/(2\pi)^{\frac{1}{2}}$ of the lowest state is independent of ϕ and corresponds to a situation in which all orientations ϕ are equally probable. Therefore the uncertainty in position is infinite and the momentum, and hence the energy, can have the precise value zero. The free particle, likewise, has no zero-point energy.

3.9 The Second Schroedinger Equation: Stationary States

To this point there has been no consideration of the time t as a variable; the amplitude wavefunctions ψ have been functions of a coordinate only. An extension of the discussion to include time requires that the amplitude functions $\psi = \psi(x)$ must be multiplied by a factor $\phi(t)$ dependent upon t but not upon x. The product

$$\psi(x)\cdot\phi(t) = \Psi(x, t) \tag{1}$$

is known as a *state wavefunction* and is denoted by the capital letter Ψ. State wavefunctions are explicit functions of both x and t. We must not, however, confuse the time-dependent part, $\phi(t)$, of Ψ with the angular coordinate ϕ, used in the discussion of the simple rotator.

The shape of the function $\phi(t)$, suggested by the wave-like aspect of particles,* is

$$\phi(t) = e^{-i2\pi\nu t} \tag{2}$$

with the supplementary assumption that ν can be expressed in the form

$$\nu = E/h. \tag{3}$$

The assumptions encompassed by eqns. (2) and (3) may be given formal recognition in a fourth (and final) postulate:

Postulate IV. Every amplitude wavefunction ψ pertaining to a dynamical system has an associated time dependent multiplier, $\phi(t)$, of the form

$$\phi(t) = e^{-i2\pi Et/h}, \tag{4}$$

where E is the energy eigenvalue belonging to the wavefunction ψ.

* The functions $\phi = \sin 2\pi\nu t$, $\phi = \cos 2\pi\nu t$, and $\phi = e^{\pm i2\pi\nu t}$ are simple representations of wave motion: of them, the third is the most convenient for our present application. The choice of $\phi = e^{-i2\pi\nu t}$, and hence of $\phi^* = e^{i2\pi\nu t}$, is a convention that could be reversed without affecting the results.

Another approach illuminates a different aspect of eqn. (4). In classical mechanics the energy E and the time t are canonically conjugate variables, and thus the operators E and t assigned to the variables E and t can be determined from Postulate II (section 3.4). If the operator $(ih/2\pi)\partial/\partial t$ is allotted to the variable E, and the operator t to the variable t, then

$$Et - tE = \frac{ih}{2\pi}\left(\frac{\partial}{\partial t}t - t\frac{\partial}{\partial t}\right) = ih/2\pi, \tag{5}$$

so that the assignments certainly obey the rule (3.4.1) of Postulate II. We now modify the first Schroedinger equation, $H\psi = E\psi$, by multiplication from the right by the time-dependent function $\phi(t)$,

$$H\psi(x)\phi(t) = E\psi(x)\phi(t). \tag{6}$$

Substituting the operator $(ih/2\pi)\partial/\partial t$ for E, and introducing the definition (1) of a state wavefunction $\Psi = \Psi(x, t)$, we find

$$H\Psi = (ih/2\pi)\frac{\partial}{\partial t}\Psi. \tag{7}$$

This is *Schroedinger's second equation*, or *Schroedinger's equation including time*. The solution of eqn. (7), as the reader should confirm is $\phi = e^{-i2\pi Et/h}$, and thus

$$\Psi = \psi e^{-i2\pi Et/h}. \tag{8}$$

Eqn. (8) embodies precisely the same conclusions as eqns. (1) and (4), and the Postulate IV is therefore not essential to the derivation of the time-dependent functions. However, if this step is avoided it becomes necessary to introduce the expression (2) as a fundamental Postulate based on the wave-like aspect of matter. In wave mechanics, as in classical mechanics, there is no unique set of fundamental postulates.

The state functions Ψ of a dynamical system are simple to write down once the amplitude wavefunctions ψ are known. Take as an example the linear oscillator: we need merely substitute into (8) the expression (3.5.12) with (3.5.3) for the amplitude function, and thus

$$\Psi_v = N_v H_v(x/r)e^{-\frac{1}{2}(x/r)^2}e^{-i2\pi E_v t/h}. \tag{9}$$

Ψ_v is then the state function of the v-th quantum state of the oscillator. We next enquire what effect, if any, the use of the state function Ψ_v in place of the amplitude function Ψ_v has on our previous conclusions. From (8) it follows quite generally that

$$\Psi_v^* \Psi_v = \psi_v^* \psi_v, \qquad (10)$$

that is, the time-dependence drops out when a state function is multiplied by its complex conjugate. Therefore the probability distribution,

$$\psi_v^* \psi_v = \Psi_v^* \Psi_v, \qquad (11)$$

the normalisation condition,

$$\int \psi_v^* \psi_v \mathrm{d}x = \int \Psi_v^* \Psi_v \mathrm{d}x, \qquad (12)$$

and the expression (3.8.1) for the average value, $\bar{q}$, of some dynamical variable q,

$$\bar{q} = \int \psi_v^* q \psi_v \mathrm{d}x = \int \Psi_v^* q \Psi_v \mathrm{d}x, \qquad (13)$$

are unaffected by the presence of the time factor $e^{-i2\pi Et/h}$ in the state functions Ψ_v. When an oscillator is in a state described by a state function Ψ_v the average values of the coordinate x, the momentum p, the energy E_v, and so forth, are *independent of time*. For this reason such states are described as *stationary states.*

3.10 Non-stationary States

We shall continue to draw upon the linear oscillator to illustrate our line of argument, although the results are equally applicable to any other dynamical system. Suppose, then, we have in our possession an oscillator known to be in the state described by the state function Ψ_v, and that we wish to enquire 'What is the probable result of an exact experimental determination of the energy of the oscillator?' Consider first the uncertainty ΔE_v associated with a single experimental measurement of the energy. In the Schroedinger scheme of things the operator associated with the total energy is the Hamiltonian operator H. Therefore, according to eqns. (3.8.1) and (3.9.13), the mean value, $\overline{E_v}$, of a large number of experimental determinations of the energy is†

$$\overline{E_v} = \int \Psi_v^* H \Psi_v \mathrm{đ}x = \int \psi_v^* H \psi_v \mathrm{đ}x. \qquad (1)$$

From the Schroedinger amplitude equation we have $H\psi_v = E_v \psi_v$, and the constant E_v commutes with ψ_v, so that

$$\int \psi_v^* H \psi_v \mathrm{đ}x = E_v \int \psi_v^* \psi_v \mathrm{đ}x. \qquad (2)$$

† Here and in subsequent equations the horizontal stroke in $\mathrm{đ}x$ indicates that integration is over the complete range of possible values of x, that is between the limits $x = \pm \infty$.

Therefore

$$\overline{E_v} = E_v \tag{3}$$

since $\int \psi_v^* \psi_v \mathrm{d}x = 1$. Likewise

$$\begin{aligned}\overline{E_v^2} &= \int \Psi_v^* H^2 \Psi_v \mathrm{d}x = \int \psi_v^* H^2 \psi_v \mathrm{d}x \\ &= E_v \int \psi_v^* H \psi_v \mathrm{d}x = E_v^2.\end{aligned} \tag{4}$$

The uncertainty ΔE_v can now be obtained from eqn. (3.8.10), yielding

$$\Delta E_v = 0. \tag{5}$$

In other words, the result of a precise experimental measurement of the energy of an oscillator in the state Ψ_v will certainly be the value E_v. The measurement of energy, unlike that of x and p (section 3.8), is sharp. The reason is apparent if we compare (4) with, say, (3.8.5): it is that Ψ_v is an eigenfunction of H.

We now relax the condition that the state of the oscillator is Ψ_v and assume instead that it is in one of two adjacent states Ψ_n or Ψ_m; that is, the oscillator is either in the state with $v = n$ or in that with $v = m$, though in advance we do not know which. (As the order of the states is immaterial we can choose the m-th to be higher in energy when, because the states are adjacent, $m = n + 1$.) Naturally the state to which the oscillator belongs could be found by experiment, since a measurement of the energy which gives the value E_m means that the oscillator is certainly in the state Ψ_m; but at present our interest is in the wavefunctions in advance of any possible measurement.

Ψ_n and Ψ_m are state functions belonging to the manifold of state functions (3.9.9). Both are possible solutions of the second Schroedinger equation (3.9.7). It follows that their sum,

$$\Psi = c_n \Psi_n + c_m \Psi_m, \tag{6}$$

with arbitrary constant coefficients, is also a solution of the Schroedinger equation. This we can demonstrate readily, for substitution in the left-hand side of (3.9.7) gives

$$H(c_n \Psi_n + c_m \Psi_m) = c_n H \Psi_n + c_m H \Psi_m = c_n E_n \Psi_n + c_m E_m \Psi_m, \tag{7}$$

while the right-hand side yields

$$\frac{ih}{2\pi} \frac{\partial}{\partial t} (c_n \Psi_n + c_m \Psi_m) = c_n E_n \Psi_n + c_m E_m \Psi_m. \tag{8}$$

Ψ in (6), then, is the state function of an oscillator which has access to the two states Ψ_n and Ψ_m. Ψ, of course, is not an eigenfunction of H [this follows from (7)]; if it were, the result of an energy measurement would be certain, and this it is not. The significance of the constants c_n and c_m can be seen by multiplying both sides of (6) from the left by Ψ_n^* and integrating between the limits of $x = \pm\infty$,

$$\int \Psi_n^* \Psi \, dx = c_n \int \Psi_n^* \Psi_n \, dx + c_m \int \Psi_n^* \Psi_m \, dx. \qquad (9)$$

The second term on the right of (9) is zero, because Ψ_n and Ψ_m have different symmetries in x when $m = n + 1$; and the first term on the right is equal to c_n since, according to (3.9.12) and (3.5.13), $\int \Psi_n^* \Psi_n \, dx = 1$. Hence

$$\int \Psi_n^* \Psi \, dx = c_n. \qquad (10)$$

A similar argument shows that

$$\int \Psi_m^* \Psi \, dx = c_m. \qquad (11)$$

Let us now evaluate the average $\overline{E} = \int \Psi^* H \Psi \, dx$ of this oscillator which divides its time between the states Ψ_n and Ψ_m. For Ψ given by (6) we have

$$\overline{E} = \int \Psi^* H \Psi \, dx = \int \Psi^* H (c_n \Psi_n + c_m \Psi_m) \, dx. \qquad (12)$$

The integral is in two parts. The first is

$$\int \Psi^* H (c_n \Psi_n) \, dx = c_n E_n \int \Psi^* \Psi_n \, dx. \qquad (13)$$

From (10) we obtain $|c_n|^2 E_n$ as the solution of the expression (13). Likewise, the second term on the right of (12) equals $|c_m|^2 E_m$, and thus

$$\overline{E} = c_n^2 E_n + c_m^2 E_m. \qquad (14)$$

Repeated measurements of the energy of the oscillator in the state Ψ therefore yield the results E_n and E_m in the proportion $|c_n|^2 : |c_m|^2$. $|c_n|^2$ and $|c_m|^2$ can be thought of as the probability that the oscillator has the energy values E_n and E_m respectively. Alternatively, if observations are made on an assembly of oscillators, all in the state defined by Ψ, $|c_n|^2$ and $|c_m|^2$ are the respective probabilities of the energies E_n and E_m.

3.11 The Absorption of Radiation: Transitions

As in section 3.10, we consider an harmonic oscillator which has access to two adjacent states Ψ_n and Ψ_m, but we assume that transitions are possible between them. The transitions of special interest are these that occur in absorption, namely,

$$\text{Oscillator in the state } \Psi_n + \text{Radiation} = \text{Oscillator in the state } \Psi_m,$$

the subscript m being used to denote the state which is higher in energy.

When an oscillator is exposed to radiation the potential energy is no longer

$$V(x) = \tfrac{1}{2}Fx^2 \tag{3.3.6}$$

though the contribution of $V(x)$ is of course present. In addition, the electric field of the radiation supplies a small potential which we designate by v, so that the Hamiltonian operator for an oscillator immersed in a bath of radiation is

$$H = -(h^2/8\pi^2 m)\,\partial^2/\partial x^2 + V(x) + v. \tag{1}$$

The second Schroedinger equation (3.9.7.) is, then

$$\left\{-(h^2/8\pi^2 m)\frac{\partial^2}{\partial x^2} + V(x) + v\right\}\Psi = (ih/2\pi)\frac{\partial}{\partial t}\Psi. \tag{2}$$

When $v = 0$, the solution of (2) is the state function Ψ given in (3.10.6) with constant coefficients. The next step is to elucidate the form of Ψ which is a solution of (2) when $v \neq 0$. It will emerge that

$$\Psi = c_n\Psi_n + c_m\Psi_m \tag{3.10.6}$$

is an acceptable solution, provided that the coefficients are functions of time.

Since the field of the radiation is very small we may assume that $V(x) \gg v$. Let us now rewrite the left-hand side of (2) as

$$\left\{-(h^2/8\pi^2 m)\frac{\partial^2}{\partial x^2} + V(x)\right\}\Psi + v\Psi. \tag{3}$$

The solution to the first term on the left of (3) was obtained in section 3.10. Substituting the result (eqn. 3.10.7), (3) becomes

$$c_n E_n \Psi_n + c_m E_m \Psi_m + v\Psi \tag{4}$$

We next deal with the right-hand side of (2). Assuming that the coefficients are functions of time, and that Ψ is the state function (3.10.6), we obtain

$$(ih/2\pi)\frac{\partial}{\partial t}(c_n\Psi_n + c_m\Psi_m)$$
$$= (ih/2\pi)\left\{\frac{dc_n}{dt}\Psi_n + \frac{dc_m}{dt}\Psi_m\right\} + c_nE_n\Psi_n + c_mE_m\Psi_m. \quad (5)$$

Gathering together the expanded forms (4) and (5) of the left and right sides of (2) yields

$$v\Psi = (ih/2\pi)\left\{\frac{dc_n}{dt}\Psi_n + \frac{dc_m}{dt}\Psi_m\right\}. \quad (6)$$

Multiplication of (6) from the left by Ψ_m^* followed by integration with respect to x between the limits of $\pm\infty$ (recalling that $\int \Psi_m^*\Psi_m\, dx = 1$ and $\int \Psi_m^*\Psi_n dx = 0$) gives

$$\frac{dc_m}{dt} = -(i2\pi/h)\ \{c_n \int \Psi_m^*\, v\Psi_n dx + c_m \int \Psi_m^*\, v\Psi_m dx\}. \quad (7)$$

This is the equation we are looking for. Suppose that we have under observation an oscillator initially in the lower state Ψ_n, when $c_m = 0$. The initial rate of transfer to the upper state, dc_m/dt, is proportional to the integral $\int \Psi_m^* v\Psi_n\, dx$. Obviously, no transition occurs unless $v \neq 0$: this confirms that the coefficients are independent of time in the absence of a perturbing potential.

We have yet to consider the form of the perturbing potential v. Up to this point the discussion has been perfectly general—except for the arbitrary restriction to two states, Ψ_n and Ψ_m—and the equations are valid for any type of perturbation; for instance, in a physically real system, that operating at the instant of collision of one molecule with another. For the present, however, we are interested in the perturbation applied when the oscillator is exposed to radiation. We denote by $\mathcal{E}_x$ the component of the electric field of the radiation along the x-coordinate of the oscillator, and represent the time-dependence of the field by

$$\mathcal{E}_x = \mathcal{E}_x^0 \cos 2\pi\nu t = \mathcal{E}_x^0(e^{i2\pi\nu t} + e^{-i2\pi\nu t})/2. \quad (8)$$

Suppose that the oscillator is an electron, or something bearing a charge. At any instant the value of the electric moment is $\mu_x = ex$, and the perturbation energy is

$$v = \mathcal{E}_x . ex. \quad (9)$$

Consider that at the time $t = 0$ the oscillator is in the lower state Ψ_n, so that $c_n(t = 0) = 1$ and $c_m(t = 0) = 0$, and at this time the system is exposed to monochromatic light of frequency ν. In this event we substitute for v in (6) the expressions (8) and (9): writing out the time-dependence of the state functions explicitly then gives, after some rearrangement

$$\frac{\mathrm{d}c_m}{\mathrm{d}t} = -(i2\pi/h)\frac{\mathcal{E}_x^0}{2}\{\int \psi_m^* ex\psi_n\, \mathrm{d}x\}\{e^{i2\pi(E_m - E_n + h\nu)t/h} + e^{i2\pi(E_m - E_n - h\nu)t/h}\} \tag{10}$$

Integration with respect to time, from $t = 0$ to t (*i.e.* for the period the perturbation is applied), yields

$$c_m = \frac{\mathcal{E}_x^0}{2}\{\int \psi_m^* ex\psi_n\, \mathrm{d}x\} \times \left[\frac{1 - e^{i2\pi(E_m - E_n + h\nu)t/h}}{E_m - E_n + h\nu} + \frac{1 - e^{i2\pi(E_m - E_n - h\nu)t/h}}{E_m - E_n - h\nu}\right]. \tag{11}$$

Of the several factors in (11), $\frac{\mathcal{E}_x^0}{2}\int \psi_m^* ex\psi_n\, \mathrm{d}x$ is always very small, and the numerator of each fraction within square brackets can vary in absolute magnitude only between 0 and 2. The whole expression (11) remains very small, therefore, unless the denominators $E_m - E_n \pm h\nu$ of the fractions happen to be very small, that is, unless $E_m - E_n \approx \pm h\nu$. When transitions take place from the lower state Ψ_n to the higher state Ψ_m, $E_m - E_n$ is positive; and hence, since h and ν are both positive, the second fraction in (11) becomes very large when the frequency of the radiation is close to that given by the *Bohr frequency condition*,

$$E_m - E_n = h\nu_{mn}. \tag{12}$$

In this special situation the right-hand side of (11) as a whole becomes large. The reader should satisfy himself that the first fraction in (11) plays a similar role in the induced emission of radiation.

For transitions induced by absorption of radiation we can neglect the contribution of the first fraction of (11) to c_m. Multiplication of both sides of the equation by the complex conjugate then yields

$$c_m^* c_m = \mathcal{E}_x^{0\,2}\{\int \psi_m^* ex\psi_n \mathrm{d}x\}^2 \, \frac{\sin^2[\pi(E_m - E_n - h\nu)t/h]}{(E_m - E_n - h\nu)^2}, \tag{13}$$

where $c_m^* c_m = |c_m|^2$ is the probability of the higher state Ψ_m at the end of the period of irradiation. The expression (13), however, is valid only for monochromatic light, and to obtain the corresponding equation for irradiation by a range of frequencies it is still necessary to integrate with respect to ν. Since the fraction appearing as the final term is very small except in one region, we may consider that $\mathcal{E}_x^0$ is constant: integration between $\nu = \pm\infty$ then gives†

$$|c_m|^2 = (\pi^2/h^2)\mathcal{E}_x^{0\,2}\{\int \psi_m^* ex\psi_n\, \mathrm{d}x\}^2 t. \tag{14}$$

† Using the standard integral $\int_{-\infty}^{\infty} (\sin^2 x/x^2)\mathrm{d}x = \pi$.

In other words, the probability, $|c_m|^2$, that an oscillator in the state Ψ_n will undergo a transition to a higher state Ψ_m is proportional to the time of irradiation t, to the square of the intensity $\mathcal{E}_x^0$ of the field of the incident light, and to the square of integral,

$$M_x = \int \psi_m^* ex\psi_n \, d\!\!\!^{-}x = e \int \psi_m^* x\psi_n \, d\!\!\!^{-}x. \tag{15}$$

M_x is known as the *transition moment*: it determines the transition probability inasmuch as the remaining factors in (14) are unchanged from one case to another.

3.12 Selection Rules for the Linear Harmonic Oscillator and the Fixed-axis Rotator

(*a*) In sections 3.10 and 3.11, Ψ_n and Ψ_m were defined to be state functions of adjacent states with m the higher, so that $m = n + 1$. In a more thorough treatment it emerges that this restriction is not necessary, and hence that the probability of a transition occurring by absorption of radiation between any pair of states of an oscillator is described by the expression (3.11.14). The oscillator wavefunctions enter this expression only in the factor $M_x = e \int \psi_m^* x\psi_n \, d\!\!\!^{-}x$ representing the transition moment. Obviously if this integral is zero, the probability of a transition from the n-th to the m-th state is also zero.

Let us establish the conditions under which M_x definitely has the value zero. Suppose that ψ_n and ψ_m are both wavefunctions of even-numbered quantum states, say $n = 0$ and $m = 2$; the integrand $\psi_m^* x\psi_n$ then comprises two terms, ψ_m^* and ψ_n, left unchanged by the operation of reflection of the coordinate at the origin, and one term, the operator x, that changes sign. Consequently the sign of the integrand, and hence of the integral $\int \psi_m^* x\psi_n \, d\!\!\!^{-}x$ changes sign under the operation of reflection, and this is only possible if the value of the integral is zero (section 3.8). Thus the probability of a transition between two even quantum states of the harmonic oscillator is zero. By the same argument this must apply also to transitions between two odd states, and so the probability of any two states having the same symmetry in x combining in a transition is always zero. Such transitions are said to be *forbidden.*

If, however, m is odd and n is even (*e.g.*, $m = 1, n = 0$), or *vice versa*, $\psi_m^* x\psi_n$ is unchanged by the symmetry operation of reflection at the origin. Symmetry now does not require the value of $\int \psi_m^* x\psi_n \, d\!\!\!^{-}x$ to be zero; but from symmetry arguments alone we cannot judge its magnitude; the integral may have any value, even—for some reason other than that of symmetry—the value zero. In fact, closer examination of the wavefunctions (Appendix II. 2) discloses that transitions between even and odd quantum states are forbidden unless the two states combining in this way are adjacent states. The energy jump from the state n can then take place by absorption of radiation if, and only if,

$$m = n + 1. \tag{1}$$

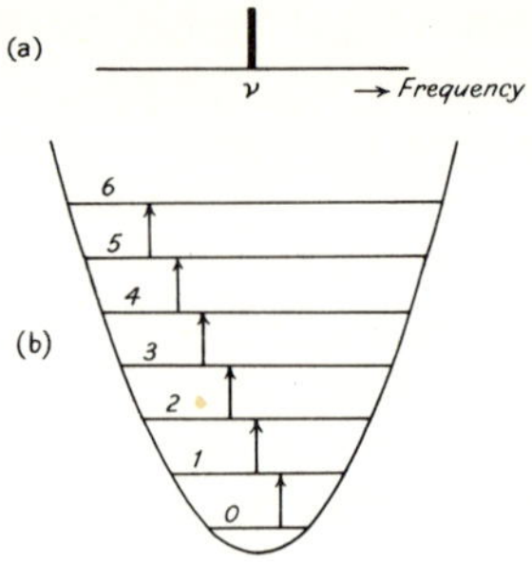

Fig. 3.12.1. Spectrum (*a*) and allowed transitions (*b*) of a simple harmonic oscillator.

This is why an adjacent pair of states was chosen for discussion in sections 3.10 and 3.11.

According to eqn. (3.5.10) the energy difference between states with quantum numbers $n = v$ and $m = v + 1$ is given by

$$E_m - E_n = h\nu \tag{2}$$

irrespective of the value of v. The frequency, ν_{mn}, of the radiation absorbed is that given by the *Bohr condition* (3.11.12). Hence the spectrum of an harmonic oscillator is a single line whose frequency ν_{mn} equals the classical oscillator frequency ν.
(*b*) The wavefunctions of the simple rotator (section 3.7) are,

$$\Phi = (2\pi)^{-\frac{1}{2}} e^{iM\phi}. \tag{3}$$

Let the electric dipole moment of the rotator have a component μ perpendicular to the axis of rotation: then the magnitude of the component along a direction fixed in space is $\mu \cos\phi$. The transition moment between two states Φ_m and Φ_n is given by

$$\int_0^{2\pi} \Phi_m^* \mu \cos\phi \Phi_n \, d\phi = \frac{\mu}{2\pi} \int_0^{2\pi} e^{-iM_m\phi} \cos\phi e^{iM_n\phi} \, d\phi. \tag{4}$$

Recalling that $\cos\phi = \frac{1}{2}(e^{i\phi} + e^{-i\phi})$, the integral (4) may be written,

$$\frac{\mu}{4\pi} \int_0^{2\pi} (e^{-i(M_m - M_n + 1)\phi} + e^{-i(M_m - M_n - 1)\phi}) d\phi. \tag{5}$$

Let $M_m - M_n \pm 1 = \gamma$, then we have for either part of (5)

$$\frac{\mu}{4\pi} \int_0^{2\pi} e^{-i\gamma\phi} d\phi = -\frac{\mu}{4\pi} \cdot \frac{e^{-i2\pi\gamma} - 1}{i\gamma}, \tag{6}$$

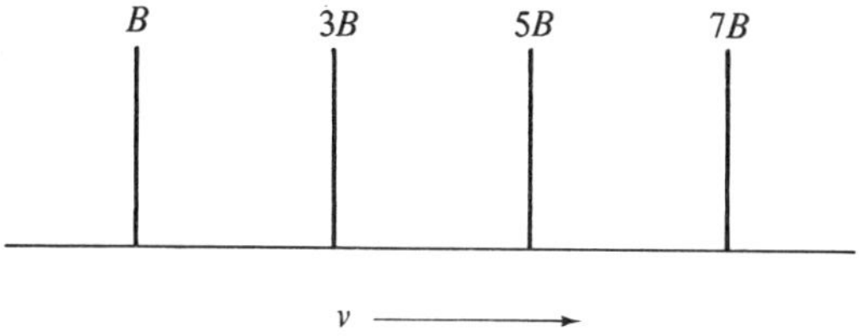

Fig. 3.12.2. Absorption spectrum of a simple, polar rotator.

which may be written in the form $\frac{\mu}{2\pi} \cdot e^{-i\pi\gamma} \frac{\sin \gamma\pi}{\gamma}$. Now $\sin \gamma\pi = 0$ and the expression vanishes for any non-zero value of the integer γ. However, if $\gamma = 0$, $e^{-i\pi\gamma} = 1$ and $(\sin \pi\gamma)/\gamma = 1$. The magnitude of the transition moment is therefore,

$$\begin{cases} \mu/2\pi \text{ if } \gamma = 0 \\ \text{zero otherwise}, \end{cases} \tag{7}$$

and the selection rule for the quantum number M is

$$\gamma = M_m - M_n \pm 1 = 0, \tag{8}$$

or

$$\Delta M = M_m - M_n = \mp 1. \tag{9}$$

Transitions are therefore allowed between adjacent states of the rotator providing $\mu \neq 0$.

Referring to eqn. (3.7.8) and Fig. 3.7.1 we see that the selection rule results in absorption of radiation at frequencies corresponding to:

$$h\nu = E_{|M|+1} - E_{|M|} = [(|M| + 1)^2 - |M|^2] \frac{h^2}{8\pi^2 I}, \tag{10}$$

or

$$\nu = (2\,|M| + 1)\frac{h}{8\pi^2 I}. \tag{11}$$

The spectrum therefore consists of a series of absorptions with a constant interval of $2B$, where we define $B = h/8\pi^2 I$.

BIBLIOGRAPHY

Pauling and Wilson, *Introduction to Quantum Mechanics,* McGraw-Hill, New York, 1935.
Rojansky, *Introductory Quantum Mechanics,* Prentice-Hall, New York, 1938.
Eyring, Walter, and Kimball, *Quantum Chemistry*, Wiley, New York, 1944.
Pitzer, *Quantum Chemistry*, Constable, London, 1953.

Chapter 4

Molecular spectroscopy: an introductory survey

Spectroscopy is a truly vast subject. It has long since broadened in scope from its origins based in purely visual phenomena and is now thought of as concerning interactions of electromagnetic radiations of all wavelengths with matter. In recent years the impact of spectroscopy on Chemistry has amounted to something of a revolution. A brief glance at the current literature reveals that there are few branches of chemistry which do not involve, in some degree, the researcher with spectroscopic considerations. The reasons for this are plain. Spectroscopic methods may provide precise, sensitive and non-destructive analytical tools; they may reveal subtleties of structure in all states of matter difficult, or impossible, to determine by other methods; they may furnish means for determining molecular energetics and a host of molecular electrical and magnetic properties. Their success stems from the fact that they operate directly at the molecular level. All spectroscopic methods are rooted in quantum theory and there is thus a uniformity of approach and there are common ideas across the whole field. In this chapter we will survey some aspects of molecular spectroscopy in an attempt to illustrate the common ground and the scope of the subject prior to the more detailed considerations of the subsequent sections.

4.1 Molecular Energy and the Electromagnetic Spectrum

A molecule consists of an assembly of positively charged nuclei and negatively charged electrons, the former accounting for the major part of the molecular mass. The energies allowed to the molecule may in principle be obtained by solving the Schroedinger equation for the system of interacting nuclei and electrons:

$$H\psi = E\psi, \tag{3.4.5}$$

but this is only possible for the very simplest of molecular examples. The nuclei and electrons in a molecule are subject to comparable forces, but as the electrons are very much lighter than the nuclei their motions are very much more rapid. We may then, to a good degree of approximation, consider the motions of the electrons separately from those of the nuclei and we can investigate the electronic problem on the assumption that the nuclei are fixed. The electronic energy and the nuclear–nuclear repulsion energy then act as an effective potential for the motions of the nuclei. This separation of the nuclear and electronic motions is embodied in the *Born–Oppenheimer approximation*. We may further divide the nuclear motions in a molecule into those corresponding to vibrations, rotations and translations. In a vibration the relative positions of the nuclei are changing, but the molecular centre of mass remains fixed in space; in a rotation the relative positions of the nuclei are constant and the motion is such that the molecular centre of mass is also fixed in space; in a translation the molecule moves as a whole in space with the nuclei remaining in constant positions relative to each other. The effective potential mentioned above influences only the vibrational motions of the nuclei; the energies associated with rotation and translation are purely kinetic in nature. In the approximation that the electronic, vibrational, rotational and translational motions are truly independent we may divide the Hamiltonian of eqn. (3.4.5) into corresponding components:

$$H = H_E + H_V + H_R + H_T, \tag{1}$$

where

$$H_E \psi_E = E_E \psi_E, \tag{2}$$

$$H_V \psi_V = E_V \psi_V, \tag{3}$$

$$H_R \psi_R = E_R \psi_R, \tag{4}$$

$$H_T \psi_T = E_T \psi_T. \tag{5}$$

The wavefunction ψ of eqn. (3.4.5) is then:

$$\psi = \psi_E . \psi_V . \psi_R . \psi_T, \tag{6}$$

and the total energy E is simply

$$E = E_E + E_V + E_R + E_T. \tag{7}$$

In these equations the subscripts E, V, R and T stand for electronic, vibrational, rotational and translational respectively.

Let us consider in turn the general nature of the energies allowed by eqns. (2)–(5). The spacing of the translational energy levels depends on the size of space in which the molecule is free to move. Consider the linear harmonic oscillator of Section 3.5 Its allowed energies were seen to be

$$E_V = (v + \tfrac{1}{2})(h/2\pi)\sqrt{F/m}. \tag{3.5.8}$$

Now suppose that $F = 0$; the parabola of Fig. 3.5.1 then becomes infinitely broad and the allowed energy levels are infinitely closely spaced, i.e. they form a continuum. Since the dimensions of a molecule are very much smaller than any normal vessel containing it we may conclude, by analogy, that the translational energy levels effectively from a continuum. The Bohr frequency condition,

$$E_m - E_n = h\nu_{mn}, \tag{3.11.12}$$

relates energy level spacings with observable spectroscopic frequencies, and since the translational levels are so closely spaced we would not expect to find interesting spectroscopic transitions associated with the translational motions of molecules.

In Section 3.7 we discussed the energies allowed to a simple, rigid, fixed-axis rotator:

$$E = M^2h^2/8\pi^2 I;\ M = 0, \pm 1, \pm 2 \ldots \tag{3.7.8}$$

A molecule freely rotating in space is not of course constrained to rotate about a fixed axis. For the simple case of a rigid diatomic molecule the allowed energy levels can be shown to be

$$E_R = J(J+1)h^2/8\pi^2 I;\ J = 0, 1, 2, 3 \ldots \tag{8}$$

where J is the rotational quantum number. As in the case of the simple, fixed-axis rotator there is no requirement for any zero-point energy here. The moment of inertia I of a diatomic molecule is given by

$$I = \mu r^2;\ \mu = m_1 m_2/(m_1 + m_2). \tag{9}$$

Here m_1 and m_2 are the atomic masses and r is the internuclear distance; μ is termed the reduced mass of the molecule. If the molecule has an electric dipole moment then arguments along the lines of those outlined in Section 3.12 show that spectroscopic transitions are allowed only between adjacent rotational levels. In other words the selection rule for J is

$$\Delta J = \pm 1. \tag{10}$$

In particular there will be absorption of energy at frequencies corresponding to

$$\nu = (E_{J+1} - E_J)/h = 2(J+1)h/8\pi^2 I. \tag{11}$$

The rotational absorption spectrum therefore consists of an evenly spaced series of transitions. For molecules with very small moments of inertia (e.g. diatomic hydrides) this series may start in the *far infrared region*, but for the majority of molecules the series starts, at least, in the *microwave region* of the electromagnetic spectrum. Rotational absorption frequencies in the microwave region may be measured very accurately and consequently molecular moments of inertia may be precisely determined. For a diatomic molecule the internuclear distance may be calculated from the moment of inertia directly with eqn. (9). The rotational energy manifolds of polyatomic molecules are generally rather more complicated than those of our simple diatomic example. Nevertheless, polar, gaseous, polyatomic molecules do display rotational spectra in the microwave region and when these are analysed precise values for moments of inertia and consequently geometric parameters may be determined.

We will now discuss the nature of molecular vibrational energy and for this purpose we will again use our simple diatomic example, though of course we must abandon our idea of it as a rigid entity. In a vibrational motion the nuclei move relative to each other under the influence of the effective potential mentioned earlier. We may intuitively deduce the general form of this potential function from our chemical experience. We know that a diatomic molecule has a more or less defined internuclear distance which must correspond to a minimum in the potential energy. We would expect the potential energy to increase if we distorted the molecule in either sense from this situation. If we compress the molecule the energy will rise, perhaps slowly at first, but then quickly to very high values as the strong nuclear–nuclear repulsion forces become dominant. On the other hand, if we stretch the molecule the energy will again rise at first, but for larger displacements will tend towards the finite limit corresponding to the dissociation of the molecule into atoms. This general behaviour is illustrated in Fig. 4.1.1. The dissociation energy D_e and the equilibrium internuclear distance r_e are also shown in this figure.

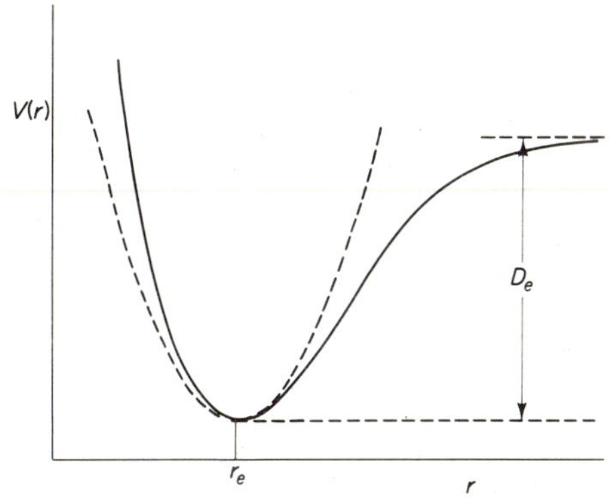

Fig. 4.1.1. Potential energy curve (solid line) for a diatomic molecule. The dotted curve represents the behaviour of the potential energy for a diatomic molecule in the harmonic approximation.

For the linear harmonic oscillator of Chapter 3 the potential function was of parbolic form with a curvature fixed by the force-constant F:

$$V(x) = \tfrac{1}{2}Fx^2. \qquad (3.3.6)$$

As we have seen this resulted in a set of evenly spaced vibrational energy levels given by

$$E_V = (v+\tfrac{1}{2})h/2\pi\sqrt{F/m}. \tag{3.5.8}$$

For small nuclear displacements we might expect our diatomic molecule to behave in the same way as the simple harmonic oscillator. The dotted curve in Fig. 4.1.1 represents the harmonic potential

$$V(r) = \tfrac{1}{2}F(r-r_e)^2, \tag{12}$$

and the resulting energy levels are

$$E_V = (v+\tfrac{1}{2})(h/2\pi)\sqrt{F/\mu} = (v+\tfrac{1}{2})h\nu. \tag{13}$$

Here, as before, μ is the reduced mass and ν is the classical vibrational frequency.

Various analytical expressions have been proposed as better approximations to the potential operating in real diatomic molecules. One of the most useful is that due to Morse:

$$V(r)=D_e\{1-\exp(-\beta(r-r_e))\}^2. \tag{14}$$

This two parameter potential function does at least include the concept of dissociation and with suitable parameters does have the general form of the solid curve of Fig. 4.1.1. The vibrational wave equation for a diatomic molecule assuming a Morse potential can be solved exactly with the result that,

$$E_V = (v+\tfrac{1}{2})h\nu - (v+\tfrac{1}{2})^2 hx\nu. \tag{15}$$

In this equation x is a small positive parameter called the *anharmonicity constant*. It is generally very much smaller than ν so that the diminution in the energy level spacings with increasing v values implied in eqn. (15) is a slow process. Nevertheless, at some finite value of v the energy level spacing becomes zero and the molecule dissociates. The energy level manifold appropriate to this model is shown in Fig. 4.1.2. We note that in both the harmonic and anharmonic approximations to the vibrational behaviour of a diatomic molecule a zero-point energy appears in accordance with the Uncertainty Principle.

For a molecule undergoing *strictly harmonic vibrations* the selection rule for v is just

$$\Delta v = \pm 1, \tag{16}$$

but greater changes in v are weakly allowed when the vibration is *anharmonic*:

$$\Delta v = \pm 1, \pm 2, \pm 3 \ldots. \tag{17}$$

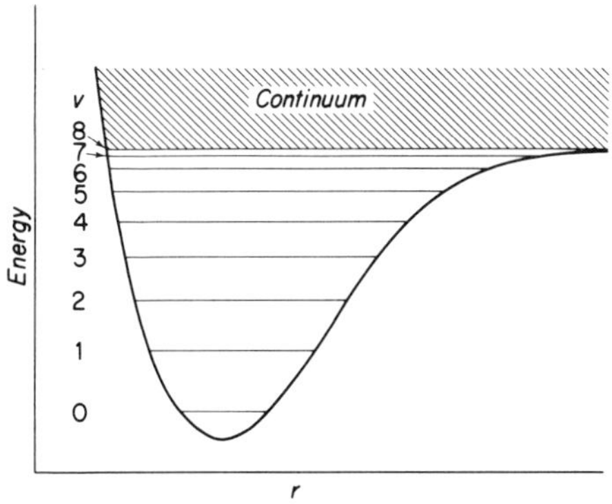

Fig. 4.1.2. Vibrational energy levels and dissociation continuum for a diatomic molecule.

Generally speaking, for a given degree of anharmonicity the intensity of a transition falls off rapidly as the change in the quantum number increases. Furthermore no spectroscopic transitions are possible unless the vibration is accompanied by a *change in the molecular dipole moment*, i.e. transitions may be observed for heteronuclear, but not homonuclear diatomics. For reasons that will be explained in the next section, at normal temperatures the majority of molecules occupy the $v = 0$, or ground vibrational level. The strongest transition observed in absorption is then that from the $v = 0$ to the $v = 1$ levels. This is called the *fundamental transition*. Much weaker transitions may be observed from the $v = 0$ to the higher levels such as those with $v = 2$ or 3. These are called *overtones* and the deviations of their observed frequencies from exact multiples of the fundamental frequency can be used to evaluate the anharmonicity constant x. The forces and masses are such in the majority of molecules that fundamental transitions, and indeed some overtones, occur in the *infrared region* of the electromagnetic spectrum.

In a diatomic molecule there is only a single mode of vibration (the internuclear stretch). A polyatomic molecule has as many independent modes of vibration as there are internal parameters required to specify its structure. For a molecule of N atoms this number is $3N-6$ when the molecule is non-linear, but $3N-5$ when it is linear. The reason for this is simple to understand. For a molecule of N atoms $3N$ coordinates are needed to describe all possible motions of the nuclei. There are then $3N$ *degrees of freedom* available to the molecule. Of these degrees of freedom we must associate three with the translational motions of the molecule and, in the case of a non-linear species, three with the rotational motions. The number remaining for the vibrational motions is therefore $3N-6$ and there must be this number of independent vibrational modes for a non-linear molecule. Rotation about the internuclear line in a linear molecule does not result in any of the $3N$ atomic coordinates changing and therefore does not correspond to a degree of freedom. There are then just two rotational degrees of freedom for a linear molecule and the number of vibrational degrees of freedom and hence modes of vibration is $3N-5$.

The energy of each of the vibrational modes of a polyatomic molecule is quantised in a similar manner to that of a diatomic molecule. Each mode has its own characteristic frequency which may appear in the infrared spectrum if the particular mode involves a dipole moment change. The form of the vibrational modes often approximates to the stretching of a particular bond or the bending of a particular valence angle in the molecule. It is thus often possible to recognise 'group frequencies' in the infrared spectra of complex molecules.

The precise calculation of the electronic energy for all but the simplest molecules presents some severe problems. In part this is because the electrons are no longer influenced by a single, central, nuclear charge as they are in an atom, but, more importantly because of the electron–electron interactions. The dynamics of systems consisting of many interacting bodies are always difficult to handle even in classical mechanics. However, tremendous strides have been made in recent years by the application of approximate methods, some of which are directly adapted from those used in problems of celestial mechanics. The feasibility of performing these complex calculations has been greatly facilitated by the development of the high speed digital computer. Whatever the difficulties may be in precisely calculating the energies allowed to the electrons in a molecule, the general picture is quite clear. A number of electronic states are available to a molecule and the lowest of these in energy is the ground electronic state. Not all electronic states are necessarily stable; when excited to some states the molecule may spontaneously dissociate. For a diatomic molecule in a stable state the potential influencing the relative motions of the nuclei

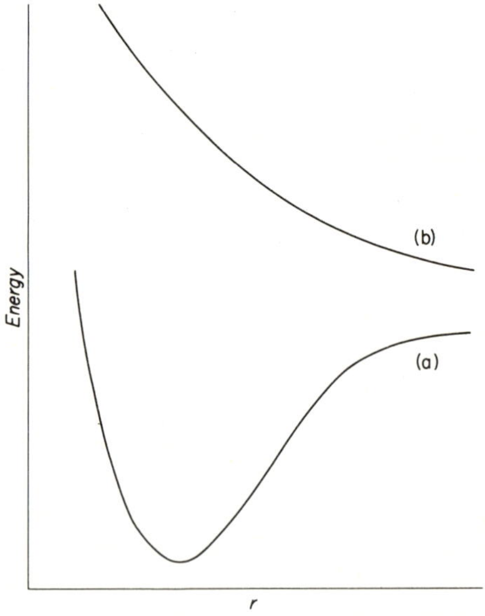

Fig. 4.1.3. Potential energy curves of a diatomic molecule in (a) a stable state, (b) an unstable or dissociative state.

would be of the Morse type function shown in the lower part of Fig. 4.1.3. At least one vibrational level must exist in the potential well for the molecule to be stable. The upper curve of Fig. 4.1.3 represents a typical potential energy function for a diatomic molecule in an unstable or dissociative state. The energy levels associated with such a state form a continuum.

The lower excited electronic states of molecules are often stable and their energy difference from the ground state typically corresponds to a quantum of radiation in the *ultraviolet* and sometimes *visible regions* of the spectrum. It is in these regions therefore that we may expect to observe molecular electronic spectra. The excitation of a molecule to a higher electronic state may result in the geometry (bond distances and angles) changing from that of the ground electronic state. Thus the internuclear distances in many diatomic molecules increase in excited electronic states from their ground state values; some polyatomic molecules such as HCN or C_2H_2 which are linear in their ground states adopt a bent configuration in some of their low lying excited states; formaldehyde and other normally planar molecules become non-planar for some electronic excitations. These effects are not unexpected since the excitation directly involves the very particles (the electrons) which bind the molecule together. It should be appreciated that one or more electrons may be completely removed from the molecule if sufficient excitation energy is provided. The resulting ion may or may not be a stable species. Some features of the electronic spectra of diatomic molecules are illustrated in Fig. 4.1.4.

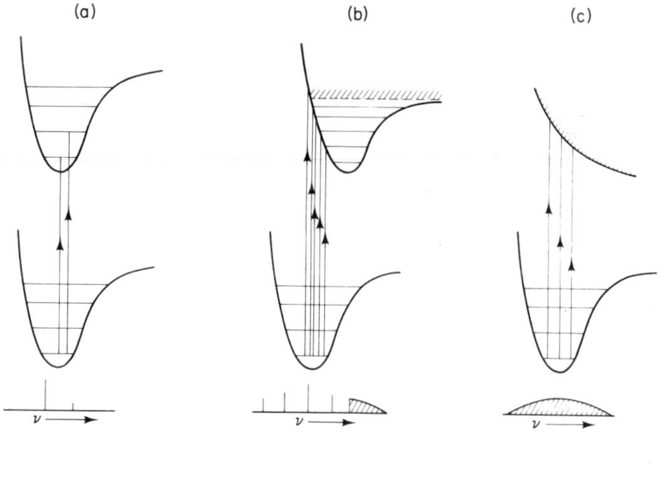

Fig. 4.1.4. Some possible types of electronic absorption transitions in diatomic molecules and the resulting spectra. In (a) the transition occurs between two stable states with similar internuclear separations and the spectrum is a short *progression* in the upper state vibrational interval. In (b) the upper state is stable, but has a larger internuclear separation than the lower state. This generally results in a long progression in the upper state interval, and may give rise to a dissociation continuum. In (c) the upper state is unstable and continuous absorption results. The intensity distributions within the spectra are in accord with the Franck–Condon Principle (Sec. 4.2.).

At this point it may be useful to summarise our ideas of the quantised energies available to a free molecule and this is done in Fig. 4.1.5. The relative magnitudes of the different energy intervals should be carefully noted.

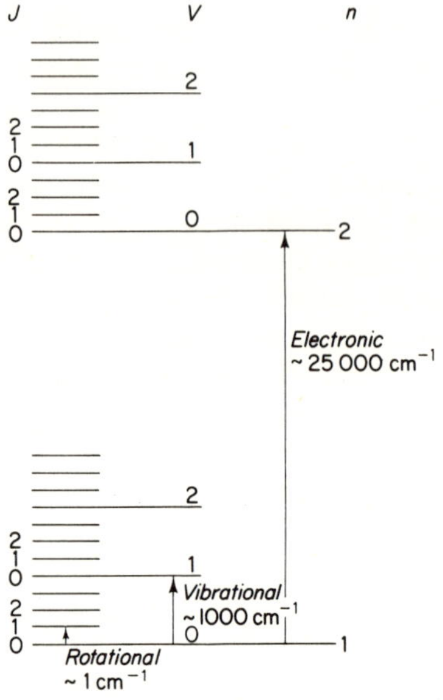

Fig. 4.1.5. Schematic representation of the rotational and vibrational energy levels for the ground and an excited electronic state of a diatomic molecule.

There are other aspects of molecular spectroscopy besides those involving rotational, vibrational and electronic transitions and some of these find wide application in chemistry. Perhaps the most important of these involve properties of nuclei and electrons which we have not mentioned previously in our discussion. Electrons and some nuclei possess intrinsic angular momentum or *spin*. Nuclear spins are characterised by a quantum number I; electron spin is characterised by a quantum number s. Individual nuclei *in their ground states* have specific values of I which may be integral, half-integral or zero. The spin quantum number s of an electron is one-half. A nucleus with non-zero spin and an electron possess magnetic moments coincident in line with the axis of the particle's spin and these magnetic moments interact with each other and with any applied magnetic field. The orientations allowed to a magnetic nucleus or an electron in an applied field are limited in number to $2I + 1$ and $2s + 1$ respectively. There are thus two possible orientations for a proton ($I = \frac{1}{2}$) and an electron. The different orientations correspond to distinct energy states which with an applied field of several thousand gauss are separated by an amount corresponding to a quantum of *radio frequency* for the proton and a quantum of *microwave* frequency for the electron. Transitions between such levels give rise to the phenomena of *nuclear magnetic resonance* and *electron spin resonance*.

A nucleus within a molecule is subject to internal magnetic fields which depend on its particular environment. Thus the exact resonance condition (radio frequency to applied field ratio) for a proton will depend on its particular chemical situation. This ability to differentiate magnetic nuclei in different chemical positions makes nuclear magnetic resonance a powerful analytical method in chemistry. Innumerable studies of proton magnetic resonances have been carried out, but work with other magnetic nuclei such as ^{19}F, ^{31}P, ^{15}N and ^{13}C is also common.

In many stable molecules the electron spins are paired and electron spin resonance may not be observed. However, radicals and other species with unpaired electron spin magnetic moments may be studied by electron spin resonance spectroscopy. Substances containing transition metal elements are frequently paramagnetic and may give rise to sharp microwave absorptions when placed in an appropriate magnetic field. However, in these materials the electron spin may not be the only contributor to the paramagnetism and the study of the resonance phenomena is then more accurately described as electron paramagnetic resonance spectroscopy. Electron resonance spectra frequently display fine structures due to the presence of magnetic nuclei in the molecule. The careful analysis of these fine structures and other details of the spectra can yield information on the distribution of the unpaired electron in the molecule, the nature of the bonding and other subtle features of the molecular electronic structure. Electron resonance spectroscopy provides a powerful method for the study of free radicals.

The charge distributions of nuclei with spin quantum numbers greater than one half are non-spherical. These nuclei possess electric quadrupole moments which can interact with non-uniform electric fields. The interaction energy depends on the orientation of the quadrupolar nucleus with respect to the electric field gradient, and just $2I + 1$ specific orientations are allowed. A nucleus within a molecule is subject to an electric field gradient due to the other charged particles in its immediate vicinity, unless these happen to have a spherical distribution. Direct transitions between the levels arising from the interactions of quadrupolar nuclei within molecules generally occur in the *radio frequency region* of the spectrum. No applied magnetic field is required to observe these transitions. Their measurement provides information on the immediate electronic environment of the quadrupolar nucleus and hence on the electronic structure of the molecule.

When formed in a radioactive decay process some nuclei occur in excited spin states and may then return to their ground states with the emission of a γ-ray photon. In an isolated atom the emission of the γ-ray is accompanied by a recoil of the nucleus so that momentum is conserved. However, for nuclei situated in small crystals effectively *recoilless emission of γ-rays* may occur. If these are allowed to impinge on a sample containing ground state nuclei identical to those emitting a resonant absorption process may occur. The precise resonance condition may not be met if the absorbing nuclei are in different environments to those emitting the

γ-rays. However, resonance may be brought about by shifting the frequency of the source radiation by moving the source relative to the absorber, i.e. by making use of the Doppler effect. Then if the relative velocity of source to absorber is varied over a range the absorption spectrum of the sample may be investigated. This type of study is known as *Mössbauer spectroscopy*. Mössbauer spectra often exhibit fine structures due to nuclear quadrupolar interactions or to the interactions of nuclear magnetic moments with internal magnetic fields. Applied magnetic fields may also produce splittings. The analysis of these effects yields similar information to that obtained by nuclear quadrupole resonance and nuclear magnetic resonance spectroscopies. In addition there are subtle shifts found in Mössbauer spectra. On excitation to a higher spin state a nucleus may change its radius. This alters the nuclear interaction with the immediate electron density. The *s* electrons are those with density at the nucleus so these 'isomer shifts' are related to the way in which these electrons are influenced on molecule formation. The scope of Mössbauer spectroscopy is somewhat limited by the relative scarcity of suitable isotopes; ^{57}Fe, ^{67}Zn, ^{119}Sn and ^{129}I are among the nuclei which do have suitable excited spin states.

A schematic representation of the electromagnetic spectrum is shown in Fig. 4.1.6. In this diagram the important spectroscopic regions we have mentioned in this section are indicated.

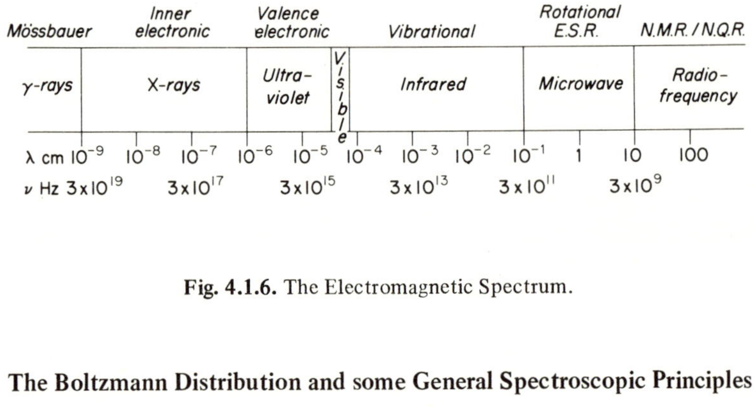

Fig. 4.1.6. The Electromagnetic Spectrum.

4.2 The Boltzmann Distribution and some General Spectroscopic Principles

The nature of spectroscopy is strongly influenced by the way in which molecules in thermal equilibrium distribute themselves among the available energy levels. Consider the simple two level system of Fig. 4.2.1. For thermal equilibrium at temperature T the number of molecules in the upper state N_m is related to the number in the lower state N_n by the Boltzmann distribution law:

$$N_m/N_n = \exp(-\Delta E/kT), \tag{1}$$

where k is the Boltzmann constant. If the upper state is g_m-fold degenerate and the lower state g_n-fold degenerate then eqn. (1) must be modified to

$$N_m/N_n = (g_m/g_n)\exp(-\Delta E/kT). \tag{2}$$

For a system of N molecules with several available levels the fraction of molecules in a given state i is

$$N_i/N = g_i \exp(-\Delta E_i/kT)/\sum_i g_i \exp(-\Delta E_i/kT), \tag{3}$$

in which ΔE_i is the energy difference between the state i and the lowest energy state in the system. The sum appearing in the denominator on the right-hand side of this expression is called the *partition function*. Neglecting the effects of degeneracies for the moment, the appearance of the energy difference as a negative exponent in these

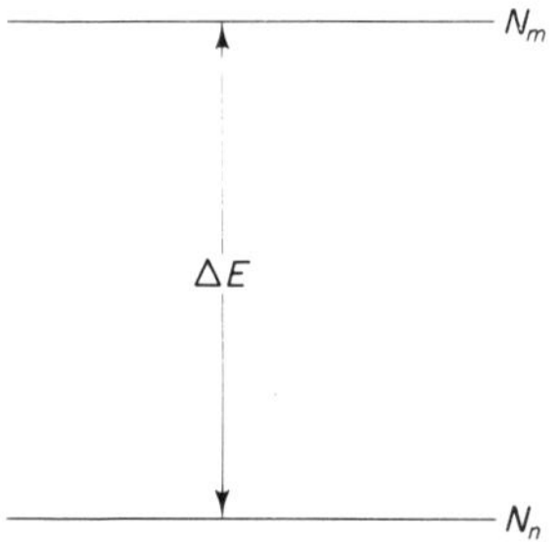

Fig. 4.2.1. A simple two level system.

expressions requires the level population to decrease as the energy level manifold is ascended. The rate of population fall-off between levels will be largest when the energy differences become large with respect to kT. At 300K kT is equivalent to about 0·596 kcal mole^{-1} or 208 cm^{-1}. Rotational, vibrational and electronic energy intervals are typically about 1 cm^{-1}, 1000 cm^{-1} and 25 000 cm^{-1} respectively. We anticipate therefore that a number of the rotational levels available to the molecules will have appreciable populations at normal temperatures, but that only the very lowest vibrational levels will be significantly occupied under these conditions. Virtually all the molecules of a substance in thermal equilibrium occupy the ground electronic state at normal temperatures.

The degeneracy factors appearing before the exponential terms in eqns. (2) and (3) may contain several contributions. Some electronic or vibrational states may be degenerate and the appropriate numerical factors must be included. The $2J + 1$ degeneracy of rotational states and any other degeneracies in the rotational energies must be allowed for. Then the $2I + 1$ orientation degeneracy of each nucleus present in a molecule will have to be taken into account unless such degeneracies are lifted by electric or magnetic interactions. As we shall see in Section 5.6 the presence of equivalent nuclei in molecules may have marked effects on the relative occupancies within the manifold of the rotational levels.

Having established these ideas on the thermal distribution of molecules among the different energy states we must return briefly to some consideration of the interaction of electromagnetic radiation with molecules.

In Section 3.11 we investigated the interaction of a charge oscillating in one direction x with one component of the electric field $\mathcal{E}_x^0$ associated with electromagnetic radiation. The result was that if the oscillator was initially in the state ψ_n it would undergo a transition to state ψ_m with a probability $|c_m|^2$ given by

$$|c_m|^2 = (\pi^2/h^2)\,\mathcal{E}_x^{0^2}\,\{\int \psi_m^* ex\psi_n d\tau\}^2 t, \tag{3.11.14}$$

or making use of eqn. (3.11.15)

$$|c_m|^2 = (\pi^2/h^2)\,\mathcal{E}_x^{0^2}\,|M_x|^2 t. \tag{4}$$

The more general problem involves the interaction of natural isotropic radiation with a three-dimensional oscillator. The electric field components of natural isotropic radiation are related by

$$\mathcal{E}_x^{0^2} = \mathcal{E}_y^{0^2} = \mathcal{E}_z^{0^2} = \mathcal{E}^{0^2}/3, \tag{5}$$

and for a general oscillating charge we have

$$M^2 = M_x^2 + M_y^2 + M_z^2. \tag{6}$$

The transition probability for this case then becomes

$$|c_m|^2 = (\pi^2/h^2)(\mathcal{E}^{0^2}/3)M^2 t. \tag{7}$$

Now the energy density associated with a sinusoidally oscillating electric field is $\mathcal{E}^{0^2}/16\pi$ and this is just half the total radiation density as an equal contribution is associated with the magnetic component. In terms of the incident radiation density ρ the transition probability may be written as

$$|c_m|^2 = (8\pi^3/3h^2)M^2 \rho t. \tag{8}$$

Let us now assume that molecules are distributed between the two states ψ_m and ψ_n in thermal equilibrium and that radiation of density ρ is present. Under these conditions let there be N_m molecules in state ψ_m and N_n in state ψ_n. Molecules may be transferred between the two states by three radiative processes: stimulated absorption, stimulated emission and spontaneous emission. The rates at which these processes occur may be expressed in terms of probability coefficients:

$$\text{Rate of stimulated absorption} = B.N_n.\rho; \tag{9}$$

$$\text{Rate of stimulated emission} = B.N_m.\rho; \tag{10}$$

$$\text{Rate of spontaneous emission} = A.N_m. \tag{11}$$

The coefficients A and B are known as *Einstein's probability coefficients* for the spontaneous and stimulated processes. From eqn. (8) we may identify B as

$$B = (8\pi^3/3h^2)M^2, \tag{12}$$

and this can in principle be calculated from a knowledge of the molecule and the transition involved. We will now deduce the relationship between A and B.

For the condition of equilibrium to be maintained the overall rate of absorption must exactly balance that of emission. We therefore have from eqns. (9), (10) and (11):

$$N_m(A + B\rho) = N_n B\rho, \tag{13}$$

or

$$A = ((N_n/N_m) - 1)B\rho. \tag{13(a)}$$

From the Boltzmann distribution law we have:

$$N_n/N_m = \exp(h\nu/kT), \tag{14}$$

where ν is the frequency equivalent to the energy difference between the two states ψ_m and ψ_n. One of Planck's revolutionary contributions to Physics about the turn of the century was to relate the energy density of radiation correctly to its frequency with the expression:

$$\rho = 8\pi h\nu^3/\{c^3(\exp(h\nu/kT) - 1)\}. \tag{15}$$

Combination of eqns. (13(a)), (14) and (15) yields the result we seek, namely,

$$A = (8\pi h\nu^3/c^3)B. \tag{16}$$

The implications of eqn. (16) should be fully appreciated. The coefficient $8\pi h/c^3$ is a very small quantity so that at low values of the frequency ν the value of A is very much less than that of B. Thus at relatively low frequencies, such as are appropriate to the radio-frequency, microwave and, to some extent, infrared regions, stimulated processes are by far the most important. Only when the frequency becomes high, as in the visible or ultraviolet region do spontaneous processes become competitive with those stimulated by the radiation. The spontaneous emission of radiation from excited atoms in the visible region will no doubt be familiar to the reader who has conducted flame tests for alkali metals and the like.

So far in this discussion we have considered the absorption and induced emission of radiation to occur through the interaction of the radiation's electric field with an electric dipole in a molecule. When it is allowed, this electric dipole mechanism gives rise to the strongest spectroscopic transitions, but other, less favourable, mechanisms are possible and these may be important. Higher order electric moments such as quadrupoles may interact with the electric field of the radiation and may enable otherwise forbidden transitions to occur weakly. Perhaps more generally important is the mechanism involving the interaction of magnetic dipoles in a molecule with the magentic field of the radiation. The transition in nuclear magnetic resonance, electron spin resonance, pure quadrupole resonance and Mössbauer spectroscopies occur essentially through a magnetic dipole mechanism and they are consequently inherently rather weakly allowed processes. To give some idea of the relative efficiencies of the various mechanisms the A coefficient for a visible transition might typically be (in s^{-1}):

for an electric dipole transition $\sim 10^8$;

for a magnetic dipole transition $\sim 10^3$;

for an electric quadrupole transition ~ 1.

If the mean excited state lifetime τ depends only on the state's ability to spontaneously radiate energy by one of the above mechanisms, its value will be just the reciprocal of the appropriate A coefficient. Thus in the above example for relaxation by emission of a visible photon:

τ electric dipole $\sim 10^{-8}$ s;

τ magnetic dipole $\sim 10^{-3}$ s;

τ electric quadrupole ~ 1 s.

The mean lifetime of a state determines the precision to which the energy of that state is defined. This is a direct requirement of the Uncertainty Principle. For short lived states the energy is only roughly defined and spectroscopic transitions involving these states will be correspondingly broad in nature; conversely long lived states will generally give rise to sharp spectral transitions. In a real situation the width of spectral features may depend on a number of factors. These include: the frequency and intensity of any radiation present; the influence of fluctuating electric and magnetic fields due to neighbouring molecules on the particular energy levels; the general frequency and nature of molecular collisions; Doppler effects arising from the spread in molecular velocities relative to the source of radiation or the detecting system.

There is one further aspect of spectral transitions that merits mention here: the time involved in the actual transition itself. For most purposes we may consider that a transition occurs in a very short time period indeed, in fact virtually instantaneously, at least as far as the nuclear motions are concerned. This idea is embodied in an important spectroscopic concept known as the *Franck–Condon principle*.

According to this principle spectroscopic transitions, such as those concerned with the electronic excitations of a diatomic molecule shown in Fig. 4.1.4, take place without alteration of the relative nuclear positions. That is the transitions may be considered to be 'vertical' in the sense that they can be represented by vertical lines on a diagram such as Fig. 4.1.4. The Franck–Condon principle finds wide application in the interpretation of molecular electronic spectra, but as we will not be considering these spectra in any detail elsewhere in this book we need go no further here.

By this time the reader should have formed some impression of the many different aspects of molecular spectroscopy and he will recognise the basic unity of the subject. In the subsequent chapters we will discuss in more detail some of the areas mentioned here. We cannot hope to be comprehensive in our coverage of topics or their details in a volume of this size. The interested reader is therefore encouraged to make use of the Bibliographies given at the end of each chapter.

Chapter 5

The rotation of molecules

In this chapter we will develop our ideas of molecular rotation in some detail, but let us first consider this motion in the context of the three states of matter.

In the vapour state at low pressure the effect of molecular collisions is quite small, essentially because the period of rotation is short in comparison with the interval between collisions; therefore a molecule may execute a substantial number of revolutions before it is disturbed, and perhaps transferred to a different rotational energy level, by colliding with another molecule. A molecule in the gaseous state is described as "isolated", to emphasise the degree to which its interaction with the other molecules can be neglected. The pressure at which interaction becomes appreciable depends upon the intermolecular forces and so varies from one molecule to another; but as a very rough guide, 0·01 mm of mercury is an upper limit to the region of pressure where the effect of collisions upon rotation can be ignored. However, the perturbing influence of collisions upon the rotational energy levels increases only gradually with pressure and the formulae of Section 5.2 may remain useful even at pressures in excess of one atmosphere.

When a molecule is closely surrounded by others, as in a liquid or solid, the character of its motion changes radically. In a crystal the degrees of freedom allotted to the rotation of an isolated molecule are expended in modes of vibration about an equilibrium *orientation*, just as the translation of a free molecule is transformed in the solid state into modes of vibration about an equilibrium position. These forms of vibratory motion in a crystal, known as the *lattice modes*, must be taken into account in the refinement of molecular structures determined by diffraction methods (Chapter 9). In a liquid the situation is still more disorganised because there is no preferred orientation, and rotation degenerates into a Brownian motion, repeatedly changing speed or direction under the buffeting that a molecule receives from its neighbours. Under these conditions the quantised nature of rotational energy disappears. Rotation in solids and liquids is sometimes examined to obtain information about intermolecular forces, or about some specific property such as the dipole moment; but if we wish to study rotational energy levels we must limit our attention to molecules in the gas phase.

5.1 Moments of Inertia of a Rigid Molecule

The moment of inertia of a rigid molecule about any axis passing through the centre of mass is defined by

$$I = \sum_i m_i r_i^2, \tag{1}$$

in which r_i is the perpendicular distance from the axis and m_i the mass of the i-th nucleus. It is a theorem of mechanics that the locus of points formed by plotting $I^{-\frac{1}{2}}$ radially from the centre of mass in the direction of the axis of rotation is the surface of a triaxial ellipsoid, known as the *momental ellipsoid*. The three mutually perpendicular axes of the momental ellipsoid coincide with the *principal axes of inertia* of the molecule, about which rotation is dynamically balanced. It is customary to label the axes of the ellipsoid a, b and c in order of decreasing length, so that a is the major axis, b the intermediate, and c the minor axis of the figure. Since the axial lengths are proportional to $I^{-\frac{1}{2}}$ it follows that $I_a \leqslant I_b \leqslant I_c$ always. Sections through the momental ellipsoid of a simple molecule, CH_2O, are shown in Fig. 5.1.1.

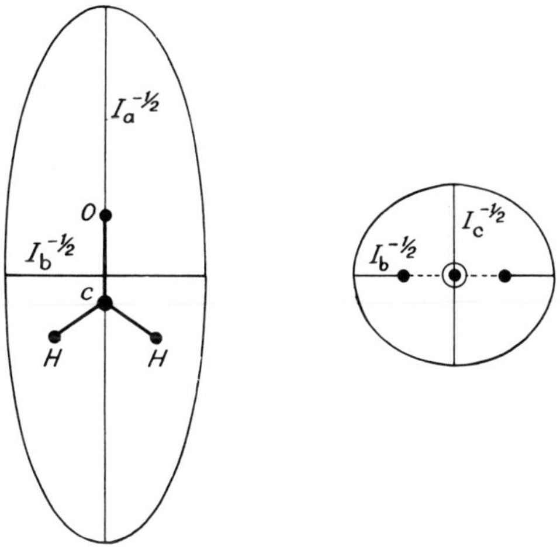

Fig. 5.1.1. Principal sections of the momental ellipsoid of CH_2O.

In symmetrical molecules the direction of the principal axes of inertia can often be determined by inspection. Thus the fact that a, b, and c are symmetry axes of the momental ellipsoid means that a symmetry axis present in the molecule must coincide with one of these axes. Likewise a plane of symmetry in the molecule must necessarily be a principal section of the ellipsoid; that is, it must contain two of the principal axes of inertia and be perpendicular to the third. Let us use these ideas to find the orientation of the principal axes of inertia in the molecule CH_2O. The elements of symmetry in the molecule are the two-fold axis C_2, and two mutually perpendicular planes of symmetry σ_v containing the C_2 axis. One of the axes of inertia must then coincide with the C_2 axis; while, of the second and third axes, one must lie in each of the planes σ_v. Which of the axes of inertia is the least (I_a) and which the greatest (I_c) is a decision to be made by calculation rather than from symmetry considerations alone. It turns out that I_a is coincident with C_2, that I_b lies in the molecular plane (normal to the C_2 axis), and that I_c is normal to the molecular plane. Indeed the direction of I_a is self-evident, for only the light hydrogen atoms contribute to the moment for rotation about C_2.

When the orientation of the principal axes can be established in this way, it is relatively easy to calculate the corresponding moments of inertia using eqn. (1). However, if the molecule lacks the necessary symmetry, so that the positions of the axes are not obvious, the moments are obtained from the determinantal equation,*

$$\begin{vmatrix} I_{xx}-I & -I_{xy} & -I_{xz} \\ -I_{xy} & I_{yy}-I & -I_{yz} \\ -I_{xz} & -I_{yz} & I_{zz}-I \end{vmatrix} = 0. \tag{2}$$

Here $I_{xx}, I_{xy} \ldots$ are moments and products of inertia with respect to any convenient frame of cartesian coordinates. Thus

$$I_{xx} = \sum_i m_i(y_i^2 + z_i^2) - \frac{(\sum_i m_i y_i)^2}{\sum_i m_i} - \frac{(\sum_i m_i z_i)^2}{\sum_i m_i}, \tag{3}$$

and

$$I_{xy} = \sum_i m_i x_i y_i - \frac{\sum_i m_i x_i . \sum_i m_i y_i}{\sum_i m_i}, \tag{4}$$

where x_i, y_i, and z_i are the coordinates of the i-th atom of mass m_i, and the sum is taken over all atoms present in the molecule. The three roots of eqn. (2) are the principal moments of inertia. Usually some simplification is possible. For instance, if a molecule possesses a symmetry plane, one of the principal axes must be perpendicular to the plane. Let this be the x direction, then the products of inertia I_{xy} and I_{xz} are zero and the determinant factorises into one linear and one quadratic expression. In the event that x, y and z coincide with the principal axes of inertia, all the products of inertia are zero, and the determinant factorises into three linear equations each precisely like (1).

* Hirschfelder, *J. Chem. Phys.*, 1940, **8**, 431.

5.2 Rotational Energy Levels

In classical mechanics the total Hamiltonian energy of a rotator with one degree of freedom is $(I\omega)^2/2I$ (eqn. 3.7.1). The generalised rotation of a molecule has components along three perpendicular directions (*i.e.*, it has three degrees of freedom) and hence the total energy of rotation is

$$E_R = \frac{P_a^2}{2I_a} + \frac{P_b^2}{2I_b} + \frac{P_c^2}{2I_c}, \tag{1}$$

in which $P_a (= I_a\omega_a)$, P_b, and P_c are the components of the total angular momentum P along the principal axes a, b and c. The total angular momentum is given by

$$P^2 = P_a^2 + P_b^2 + P_c^2. \tag{2}$$

Before we consider the quantisation of the energy, let us describe the possible simplification of the classical expression (1) for molecules that possess some degree of symmetry. The basis of the simplification is the symmetry of the momental ellipsoid, and for real molecules there are four separate classes:

(*a*) The highest symmetry of the momental ellipsoid is attained when the principal axes are of equal lengths, $a = b = c$: the figure is then a sphere. The moment of inertia is the same about any axis through the centre of mass, so that we may drop the subscripts formerly attached to the moments. Hence (1) becomes

$$E_R = (P_a^2 + P_b^2 + P_c^2)/2I. \tag{3}$$

Introducing (2) we find

$$E_R = P^2/2I. \tag{4}$$

Molecules in this class are known as *spherical tops*. Examples are methane and carbon tetrachloride.

(*b*) For a *linear molecule* $I_a = 0$ (corresponding to zero moment about the internuclear axis) and $I_b = I_c$. Therefore the principal axes of inertia are mutually perpendicular, and perpendicular to the internuclear axis: the momental ellipsoid is a circular cylinder of infinite length. Because I_a is zero, the component of momentum $P_a (= I_a\omega_a)$ along a is zero also. Thus (2) reduces to $P^2 = P_b^2 + P_c^2$, and eqn. (1) gives

$$E_R = P^2/2I_b. \tag{5}$$

(*c*) A *symmetric top* has two moments of inertia equal to one another but different from the third. The top is described as *prolate* if $I_a < I_b = I_c$, or *oblate* if $I_a = I_b < I_c$. A prolate top is shaped like a shuttlecock (examples are CH_3Cl and CH_3CN), whilst an oblate top resembles a discus (*e.g.*, BF_3 and C_6H_6). Any molecule with a single three-fold or higher principal axis of symmetry is necessarily a symmetric top, the principal axis of the molecule being coincident with the unique axis of the top. In the examples mentioned above the principal axis is C_3, except for the benzene molecule where it is C_6. The momental ellipsoid of a symmetric top is an ellipsoid of revolution.

For a prolate symmetric top ($I_b = I_c$) the expression (1) for the energy can be rearranged to

$$E_R = \frac{P^2}{2I_b} + \frac{P_a^{\,2}}{2}\left\{\frac{1}{I_a} - \frac{1}{I_b}\right\}. \tag{6}$$

This equation can be verified quite easily from the vector diagram in Fig. 5.2.1. If the top is oblate the only effect is that the subscript c replaces a in (6), as the reader should check for himself.

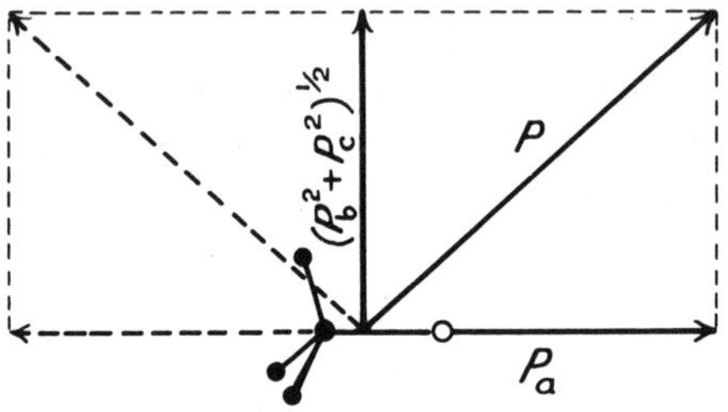

Fig. 5.2.1. Vector diagram for a prolate symmetric top. The dotted lines show the effect of reversing the sign of rotation about the top axis. In wave mechanics, $P = J$ and $P_a = K$.

(*d*) The *asymmetric top* class of molecules has all three principal moments unequal. Therefore the momental ellipsoid has three unequal axes, and the classical expression (1) cannot be reduced to a simpler form.

A molecule is an asymmetric top unless it possesses at least one three-fold or higher symmetry axis. A majority of molecules found in nature are asymmetric tops.

Quantisation of Rotation. The quantisation of rotation restricts the rotational energy of molecules to the sharply-defined values allowed by wave mechanics. In a rigorous approach one must determine the eigenvalues of the operator corresponding to the classical Hamiltonian energy for rotation, given by the expression (1). This treatment can be found in standard texts on wave mechanics: it leads to a number of rules which may be used retrospectively to adapt the classical equations to quantised conditions. This is the procedure we shall follow.

The first rule relates to the total angular momentum of a molecule. For quantised rotation we use the letter $\boldsymbol{J}$ to denote the total angular momentum vector, in place of the classical symbol $\boldsymbol{P}$. The quantum rule is that the permitted values of $\boldsymbol{J}$ are multiples of $h/2\pi$ given by the expression

$$\boldsymbol{J} = \sqrt{J(J+1)}\,h/2\pi, \qquad J = 0, 1, 2, \ldots \tag{7}$$

where J is a quantum number that may be any positive integer, zero included.

The second rule refers to the component $\boldsymbol{K}$ of the total angular momentum $\boldsymbol{J}$ along the axis of a symmetric top. We recall the discussion in Section 3.7, where it was found that the angular momentum P_z about an axis z fixed in space is given by

$$P_z = Mh/2\pi, \qquad M = 0, \pm 1, \pm 2, \ldots \tag{3.7.10}$$

The substance of the rule is that, for symmetric top molecules, the component of $\boldsymbol{J}$ parallel to the top axis obeys an expression of the same form as (3.7.10). Therefore the axial component of $\boldsymbol{J}$ may be written

$$\boldsymbol{K} = Kh/2\pi, \tag{8}$$

wherein K is a second quantum number which may assume integral values, including the value zero. However, since $\boldsymbol{K}$ is the axial component of $\boldsymbol{J}$, the quantum number K cannot be greater than J; therefore the spectrum of possible values of K is

$$K = 0, \pm 1, \pm 2, \ldots \pm J. \tag{9}$$

The simultaneous quantisation of the total angular momentum and of the component parallel to the top axis means that the orientation θ of the total angular momentum with respect to the axis is also quantised. The angle θ can assume only the discrete values given by

$$\cos\theta = \boldsymbol{K}/\boldsymbol{J} = K/\sqrt{J(J+1)}. \tag{10}$$

The spatial quantisation is illustrated for $J = 3$ in Fig. 5.2.2. Since the magnitude $\sqrt{J(J+1)}h/2\pi$ of the vector $\boldsymbol{J}$ exceeds the maximum value $Jh/2\pi$ of its component $\boldsymbol{K}$, it is clear that $\boldsymbol{J}$ can never point exactly in the direction of the top axis.

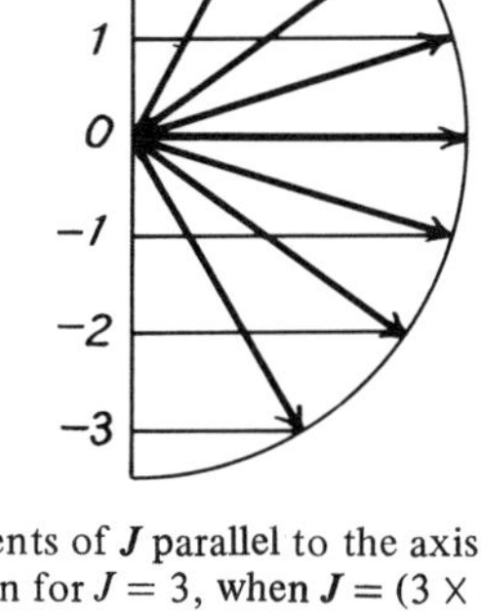

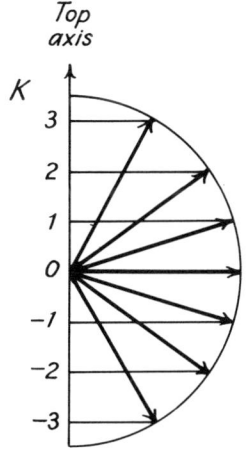

Fig. 5.2.2. Components of $\boldsymbol{J}$ parallel to the axis of a symmetric top. The diagram is drawn for $J = 3$, when $\boldsymbol{J} = (3 \times 4)^{\frac{1}{2}} = 3{\cdot}464$ in units $h/2\pi$.

Let us determine the quantised rotational energy levels of a prolate symmetric top. Eqn. (6) gives the classical energy: merely substituting $\boldsymbol{J}$ and $\boldsymbol{K}$ for the classical symbols P and P_a we have

$$E_R = \frac{\boldsymbol{J}^2}{2I_b} + \frac{\boldsymbol{K}^2}{2}\left\{\frac{1}{I_a} - \frac{1}{I_b}\right\}. \qquad (11)$$

Introducing the quantisation of momentum given by eqns. (7) and (8) then yields

$$E_R = J(J+1)\frac{h^2}{8\pi^2 I_b} + K^2\left\{\frac{1}{I_a} - \frac{1}{I_b}\right\}\frac{h^2}{8\pi^2}. \qquad (12)$$

This is the equation we require. If both sides are divided by h the equation simplifies in form to

$$E_R/h = J(J+1)B + K^2(A-B) \qquad (13)$$

in which

$$A = h/8\pi^2 I_a \text{ and } B = h/8\pi^2 I_b. \qquad (14)$$

A and B are known as *rotational constants*; they are proportional to reciprocal moments of inertia, the factor of proportionality being $h/8\pi^2$. This factor may be expressed in any convenient units. For rotational constants in MHz (10^6 Hz) and moments of inertia in u.Å^2, $h/8\pi^2$ has the value 505 376; for rotational constants given as wavenumbers (cm^{-1}) and moments of inertia in u.Å^2, the factor is $h/8\pi^2 c$ which has the value 16.8575.

The rotational energy levels of a prolate top are shown diagrammatically in Fig. 5.2.3. When the top is oblate the subscript c replaces a in (6), and therefore $C = h/8\pi^2 I_c$ replaces A in (13). The formula for the energy levels of an oblate top then is

$$E_R/h = J(J+1)B + K^2(C-B). \qquad (15)$$

From the convention that $I_a \leqslant I_b \leqslant I_c$ it follows that $A \geqslant B \geqslant C$, and hence that the quantity $(C-B)$ in (15) is always negative.

If we refer to Fig. 5.2.2 we see that the axial component $\boldsymbol{K}$ of the angular momentum of a symmetric top may be positive or negative. However, it follows from eqn. (11) that E_R depends upon the magnitude of $\boldsymbol{K}$ but not upon its sign: thus every energy level for which $K \neq 0$ is doubly degenerate. The degeneracy corresponds physically to opposite directions of rotation about the top axis. The total number of sub-levels belonging to any value of J is therefore $2J+1$, of which $2J$ sub-levels occur as degenerate pairs and one is non-degenerate.

For a linear molecule we replace P^2 in the classical expression (5) by $\boldsymbol{J}^2$ and then introduce the quantisation condition (7) to obtain

$$E_R = J(J+1)h^2/8\pi^2 I_b \tag{16}$$

whence

$$E_R/h = J(J+1)B. \tag{17}$$

The energy levels are the same as those in the $K = 0$ manifold of sub-levels of a symmetric top. The formulae (16) and (17) apply also to spherical top molecules.

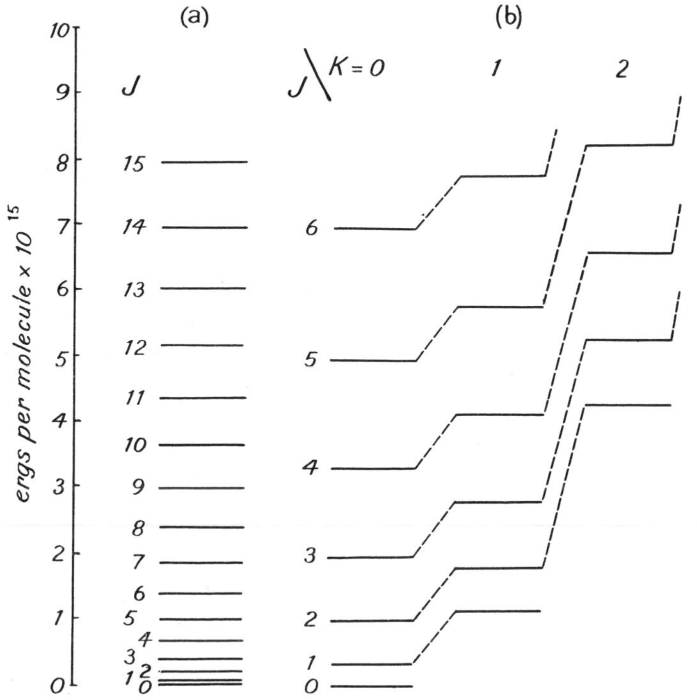

Fig. 5.2.3. Rotational energy levels for a linear molecule (a) and a prolate symmetric top (b). The energies are calculated for $B = 5000$ MHz (a) and $A = 150\,000$, $B = 25\,000$ MHz (b); *i.e.*, they are roughly to scale for carbon oxysulphide and methyl fluoride, respectively.

In asymmetric top molecules the K-degeneracy of the symmetric top is lifted and there are $2J + 1$ levels of different energy for each value of J. The quantisation of the classical energy is too involved to be treated here; but explicit solutions have been obtained for small values of J, of which those for $J = 0$ to 2 are given in

Table 5.2.1. The lowest energy sub-level belonging to a particular value of J is denoted J_{-J}, the next J_{-J+1}, and so on up to J_{+J}. Thus the sub-level 1_{-1} is lower in energy than 1_0 which in turn is lower than 1_{+1}, as the reader should check for himself with the aid of the table, recalling that $A > B > C$.

Table 5.2.1. Rotational Energy Levels ($J = 0$ to 2) of a Rigid Asymmetric Top

J_τ	E_R/h
0_0	0
1_{-1}	$B + C$
1_0	$A + C$
1_{+1}	$A + B$
2_{-2}	$2A + 2B + 2C - 2[(B - C)^2 + (A - C)(A - B)]^{\frac{1}{2}}$
2_{-1}	$A + B + 4C$
2_0	$A + 4B + C$
2_{+1}	$4A + B + C$
2_{+2}	$2A + 2B + 2C + 2[(B - C)^2 + (A - C)(A - B)]^{\frac{1}{2}}$

It is useful to have a quantitative measure of the degree of asymmetry of a molecule. Of the several possible relationships between the rotational constants used for this purpose we will consider just one, κ, known as Ray's asymmetry parameter. This is defined as

$$\kappa = \frac{2B - A - C}{A - C}. \tag{18}$$

For a prolate symmetric top $B = C$ and $\kappa = -1$; for an oblate symmetric top $A = B$ and $\kappa = +1$; for any other situation κ lies between these limits and the molecule is an asymmetric top. If κ is close to either of the limits -1 or $+1$ we describe the rotor as *near prolate* or *near oblate* respectively, while a situation with $\kappa \approx 0$ represents the most asymmetric case.

A schematic representation of the first few levels of an asymmetric top and the way they correlate with the levels of the limiting prolate and oblate symmetric top cases is given in Fig. 5.2.4. The sub-levels belonging to a given J value are in the specific order mentioned above and may never cross. There is no such restriction on levels of different J values and they may and often do cross, although this feature is not depicted in Fig. 5.2.4 for reasons of clarity. Quite generally the rotational energy of an asymmetric top may be written

$$E_R/h = J(J + 1)\frac{A + C}{2} + \frac{A - C}{2}E(\kappa) \tag{19}$$

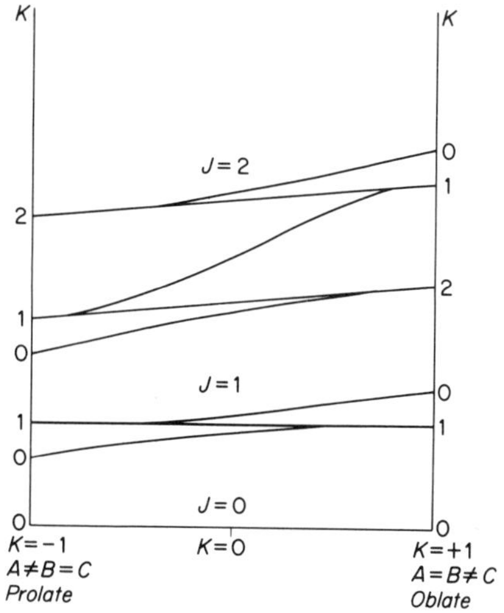

Fig. 5.2.4. Schematic representation of the rotational energy levels of an asymmetric top molecule.

where $E(\kappa)$ is the *reduced energy* for the specified level and degree of asymmetry. Formally the reduced energy is the rotational energy appropriate to the hypothetical system with $A = 1, B = \kappa, C = -1$, and it has been tabulated* at small intervals in κ up to quite high J values. These tabulations enable us to calculate, with the aid of (18) and (19), rotational energy levels for any rigid asymmetric top molecule.

5.3 The Thermal Population of Rotational Levels

In a sample of a gas or vapour we have to deal with a very large number of molecules, some of which will be in the lowest ($J = 0$) rotational energy level and others in excited levels. According to the *Boltzmann distribution law*, the number of molecules having the classical energy E_R is proportional to $e^{-E_R/kT}$, where k is Boltzmann's constant (the gas constant per molecule) and T is the absolute temperature. In quantum mechanics we have to multiply the Boltzmann factor $e^{-E_R/kT}$ of the level by the number of times the level occurs, so that the population of the level E_R is proportional to

$$ge^{-E_R/kT}. \tag{1}$$

The factor g in (1) is known as the *statistical weight* of the energy level E_R. For simplicity let us consider linear unsymmetrical molecules all of which belong to the

* See for example, Townes and Schawlow, *Microwave Spectroscopy*, McGraw-Hill, New York and London, 1955, p. 527.

point group, $C_{\infty v}$. For these molecules the statistical weight of a level of given J is just $2J+1$. This is the spatial degeneracy of the level and, indeed, is closely analogous to the $2l+1$ degeneracy of hydrogen-like orbitals (s, p, d, etc.). Substituting into (1) the statistical weight $g = 2J+1$, and the energy E_R given by eqn. (5.2.17), we find

$$N_J \propto (2J+1)e^{-hJ(J+1)B/kT}, \tag{2}$$

in which N_J denotes the number of molecules present in the level for which the quantum number is J. The total number of molecules, N, is the sum of those present in all the levels, and thus

$$N = \sum_J N_J. \tag{3}$$

The proportionality constant in (2) is the same for every level, and so we have

$$N \propto \sum_J (2J+1)e^{-hJ(J+1)B/kT}. \tag{4}$$

The expression on the right-hand side of (4) is termed the *rotational partition function*, f_R. Written out explicitly, f_R for a linear unsymmetric molecule is given by

$$f_R = 1 + 3e^{-2hB/kT} + 5e^{-6hB/kT} + \dots \tag{5}$$

If the levels are fairly close and the temperature not too low, this sum may be expressed approximately by the integral

$$\int_0^\infty (2J+1)e^{-hJ(J+1)B/kT}dJ. \tag{6}$$

Let $J(J+1) = y$, then the integral becomes

$$\int_0^\infty e^{-hBy/kT}dy = \frac{kT}{hB}. \tag{7}$$

Hence for this simple case, $f_R = kT/hB$. As an example, f_R for the molecule HCN ($B = 44\,316$ MHz) at 300 K is, according to (7), 141·06. The exact value calculated from the summation (5) is 141·39.

The next step in evaluating the relative population of the levels is to divide (2) by (4), when the proportionality constant cancels and we obtain

$$\frac{N_J}{N} = \{(2J+1)e^{-hJ(J+1)B/kT}\}/f_R. \tag{8}$$

Introducing the value of f_R given by (7), we have

$$\frac{N_J}{N} = \frac{h(2J+1)B}{kT} e^{-hJ(J+1)B/kT}. \tag{9}$$

N_J/N is the fractional number of molecules in the J-th rotational level. Fig. 5.3.1 illustrates the variation of N_J/N with J for the molecule HCN at 300 K. The population of the levels passes through a maximum owing to the linearly-increasing degeneracy with ascending J. The location of the maximum can be obtained by differentiating eqn. (9), when we find

$$J_{max} + \tfrac{1}{2} = (kT/2Bh)^{\frac{1}{2}}. \tag{10}$$

Hence the energy of the level which has the greatest population of molecules is

$$E_R(max) = hJ_{max}(J_{max}+1)B = \tfrac{1}{2}kT - \tfrac{1}{4}hB. \tag{11}$$

Except at very low temperature, $\frac{1}{2}kT \gg \frac{1}{4}hB$; the energy of the most populous level therefore is $\approx \frac{1}{2}kT$.

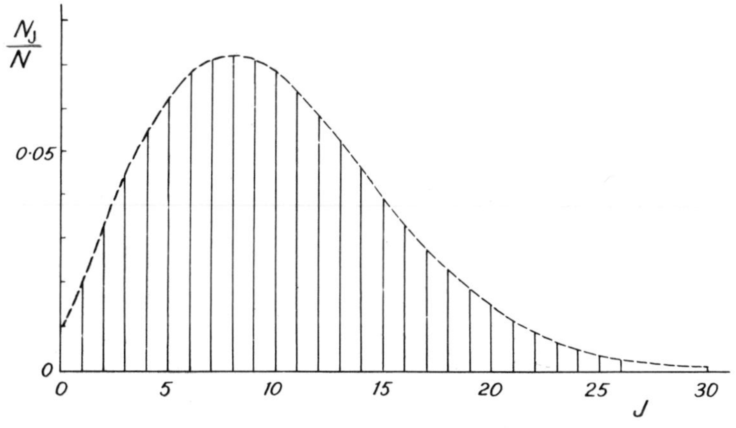

Fig. 5.3.1. Thermal population of rotational states (HCN at 300°K).

The mean rotational energy $\bar{E}_R$ is given by

$$\bar{E}_R = \frac{\sum_J N_J E_J}{\sum_J N_J} \tag{12}$$

The student should show that, in the high temperature approximation, $\bar{E}_R$ is just kT corresponding to the classical equi-partition of energy for two degrees of freedom.

5.4 Rotation Spectra of Gases

Except under the influence of collisions, a molecule can transfer from one rotational energy state to another only with absorption or emission of radiation. These transitions between rotational states comprise the *pure rotational spectrum* of the molecule. As was mentioned in Section 4.1 the frequencies necessary to excite rotational transitions for a majority of molecules fall in an accessible part of the spectrum, between about 0·1 and 10 cm wavelength known as the *microwave region*. In this region several electronic devices may be used as radiation sources. One typical device is the Klystron tube. The oscillation frequency of a Klystron is readily variable and is virtually monochromatic, thus enabling the realisation of a spectrometer containing no prism or other dispersing element. Microwave radiation is conveniently propagated through hollow metal tubes or *waveguides*. These are often of a rectangular cross-section and also serve to contain the vapour under investigation. Radiation frequencies are generally expressed in MHz and can be measured with great accuracy. It is useful to remember that a wavelength of 1 cm corresponds to a frequency of approximately 30 000 MHz.

Transitions between rotational states are not unrestricted. It is found that the transition probability is zero for many combinations, i.e. that the transition is a *forbidden* one. In reaching a preliminary understanding of a spectrum it is clearly of the greatest importance to know which transitions are forbidden and which are not. This information is obtained from a consideration of the wavefunctions for rotation. The principle was outlined for the case of the rigid fixed axis rotator in Section 3.12 where the selection rule $\Delta M = \pm 1$ was derived.

For a freely rotating molecule the general selection rule is

$$\Delta J = 0, \pm 1 \qquad (1)$$

although additional restrictions may apply in certain cases. A rationalisation of the general rule may be obtained if we ascribe to a quantum of radiation or *photon* a spin angular momentum quantum number of unity. We may then think of the interaction of a photon with a rotating molecule as a process involving the coupling of angular momenta according to well established quantum rules and eqn. (1) immediately follows.

The transition moment for the fixed-axis rotator was shown to be *proportional to the permanent dipole moment* of the system (Section 3.12), and the same result is true for a freely rotating molecule. Consequently molecules which lack a permanent dipole moment cannot give rise to a pure rotation spectrum. Conversely, the intensity of absorption in the pure rotation spectrum can be used to calculate the permanent dipole moment, but accurate intensity measurements are difficult and the method is applied only infrequently.

The conditions imposed upon the rotational transitions of molecules are therefore very restrictive indeed. For polar molecules we shall expect to observe absorption in the microwave region at frequencies controlled by the selection rule (1). The rotation of non-polar molecules cannot be excited by absorption of radiation; thus molecules with a centre of symmetry or possessing tetrahedral symmetry (*e.g.* N_2, CO_2, C_2H_2, C_6H_6 and CH_4) have no microwave spectra.

Linear Molecules

For this class of molecules the general pure rotational selection rule (1) is further restricted to

$$\Delta J = \pm 1. \tag{2}$$

$\Delta J = 0$ would be of little significance in the *pure rotational spectrum* of a linear molecule as it involves no energy change, but it is generally forbidden for the following reason. The intensity of a transition depends on the magnitude of the transition moment integral which, for $\Delta J = 0$, we write as

$$M = \int \psi_J^* \mu \psi_J d\tau. \tag{3}$$

This is no more than the *average value* of the dipole moment in *space* for the state described by the quantum number J. Since for a linear molecule the dipole moment lies along the linear axis and the rotational angular momentum is always perpendicular to it, M is clearly zero. Only $\Delta J = +1$ corresponds to the absorption of radiation. Substituting $J'' = J$ and $J' = J + 1$ in eqn. (5.2.17)* we obtain the frequencies of absorption in the pure rotation spectrum.

$$h\nu = E_{R'} - E_{R''} = E_R(J+1) - E_R(J) = 2Bh(J+1). \tag{4}$$

The spectrum consists of a set of uniformly spaced lines with frequencies in the ratio 1 : 2 : 3 . . . corresponding to the J values 0, 1, 2 . . . The uniform spacing is strikingly verified by experiment, as for instance by the frequencies of the four transitions of carbon oxysulphide given in Table 5.4.1. The tendency for the lines to converge very slightly with ascending values of J is due to centrifugal stretching of the molecule which causes the moment of inertia to increase with J.

Under favourable conditions the values of the rotational constant B obtained from measurements in the microwave region are accurate to one part in 10^6. The moments of inertia, $I_b = h/8\pi^2 B$, are subject to rather more uncertainty in the absolute sense because Planck's constant is not known to the same degree of accuracy, but this is of little practical importance. The calculation of bond lengths from

* An accepted spectroscopic convention is to denote upper state quantities by a single prime and lower state quantities by a double prime.

Table 5.4.1. Pure Rotation Spectrum of Carbon Oxysulphide ($^{16}O^{12}C^{32}S$)

$J \rightarrow J+1$	$1 \rightarrow 2$	$2 \rightarrow 3$	$3 \rightarrow 4$	$4 \rightarrow 5$
ν, MHz	24 325·92	36 488·82	48 651·64	60 814·08
$B = \nu/2(J+1)$	6 081·48	6 081·47	6 081·45	6 081·41

the moments of inertia introduces further complications, arising from the fact that the nuclei are not motionless even in the lowest, zero-point vibrational level. The pure rotational line frequencies depend on the B rotational constant averaged over the nuclear motions. For the simple case of a diatomic molecule the line frequencies thus depend on $\langle\frac{1}{r^2}\rangle$, where the brackets indicate the average value. Even for strictly harmonic vibrations of the nuclei $\langle\frac{1}{r^2}\rangle \neq \frac{1}{r_e{}^2}$, where r_e is the equilibrium internuclear distance which we seek to determine. Therefore in a precise determination of internuclear distances it is necessary to allow for the effects of the residual amplitude of vibration.

The moment of inertia of a rigid linear molecule is given by

$$I_b = \sum_i m_i z_i^2, \tag{5}$$

in which z_i is the coordinate of the i-th mass along the internuclear axis z and with the molecular centre of mass as origin. From the definition of the centre of mass we also have

$$\sum_i m_i z_i = 0. \tag{6}$$

For a diatomic molecule, XY, there are just two coordinates, z_X and z_Y to be determined and the structural problem is directly soluble if we have a value for the moment of inertia. Remembering that $z_Y - z_X = r_{XY}$ the internuclear distance, eqns. (5) and (6) reduce to

$$I_b = \left(\frac{m_X m_Y}{m_X + m_Y}\right) r_{XY}{}^2 = \mu r_{XY}{}^2, \tag{7}$$

in which μ is the reduced mass for the diatomic molecule.

For a linear triatomic molecule XYZ there are three coordinates z_X, z_Y and z_Z to be determined, but we have only the two relationships between them. To solve this problem we must make use of data for isotopic molecules. If one atom in the molecule XYZ is replaced by an isotope and there are no changes in the internuclear distances, the accompanying mass change, Δm_i, will result in a change in the moment of inertia given by

$$\Delta I_b = \frac{M \cdot \Delta m_i}{M + \Delta m_i} \cdot z_i^2. \tag{8}$$

Here M is the mass of the molecule before isotopic replacement. Eqn. (8) enables us to directly determine the coordinate of the substituted atom and the remaining two coordinates can be calculated from eqns. (5) and (6). The nature of eqns. (5) and (8) gives rise to ambiguities in the signs of the atomic coordinates, but such problems are generally easily resolved by simple intuitive considerations as to the reasonableness of the resulting parameters.

Fortunately the microwave spectrometer is sufficiently sensitive to detect absorption lines due to isotopic molecules in their natural abundance or only slightly enriched, so that, unless the molecule contains atoms with no convenient isotopes (*e.g.* ^{19}F, ^{31}P), the evaluation of the moments of inertia of isotopic species presents no special problem. The moments of two isotopic forms of carbonyl sulphide are given in Table 5.4.2, together with values calculated from the atomic masses and internuclear distances r(C—O) = 1·1637 and r(C—S) = 1·5584 Å.

Table 5.4.2 Moments of Inertia of Carbon Oxysulphide (u.Å^2)

	$^{16}O^{12}C^{32}S$	$^{18}O^{12}C^{32}S$
observed	83·0626	88·5637
calculated	83·0626	88·5637

The assumption that isotopic molecules have identical bond lengths is not one to be made lightly. It is not strictly valid unless the nuclei are motionless: otherwise the apparent moments of inertia (and hence the internuclear distances) depend upon the amplitude of vibration and consequently upon the isotope employed, for the vibrational amplitude of an atom changes with its mass. To obtain really good internuclear distances it is desirable to correct the moments of inertia for the residual amplitude. The details of how the correction is applied need not concern us here, though it should be noted that it can be determined rigorously for a limited number of small molecules only. The moments in Table 5.4.2 are corrected in this way and thus the bond lengths in carbon oxysulphide are reliable to ± 0·0005 Å. Uncorrected moments may lead to uncertainties in the bond lengths as high as 0·01 Å.

Symmetric Tops

When a molecule is a symmetric top for reasons of symmetry, the permanent dipole moment must coincide with the axis of the top. It follows there is no component of the dipole moment perpendicular to the top axis; and consequently that the absorption of radiation cannot excite rotation about the top axis, or add to it if it is already present. Therefore we have the selection rule

$$\Delta K = 0. \tag{9}$$

The permanent moment does however have a component perpendicular to the generalised axis of rotation (the axis of the total angular momentum vector in Fig. 5.2.1), and thus rotational transitions will be observed in the microwave region, provided the molecule possesses a permanent moment.

The overall selection rule is $\Delta J = 0, \pm 1$ except when $K = 0$ and then $\Delta J = \pm 1$ applies. We recall that the transition moment integral for $\Delta J = 0$ is the average value of the dipole moment in space and this is not zero for a symmetric top except when $K = 0$ and then the dipole moment and the total angular momentum are mutually perpendicular. In the present discussion we need only consider $\Delta J = +1$ as we did for the linear case, but we must remember the full selection rules when we consider rotational energy changes accompanying vibrational transitions in the next chapter.

The rotational energy levels of a rigid symmetric top are given by the expressions (5.2.13) and (5.2.15). With $K' = K''$, $J'' = J$, and $J' = J + 1$ we obtain for the frequency of the absorption lines,

$$h\nu = E_R(J+1,K) - E_R(J,K) = 2Bh(J+1). \tag{10}$$

The pure rotation spectrum thus consists of a set of evenly spaced lines which obey the same formula as a linear molecule. It is important to note, however, that each line contains $2J + 1$ components corresponding to the possible K values for each J value. Centrifugal distortion frequently effects a partial separation of these K components and thus distinguishes the symmetric top spectrum from that due to a linear molecule. Nevertheless, only one rotational constant, $B = h/8\pi^2 I_b$, can be evaluated from the frequencies. For a complete structure determination, therefore, we need at least as many isotopic moments as there are independent bond distances and interbond angles in the molecule. In a straightforward case like the pyramidal molecule PCl_3 one bond length r(P—Cl) and one interbond angle ∠ClPCl are required to establish the molecular geometry: these can be determined from I_b for $^{31}P^{35}Cl_3$ and $^{31}P^{37}Cl_3$. (A mixed isotope like $^{31}P^{35}Cl_2{}^{37}Cl$ can also be used, but is by no means a symmetric top). The apparent, uncorrected moments yield r(P—Cl) = 2·044 Å and ∠ClPCl = 99°56′*, the estimated uncertainty due to neglect of the residual vibrational amplitude being 0·003 Å and 20′ respectively.

Asymmetric Tops

Most asymmetric top molecules have a permanent dipole moment and can be expected to absorb in the microwave region. The pure rotational spectrum is often very rich in lines as the full general selection rule $\Delta J = 0, \pm 1$ applies and all three possible changes in J may give rise to *absorption* of radiation. It is usual to describe transitions with $\Delta J = 0$ as Q-branch transitions, and those with $\Delta J = +1$ and $\Delta J = -1$ as R- and P-branch transitions respectively. In principle it is possible to determine all

* Kisliuk and Townes, *J. Chem. Phys.*, 1950, **18**, 1109.

three rotational constants, A, B and C from the measured line frequencies. However, in practice, all three constants may not be obtainable to the same degree of precision; the particular situation depending on the asymmetry of the molecule, the direction of the dipole moment within it, and the available transitions.

In this class of molecules it is easier to follow the steps in a structure determination if we consider an actual example. The molecule CH_2O, seen in Fig. 5.1.1, is a fairly typical asymmetric top. Using a chemist's intuition we shall suppose that the molecule is symmetrical and planar (point group C_{2v}) so that three parameters, the distances r(C—O) and r(C—H) and the ∠HCH, are required to describe its geometry. A number of rotational transitions of H_2CO have been observed in the microwave region.* Two of them have an especially simple explanation; they are,

$J_{\tau}'' \rightarrow J_{\tau}'$	ν, MHz	*Interpretation*
$0_0 \rightarrow 1_{-1}$	72 838·14	$B + C$
$2_{-1} \rightarrow 2_0$	14 488·74	$3(B - C)$

According to Table 5.2.1 the frequency of the $0_0 \rightarrow 1_{-1}$ transition is equal to $(B + C)$, while that of the $2_{-1} \rightarrow 2_0$ transition equals $3(B - C)$. The experimental frequencies thus lead directly to $B = 38\,834$ and $C = 34\,004$ MHz. Other lines in the spectrum confirm these figures and also yield a value for the third rotational constant, A, of 282 106 MHz. The corresponding moments of inertia are

$$I_a^0 = 1{\cdot}791,\ I_b^0 = 13{\cdot}014,\ I_c^0 = 14{\cdot}862, \text{ all u.Å}^2, \tag{11}$$

whence

$$I_c^0 - (I_a^0 + I_b^0) = 0{\cdot}057 \text{ u.Å}^2. \tag{12}$$

The superscript added to the I's is a reminder that these are apparent moments of inertia for the zero-point vibrational state. In fact the influence of the vibration amplitude is easily seen, for a planar *rigid* molecule necessarily has

$$I_c - (I_a + I_b) = 0, \tag{13}$$

and this relation is not accurately obeyed by the uncorrected moments. The discrepancy is well beyond the limit of experimental error.

Since the corrected moments must conform to (13) it follows that only two of the moments are independent. Consequently we cannot solve for the three structural parameters unless we know at least one moment of inertia of an isotopic molecule. The extra datum is provided by a determination of I_b^0 for $H_2{}^{13}CO$ whose transitions are observable near those of $H_2{}^{12}CO$ at the natural abundance (1·1%) of the isotope ^{13}C. (A moment of inertia of DCHO or $H_2C^{18}O$ would have served equally well, but the natural abundance of deuterium (0·015%) and ^{18}O (0·2%) is lower than that of

* Lawrance and Strandberg, *Phys. Rev.*, 1951, **83**, 363. Erlandsson, *J. Chem. Phys.*, 1956, **25**, 579.

^{13}C and the transitions correspondingly more difficult to observe). The correction to the moments for the non-rigidity due to vibration can, in principle, be calculated from the mean vibrational amplitude of the atoms which, in turn, may be obtained from the vibration frequencies. Unfortunately, the calculation is somewhat difficult and a good deal of uncertainty attaches to the values that finally emerge. It turns out that the best set of structural parameters is

$$r(\text{C—O}) = 1{\cdot}21 \pm 0{\cdot}01\ \text{Å},\ r(\text{C—H}) = 1{\cdot}12 \pm 0{\cdot}01\ \text{Å},\ \angle\text{HCH} = 118 \pm 2°. \qquad (14)$$

The uncertainties attributed to these values arise very largely from the difficulty of taking the vibrational non-rigidity properly into account.

The determination of the C—O bond length in formaldehyde by electron-diffraction is discussed in Section 10.4.

5.5 Some Nuclear Properties

So far in discussing the rotation of molecules it has only been necessary to envisage the atomic nucleus as a small entity accounting for the major part of the atomic mass. There are however two further properties of nuclei which influence the rotational spectra of molecules. These are the *nuclear spin, $\boldsymbol{I}$*, and the *electric quadrupole moment, Q*. A third property, the *magnetic moment*, μ, will be discussed in another context in Chapter 8; its influence on rotational spectra is generally only slight.

The substance of the idea of nuclear spin is that the nucleus is capable of rotation about an axis. The spin generates a quantised angular momentum $\boldsymbol{I}$ whose magnitude is given by

$$\boldsymbol{I} = \sqrt{I(I+1)}\,h/2\pi. \qquad (1)$$

Here I, the quantum number for nuclear spin, can assume integral, half-integral, or zero values. (1) is a typical wave-mechanical equation for the quantisation of spin angular momentum and may be compared with eqn. (5.2.7) for the angular momentum $\boldsymbol{J}$ associated with molecular rotation. Values of I for some common nuclei are given in Table 5.5.1.

It should be emphasised that in chemical applications we are interested in a single value of I (that given in the table) for each nucleus. To understand why this is so requires some explanation of nuclear structure. The angular momentum $\boldsymbol{I}$ can be thought of as the resultant of the angular momenta of the protons and neutrons of which the nucleus is composed. Each of these fundamental particles is considered to have a spin quantum number $I = \frac{1}{2}$, the value for the proton being experimentally observable. As the deuterium nucleus, or deuteron, possesses one proton and one neutron the resultant spin must be 1 or 0, depending on whether the individual spins are aligned parallel or antiparallel. Experiment shows that the deuteron in its ground state has $I = 1$, corresponding to the parallel alignment. $I = 0$ then represents an excited state of the deuteron, but the excitation energy is so high that it actually exceeds the dissociation energy of the nucleus. Many heavier nuclei do however possess excited spin states which are stable to dissociation.

The assignment of a spin quantum number $I = \frac{1}{2}$ to protons and neutrons has one obvious consequence, that nuclei with even mass number—hence with an even total number of protons and neutrons—must have zero or integral values of I, whereas nuclei of odd mass must have half-integral values.

Table 5.5.1. Physical Constants of some Common Nuclei*

Isotope	*Natural abundance %*	*Spin Quantum number I*	*Magnetic Moment gI (in units of the nuclear magneton)*	*Electric Quadrupole moment, Q cm² × 10²⁴ (in units of e)*
^{1}n		$\frac{1}{2}$	−1·913	
^{1}H	99·985	$\frac{1}{2}$	+2·793	
$^{2}H(D)$	0·015	1	+0·857	+0·0027
^{12}C	98·89	0		
^{13}C	1·11	$\frac{1}{2}$	+0·702	
^{14}N	99·63	1	+0·404	+0·02
^{15}N	0·37	$\frac{1}{2}$	−0·283	
^{16}O	99·76	0		
^{17}O	0·04	$\frac{5}{2}$	−1·893	−0·005
^{18}O	0·20	0		
^{19}F	100	$\frac{1}{2}$	+2·627	
^{31}P	100	$\frac{1}{2}$	+1·131	
^{32}S	95·0	0		
^{33}S	0·75	$\frac{3}{2}$	+0·643	−0·067
^{34}S	4·2	0		
^{35}Cl	75·4	$\frac{3}{2}$	+0·821	−0·085
^{37}Cl	24·6	$\frac{3}{2}$	+0·683	−0·067
^{79}Br	50·52	$\frac{3}{2}$	+2·099	+0·33
^{81}Br	49·48	$\frac{3}{2}$	+2·263	+0·28
^{127}I	100	$\frac{5}{2}$	+2·794	−0·75

* Townes and Schawlow, *Microwave Spectroscopy*, McGraw-Hill, 1955.

A glance at the table will show that this is true. A useful rule, more difficult to understand theoretically, is that nuclei with even charge and even mass have zero spin. A theory of nuclear structure should predict the numerical value of I for nuclei of any mass, but at present the situation is mysterious and nuclear spins are accessible only by experiment.

To make plain the nature of a quadrupole, consider the arrangement of charges in Fig. 5.5.1 (a). This quadrupole system has no dipole moment ($\Sigma e_i z_i = 0$) yet it gives an external field, which however falls off more rapidly than that of a dipole.

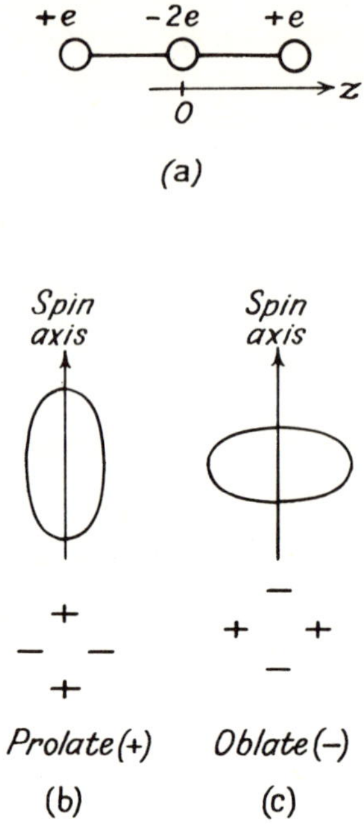

Fig. 5.5.1. A simple quadrupole (*a*) and nuclear quadrupoles (*b* and *c*). The nucleus has a net positive charge, and thus the —ve signs in (*b*) and (*c*) merely indicate regions of low density of positive charge.

Its action is characterised by a *quadrupole moment* which has the dimensions of charge × (length)^2 and in this example is simply $\Sigma e_i z_i^2$; where z is measured along the axis on which the charges are located. Nuclear quadrupoles do not have the linear form shown in (*a*): they are more complex shapes which arise when distribution of nuclear charge is not spherical. In such cases the nucleus can be thought of as elongated (case *b*) or flattened (case *c*) with respect to the axis of spin (about which the nucleus has axial symmetry). The charge distribution can then be regarded as equivalent to a quadrupole, indicated in the figure by the +, — charges, superposed on a spherical charge Ze. The analytical expression for the quadrupole moment need not concern us, for the nuclear electric quadrupole moments, like the nuclear magnetic moments, must be found by experiment. It emerges that nuclei with $I = 0$ or $\frac{1}{2}$ have zero quadrupole moments: such nuclei must have a perfectly spherical charge distribution. When $I \geqslant 1$ the quadrupole moment is always different from zero, though its variation from one nucleus to another is seemingly irregular.

The nuclear quadrupole moment is designated eQ, where e is the proton charge and Q has the dimensions of cm^2. Values of Q in units of 10^{-24} cm^2 are given for some common nuclei in Table 5.5.1. To illustrate the use of the data consider the nucleus ^{127}I; here $Q = -0{\cdot}75 \times 10^{-24}$ cm^2, hence the quadrupole moment $eQ = -4{\cdot}80 \times 0{\cdot}75 \times 10^{-34} = -3{\cdot}60 \times 10^{-34}$ esu, the negative sign signifying that the nuclear charge distribution is oblate (Fig. 5.5.1(*c*)).

5.6 Nuclear Spin Effects in Rotational Spectra

Particles with zero or integral spin quantum numbers obey Bose–Einstein statistics; the total wavefunction for a molecule containing an equivalent pair of such particles in symmetric (remains unaltered) with respect to exchanging these.

Particles with half-integral spins obey Fermi–Dirac statistics; the total wavefunction in this case is antisymmetric (changes signs) when an equivalent pair is exchanged.

These two statements must be taken into account when we consider the populations of the rotational energy levels for those molecules which contain equivalent nuclei, *e.g.* H_2, N_2, O_2, CO_2, CH_2O, CH_3CN, etc. Consider the case with two equivalent nuclei of spin I. For each nucleus there are $2I + 1$ orientations of I with respect to a space fixed axis, z, each of which is characterised by a projection quantum number, M_I. M_I can take the values

$$I, I-1, \ldots -I \tag{1}$$

Writing spin wavefunctions for each nucleus as $\phi_{M_I}(1)$ and $\phi_{M_I'}(2)$, there are $(2I + 1)^2$ spin functions for the two nuclei of the type $\phi_{M_I}(1).\phi_{M_I'}(2)$. For the $2I + 1$ cases when $M_I = M_I'$ the two nuclei spin functions are clearly symmetric with respect to interchanging (1) and (2). When $M_I \neq M_I'$ these functions are neither symmetric nor antisymmetric, but there are equal numbers ($[(2I + 1)^2 - (2I + 1)]/2$) of symmetric and antisymmetric combinations of the type,

$$\phi_{M_I}(1)\,\phi_{M_I'}(2) + \phi_{M_I'}(1)\,\phi_{M_I}(2) \qquad \text{(symmetric)},$$

$$\phi_{M_I}(1)\,\phi_{M_I'}(2) - \phi_{M_I'}(1)\,\phi_{M_I}(2) \qquad \text{(antisymmetric)}.$$

The total number of symmetric functions is therefore

$$n_s = (2I + 1)(I + 1), \tag{1}$$

and the antisymmetric function number

$$n_a = (2I + 1)I. \tag{2}$$

For many situations we may write the overall wavefunction of a molecule as a product of its component parts, i.e.

$$\Phi_{total} = \Phi_E.\Phi_V.\Phi_R.\Phi_N, \tag{3}$$

where the subscripts E, V, R and N stand for electronic, vibrational, rotational and nuclear respectively. The symmetry of Φ total depends on the individual symmetries of its factors. Here, for simplicity, we will limit our discussion to linear molecules with equivalent nuclei, but the ideas developed can be readily extended to other molecular types.

The ground state electronic wavefunction, for the majority of molecules, is symmetric with respect to interchanging nuclei as is the wavefunction for the lowest vibrational state. The rotational wavefunction is symmetric when J is even and antisymmetric when J is odd. If the equivalent nuclei obey Fermi–Dirac statistics the symmetric nuclear spin functions must be taken with the odd J values, and the antisymmetric spin functions with the even J values in order that the overall product be antisymmetric. The inverse state of affairs applies if the equivalent nuclei obey Bose–Einstein statistics.

Some specific examples should make these ideas plain. First consider the H_2 molecule which has two equivalent nuclei of spin, $I = \frac{1}{2}$, obeying Fermi–Dirac statistics. The ratio of symmetric to antisymmetric spin states is $(I + 1)/I = 3$. The statistical weights of rotational levels with odd J values will therefore be three times those of rotational levels with even J values and this will have a marked influence on the populations of the respective levels in H_2 (Fig. 5.6.1(*a*)). In D_2 the two nuclei have $I = 1$ and hence follow Bose–Einstein statistics. In this case $(I + 1)/I = 2$ and the even J values will have twice the statistical weight of the odd J values (Fig. 5.6.1(*b*)). No such relative effects are present in HD as there are no identical nuclei.

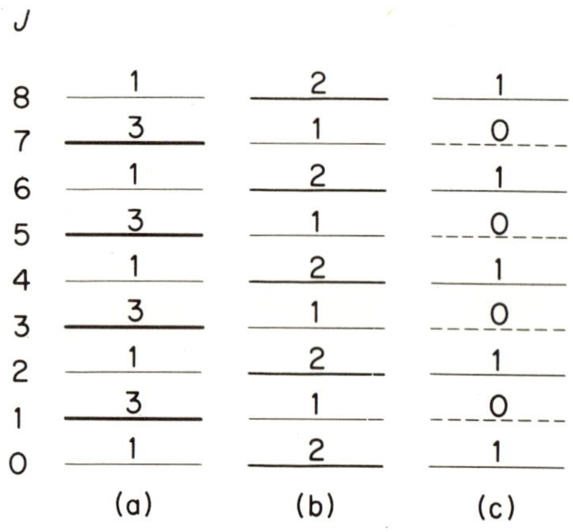

Fig. 5.6.1. Nuclear statistical weightings for the rotational levels in the ground electronic and vibrational states of some simple molecules (*a*) H_2 (*b*) D_2 (*c*) CO_2.

For a second example we will take a molecule in which the identical nuclei have $I = 0$. CO_2 is such a case. Here the ratio $(I + 1)/I$ is infinite; there are no antisymmetric spin wavefunctions and since the overall state must be symmetric with respect to nuclear exchange (Bose–Einstein statistics) the *levels of odd J are entirely missing* (Fig. 5.6.1(*c*)). This situation must clearly be taken into account in any detailed spectroscopic considerations of CO_2.

A variation of the CO_2 case is afforded by the O_2 molecule. The ground electronic state of $CO_2\,(^1\Sigma g^+)$ is typical of the majority of stable, centrosymmetric, linear molecules in that its wavefunction is symmetric with respect to nuclear exchange. The ground electronic state of $O_2\,(^3\Sigma g^-)$ is somewhat of an exception in that it is antisymmetric with respect to exchanging the nuclei, and for this reason the *even numbered rotational levels are the missing ones*.

It may happen that a molecule contains more than one pair of equivalent nuclei, *e.g.*, C_2H_2. The overall symmetry with respect to nuclear exchange is simply the product of the symmetries for the individual pairs. Thus, in the case of acetylene we have a pair of protons ($I = \frac{1}{2}$, antisymmetric) and a pair of carbon nuclei ($I = 0$, symmetric); the *overall symmetry is antisymmetric*. For a molecule with two pairs of Fermi-Dirac nuclei (antisymmetric) the *overall symmetry is symmetric*. Quite generally, when there are n pairs of identical nuclei present, the numbers of symmetric (n_s) and antisymmetric (n_a) spin wavefunctions are given by

$$n_s = \tfrac{1}{2}\left[\prod_{i=1}^{n}(2I_i + 1)\right]\left[\prod_{i=1}^{n}(2I_i + 1) + 1\right], \tag{4}$$

$$n_a = \tfrac{1}{2}\left[\prod_{i=1}^{n}(2I_i + 1)\right]\left[\prod_{i=1}^{n}(2I_i + 1) - 1\right], \tag{5}$$

5.7 Nuclear Quadrupole Coupling in Pure Rotation Spectra

The rotation spectra of molecules containing quadrupolar nuclei frequently exhibit fine structures. Analysis of these can yield quite detailed information on the electronic structures of molecules. In order to gain some insight into this type of nuclear effect let us first consider the classical energy of a simple quadrupole in an electric field. The linear quadrupole, Fig. 5.5.1(*a*), can be thought of as forming two electric

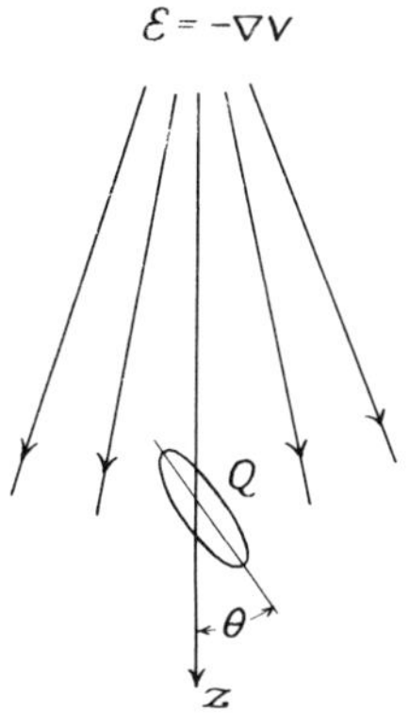

Fig. 5.7.1. Nucleus of quadrupole moment Q in an inhomogeneous electric field.

dipoles set back to back. In a uniform electric field of intensity $\mathcal{E}$ each dipole has the energy $-\mu\mathcal{E}\cos\theta$ and hence the energy of the quadrupole is $-\mu\mathcal{E}[\cos\theta + \cos(\theta + 180°)] = 0$, *i.e.*, all orientations have equal energy. However, in a non-uniform field, Fig. 5.7.1, the field strength differs slightly at each dipole so that the energies no longer cancel. In this case the potential energy of the quadrupole varies according to its orientation in the field, the extent of the variation depending upon the degree of inhomogeneity of the field.

Secondly, then, we need to characterise the inhomogeneity of the electric field. Consider the field due to a single particle of charge e. The field is radial and therefore non-uniform at any point outside the particle. Let the charge be at the origin of a set of cartesian coordinates, as shown in Fig. 5.7.2.

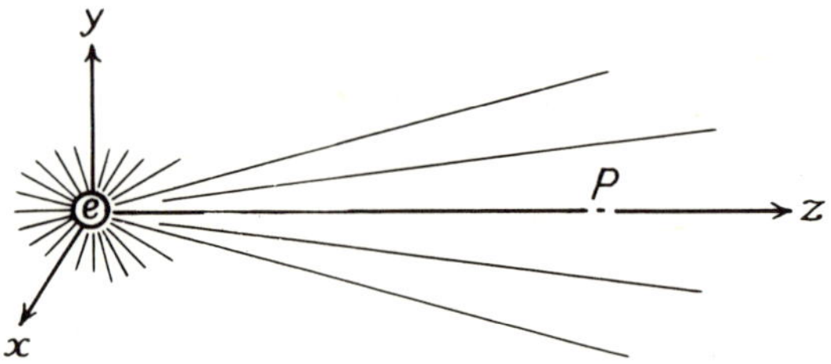

Fig. 5.7.2. Electric field of a charge e.

At a point P on the z-axis the electric potential is

$$V = e/z, \tag{1}$$

the z-component of the electric field intensity is therefore

$$-\partial V/\partial z = e/z^2, \tag{2}$$

and the *gradient* of the z-component of the field intensity is

$$\partial^2 V/\partial z^2 = 2e/z^2. \tag{3}$$

Actually the field is non-uniform not merely along z but in directions perpendicular to z also; this means that $\partial^2 V/\partial x^2$ and $\partial^2 V/\partial y^2$, the field gradients at P along the x-

and y-axes, are likewise different from zero. However, the field due to the charge e must be symmetrical about any axis which passes through it and, in particular, must be symmetrical about z. Thus for the components perpendicular to z we have

$$\partial^2 V/\partial x^2 = \partial^2 V/\partial y^2, \tag{4}$$

since each can be transformed into the other by a simple rotation of the axes. Further, we may apply the Laplace equation,

$$\partial^2 V/\partial x^2 + \partial^2 V/\partial y^2 + \partial^2 V/\partial z^2 = 0, \tag{5}$$

at the point P, whence

$$\partial^2 V/\partial z^2 = -2\partial^2 V/\partial x^2 = -2\partial^2 V/\partial y^2. \tag{6}$$

We see that, in the chosen example of an axially symmetric field, the three terms $\partial^2 V/\partial x^2$, $\partial^2 V/\partial y^2$ and $\partial^2 V/\partial z^2$ which describe the field inhomogeneity at P are interrelated in the sense that any one of them determines the value of the other two; or, in other words, that only one parameter is required to define the inhomogeneity. As a rule the parameter chosen is $\partial^2 V/\partial z^2$: it is known as the field gradient and is denoted by the symbol q. In terms of q the expression for the classical electrostatic energy E_Q of a quadrupole at P is

$$E_Q = \tfrac{1}{8} eQq\,(3\cos^2\theta - 1). \tag{7}$$

This is the form expected, for E_Q then is proportional to the quadrupole moment eQ, to the field inhomogeneity q, and to a function of the angle θ between the quadrupole and the symmetry axis z of the electric field.

Next we apply these ideas to a nucleus present in a molecule. At a nucleus there is an electric field due to all the electrons in the molecule, and to all the nuclei except the one under consideration: but because the field gradient varies inversely with the cube of distance the most important contribution comes from electrons immediately surrounding the given nucleus. The form of this contribution is discussed later (Section 8.7). In order that eqn. (7) shall be applicable, we suppose that the given nucleus is on a three-fold or higher axis of symmetry (as in a linear molecule, or on the axis of a symmetric top), to ensure that the electric field at the nucleus has axial symmetry.* The axis of the electric field then is coincident with the symmetry axis of the molecule.

In the present connection we are concerned with the quadrupole energy appropriate to a molecule rotating freely in the gas phase. The quadrupolar interaction provides a mechanism by which a nuclear spin, $\boldsymbol{I}$, and the overall rotation, $\boldsymbol{J}$, are coupled to produce a resultant angular momentum, $\boldsymbol{F}$. The magnitude of $\boldsymbol{F}$ is given by

$$\boldsymbol{F} = \sqrt{F(F+1)}\ \ h/2\pi \tag{8}$$

wherein the quantum number F may assume any of the values

$$F = (J+I), \quad (J+I-1), \ldots, |J-I|. \tag{9}$$

Thus F may have either $2I + 1$ or $2J + 1$ values, whichever is less, for each value of J.

The coupling may cause the $(2J + 1)$-fold degeneracy of the pure rotational energy level to lift in favour of $2I + 1$ (or $2J + 1$) sub-levels of different energy, corresponding to different relative orientations of $\boldsymbol{I}$ and $\boldsymbol{J}$. The physical reason for the splitting (if splitting occurs) is that each eigenvalue $\boldsymbol{F}$ imposes a different orientation upon the nucleus whose spin angular momentum $\boldsymbol{I}$ is coupled with $\boldsymbol{J}$. If the nucleus has a quadrupole moment, the electrostatic energy of the quadrupole in the electric field of other charges in the molecule must be added to the pure rotational

* Strictly, for the electric field to have axial symmetry, the distribution of nuclei *and electrons* must be axially symmetric. In nearly all stable polyatomic molecules the electrons have as much symmetry as the nuclei; but a few diatomic molecules are exceptions to this rule, as are many paramagnetic molecules.

Moreover a nucleus is invariably surrounded in an atom (or molecule) by electrons which give rise to a charge density ρ at the position of the nucleus. (Only the s-electrons contribute to ρ, however, for the $p, d, \ldots$ electronic wavefunctions have zero amplitude at the nucleus.) The components of the field gradient are then related by the Poisson equation,

$$\partial^2 V/\partial x^2 + \partial^2 V/\partial y^2 + \partial^2 V/\partial z^2 = 4\pi\rho$$

rather than by the Laplace equation (5). When the charge distribution is spherically symmetrical we have

$$\partial^2 V/\partial x^2 = \partial^2 V/\partial y^2 = \partial^2 V/\partial z^2 = -\frac{4}{3}\pi\rho.$$

In an axially symmetrical electric field $\partial^2 V/\partial z^2$ has some value other than $-\frac{4}{3}\pi\rho$. The difference between the values of $\partial^2 V/\partial z^2$ for the axial field and for the spherically symmetrical field is taken to be the quantity q: thus

$$q = \frac{\partial^2 V}{\partial z^2} + \frac{4}{3}\pi\rho.$$

This is equivalent to the definition given above: inclusion of the charge density due to the s electrons merely adds the constant term $\frac{4}{3}\pi\rho$.

energy E_R. This electrostatic energy depends upon the orientation of the nucleus and so is different for each allowed value of F. Take as an example the molecule HCN. The nuclei of ^1H and ^{12}C have no quadrupole moment, consequently their contribution to the potential energy is constant and we can disregard them. The nucleus ^{14}N has a spin $I = 1$ and a small quadrupole moment (Table 5.5.1). In the $J = 1$ rotational energy level the quantum number F, by eqn. (9), may assume the values $F = 2, 1$, and 0; hence there are three sub-levels of different energy. When $J = 2$ we have $F = 3, 2$, and 1. In general, then, each rotational energy level is resolved into $2I + 1$ sub-levels of different energy; except that the maximum number of sub-levels can in no case exceed $2J + 1$. The possible sub-levels of the rotational levels with $J = 0$ to 2 arising from the coupling of a single nucleus with spin $I = 1, \frac{3}{2}$, and $\frac{5}{2}$ are indicated in Table 5.7.1.

Table 5.7.1

J	F		
	$I = 1$	$I = \frac{3}{2}$	$I = \frac{5}{2}$
0	1	$\frac{3}{2}$	$\frac{5}{2}$
1	2 1 0	$\frac{5}{2}$ $\frac{3}{2}$ $\frac{1}{2}$	$\frac{7}{2}$ $\frac{5}{2}$ $\frac{3}{2}$
2	3 2 1	$\frac{7}{2}$ $\frac{5}{2}$ $\frac{3}{2}$ $\frac{1}{2}$	$\frac{9}{2}$ $\frac{7}{2}$ $\frac{5}{2}$ $\frac{3}{2}$ $\frac{1}{2}$

The calculation of the energy of the sub-levels is complex and we shall here merely state the result. For a *linear molecule* containing a single quadrupolar nucleus the energy correction $E_{Q'}$ is

$$E_{Q'} = -(eQq)f(I, J, F), \tag{10}$$

where

$$f(I, J, F) = \frac{\frac{3}{4}C(C+1) - I(I+1)J(J+1)}{(2J+3)(2J-1)2I(2I-1)} \tag{11}$$

and

$$C = F(F+1) - I(I+1) - J(J+1). \tag{12}$$

Thus $E_{Q'}$ may be zero for any one of three reasons: (*i*) if the molecule lacks a nucleus with a quadrupole moment, for then $Q = 0$; (*ii*) if the field gradient q is zero at the quadrupolar nucleus; or (*iii*) if $J = 0$, for then $f(I, J, F) = 0$. Otherwise one must add the quadrupole interaction energy $E_{Q'}$ to the pure rotational energy E_R in order to obtain the energy of the sub-levels arising from the coupling.

The effect of quadrupole coupling upon the pure rotation spectrum is best illustrated by an example. Let us investigate the theoretical appearance of the $J_{0\to1}$ rotational transition of $^1H^{12}C^{14}N$. The $J = 0$ energy level is non-degenerate and so cannot be split by the quadrupole coupling; but the $J = 1$ level, as we have seen, splits into the group of three sub-levels shown in Fig. 5.7.3. The values of the function $f(I,J,F)$ obtained from eqns. (11) and (12) are:

J	0		1	
F	1	2	1	0
$f(I,J,F)$	0	+0·05	−0·25	+0·50

The selection rule ($\Delta F = 0, \pm 1$) allows transitions from the lower state to all three sub-levels of the upper state. Consequently in place of the single line expected for a pure rotational transition, a group of three lines is observed corresponding to the transitions $F_{1\to1}, F_{1\to2}$, and $F_{1\to0}$. The spectrum seen experimentally* has precisely this structure of three peaks, clearly resolved from one another. The calculated and observed frequencies are:

$F_{F''\to F'}$	$(\Delta E_R + \Delta E_{Q'})/h$	ν, MHz
$F_{1\to0}$	$(\Delta E_R - 0{\cdot}50\ eQq)/h$	88 633·56
$F_{1\to2}$	$(\Delta E_R - 0{\cdot}05\ eQq)/h$	88 631·49
$F_{1\to1}$	$(\Delta E_R + 0{\cdot}25\ eQq)/h$	88 630·11

The solution to these linear equations is $\Delta E_R/h = 88\,631.26$ and $eQq = -4.6_0$ MHz. Both the magnitude and sign of the coupling constant, eQq, are determined in this

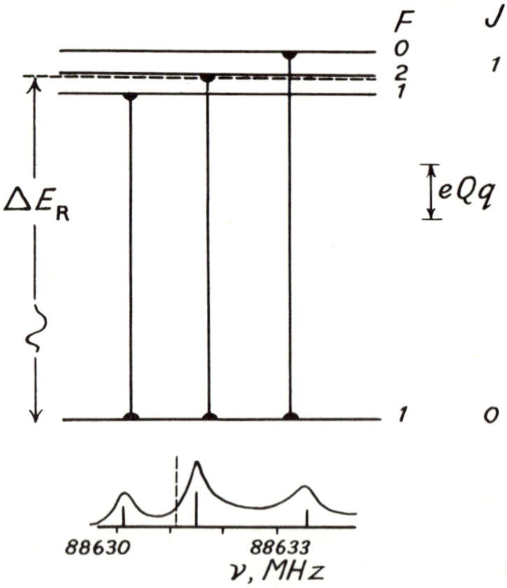

Fig. 5.7.3. Hyperfine structure of the $J = 0 \to 1$ transition of HCN. In the lower portion of the diagram the dotted line shows the position of the (hypothetical) pure rotational transition.

* Simmons, Anderson and Gordy, *Phys. Rev.*, 1950, **77**; 1952, **86**, 1055.

case. The nitrogen nucleus is prolate (eQ positive) and hence the sign of eQq indicates that the field gradient at the nitrogen nucleus in HCN is negative. Evidently when nuclear quadrupole coupling exists it is necessary to analyse the resulting hyperfine structure in order to evaluate the frequency $\Delta E_R/h$ of the pure rotational jump, and hence the moment of inertia of the molecule.

Some eQq constants obtained from rotation spectra are listed in Table 5.7.2; the wide range of values should be noted. The interpretation of quadrupole coupling constants, in terms of the electronic environments of nuclei within molecules, will be discussed in Chapter 8, following the section on pure quadrupole resonance in solids.

Table 5.7.2. Coupling Constants from Rotational Hyperfine Structure

Nucleus	*Substance*	*eQq* (MHz)
^{14}N	HCN	−4·58
	ClCN	−3·63
	FCN	−2·67
	CH_3CN	−4·35
	CH_3NC	$<$0·5
^{35}Cl	ClCN	−83·33
	CH_3Cl	−74·77
	HCCCl	−79·67
	KCl	$<$0·04
^{37}Cl	ClCN	−65·7
	CH_3Cl	−58·93
	HCCCl	−62·75
^{79}Br	BrCN	685·9
	CH_3Br	577·3
	KBr	10·24
^{81}Br	BrCN	572·5
	CH_3Br	482·4
	KBr	8·56
^{127}I	ICN	−2420
	CH_3I	−1934
	KI	−60
^{17}O	OCS	−1·32
^{33}S	OCS	−29·07
	CS	12·84
^{10}B	BH_3CO	3·4
^{11}B	BH_3CO	1·55

5.8 The Stark Effect in Rotation Spectra

The rotational absorption of a molecule in an electric field is usually somewhat different from its normal spectrum. Each line observed when the external field is zero may separate into several components when an electric field is switched on, the splitting of the original line under the influence of the field being known as the *Stark effect*. The origin of the Stark effect lies in the interaction of the electric dipole moment of a molecule with the external field.

The electric lines of force of an external field establish a direction fixed in space. The principles governing the quantisation of angular momentum about a space-fixed axis were described in Section 3.7. Theory allows the angular momentum about an axis of this kind to assume any one of the values,

$$Mh/2\pi, \tag{3.7.10}$$

in which M is some positive or negative integer, or zero. For a rotating molecule, the angular momentum parallel to the external field is the component of the total angular momentum vector $J = \sqrt{J(J+1)}h/2\pi$ along the axis of the electric lines of force. Since this space-fixed component of J obviously cannot exceed J itself, the spectrum of permitted values of the quantum number M is

$$M = J,\ J-1, \ldots, -J, \tag{1}$$

so that the total number of possible values of M is $2J + 1$. The simultaneous quantisation of the total angular momentum and of its component parallel to the field means that the orientation of the total angular momentum with respect to the axis of the field is restricted to the discrete values θ_i given by

$$\cos\theta_i = M/\sqrt{J(J+1)}. \tag{2}$$

Thus θ_i may assume only $2J + 1$ different values for each value of J.

The $2J + 1$ orientations of the rotating molecule may differ in energy for the following reason. Let the component of the permanent electric dipole moment along the axis of the vector J be μ_J; then the inclination θ_i of μ_J with respect to the electric lines of force of an external field is given by (2). Electrostatic theory gives for the potential energy $E_{\mathcal{E}(1)}$ of a dipole μ_J inclined at θ_i to a uniform field of intensity $\mathcal{E}$ the expression

$$E_{\mathcal{E}(1)} = -\mu_J \mathcal{E} \cos\theta_i \tag{3}$$

and hence, from (2),

$$E_{\mathcal{E}(1)} = -\mu_J \mathcal{E} M/\sqrt{J(J+1)}. \tag{4}$$

Consider first a symmetric top molecule. The permanent electric moment coincides with the top axis. The inclination θ of the top axis to the axis of J depends upon the quantum numbers J and K and is given by

$$\cos\theta = K/\sqrt{J(J+1)}. \tag{5.2.10}$$

Therefore the magnitude of the component μ_J is

$$\mu_J = \mu\cos\theta = \mu K/\sqrt{J(J+1)}. \tag{5}$$

The expression for the correction to the potential energy of the rotating symmetric top molecule is obtained by substituting (5) into (4). We then have

$$E_{\mathcal{E}(1)} = -\mu\mathcal{E}KM/J(J+1). \tag{6}$$

The energy correction $E_{\mathcal{E}(1)}$ depends upon the three quantum numbers J, K, and M, and is zero whenever K or M is zero; otherwise the correction varies linearly with the product, $\mathcal{E}\mu$, of the electric field intensity and the permanent dipole moment, and is described as a *first-order* Stark correction.

The frequency of the displaced lines that appear in the Stark effect depends upon the selection rule for the quantum number M. It is usual to arrange that the external electric field is parallel to the electric field of the incident microwave radiation, and the selection rule then is

$$\Delta M = 0. \tag{7}$$

This condition, together with the usual selection rules for a symmetric top, $\Delta J = 1$ and $\Delta K = 0$, can be used to calculate the frequency of the displaced lines in the Stark effect. Introducing $M'' = M' = M$, $K'' = K' = K$, $J'' = J$, and $J' = J + 1$ into (6), we obtain for the frequency displacements $\Delta\nu$,

$$h\Delta\nu = E_{\mathcal{E}(1)}(J+1, K, M) - E_{\mathcal{E}(1)}(J, K, M) = -\mu\mathcal{E}\left\{\frac{KM}{(J+1)(J+2)} - \frac{KM}{J(J+1)}\right\}$$

$$= 2\mu\mathcal{E}\frac{KM}{J(J+1)(J+2)}. \tag{8}$$

Let us enumerate the Stark displacements of the $J = 1 \rightarrow 2$ transition of a symmetric top. When $J(\equiv J'') = 1$, K may assume the values 0 and ± 1, while M may take the values 0 and ± 1. The frequency displacements are given in Table 5.8.1. Of a total of six transitions, four occur at the frequency of the undisplaced line (the line seen in

Table 5.8.1. Stark Displacements for the $J = 1 \rightarrow 2$ Transition of a Symmetric Top

K		0			± 1	
M	-1	0	$+1$	∓ 1	0	± 1
$\Delta\nu$	0	0	0	$-(\mu\mathcal{E}/3h)$	0	$\mu\mathcal{E}/3h$

the absence of the external electric field), and two are disposed symmetrically on either side. Typical values of $\mu = 10^{-18}$ esu. cm (1 Debye unit) and $\mathcal{E} = 300$ volts cm^{-1} (1 esu) give a displacement $\mu\mathcal{E}/3h$ of about 50 MHz which is easily observable.

In the $K = 0$ sub-levels of a symmetric top the total angular momentum vector $\boldsymbol{J}$ is perpendicular to the top axis: the component μ_J of the permanent electric moment is thus zero. This is the reason why there is no first-order Stark correction to the potential energy when $K = 0$. Likewise in the rotation of a linear molecule the vector $\boldsymbol{J}$ is perpendicular to the internuclear axis, and hence to the axis of the electric dipole; therefore no first-order Stark effect is expected for linear molecules either. Nevertheless, linear molecules do have an observable Stark effect in strong electric fields. Fig. 5.8.1 represents the rotation of a linear molecule in an external electric field. The field tends to twist the dipole and hence to speed up the rotation of the molecule

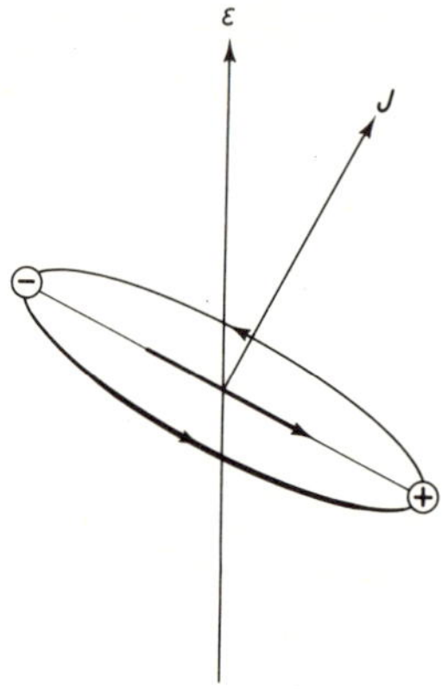

Fig. 5.8.1

when the dipole is pointing with the field, and to retard it when the dipole is pointing oppositely to the field. Therefore the dipole spends more time orientated away from the field than with it, contrary to what one would expect if there were no rotation. In consequence the energy of the dipole does not average to zero, as it would if the speed of rotation were uniform. The degree of non-uniformity in the rotation is proportional to the dipole–field interaction and inversely proportional to the rotational energy. It is thus proportional to $\mu\mathcal{E}/B$, and the energy correction to the rotational energy is proportional to this factor multiplied by $\mu\mathcal{E}$. This energy change is known as the *second-order* Stark correction, $E_{\mathcal{E}(2)}$, and we deduce

$$E_{\mathcal{E}(2)} \propto (\mu\mathcal{E})^2/B. \tag{9}$$

The exact quantum mechanical expression for $E_{\mathcal{E}(2)}$ for a linear molecule is

$$E_{\mathcal{E}(2)} = \frac{\mu^2 \mathcal{E}^2 [J(J+1) - 3M^2]}{2hBJ(J+1)(2J-1)(2J+3)} \tag{10}$$

and when $J = 0$

$$E_{\mathcal{E}(2)} = -\frac{\mu^2 \mathcal{E}^2}{6hB} \tag{10(a)}$$

In general $E_{\mathcal{E}(2)}$ is small compared with $E_{\mathcal{E}(1)}$ and can often be neglected for cases when $E_{\mathcal{E}(1)}$ does not vanish, i.e. for a symmetric top when $K \neq 0$. For linear molecules and the $K = 0$ levels of symmetric tops, $E_{\mathcal{E}(1)}$ is identically zero and the whole Stark energy is just $E_{\mathcal{E}(2)}$. An electric field of the order of a thousand volts per centimetre may then be necessary to record the Stark displacements.

First-order Stark effects are characteristic of systems with degenerate energy levels. In asymmetric tops the K-degeneracy of the symmetric rotor is lifted and the Stark effects are usually second-order; the Stark displacements being proportional to $\mathcal{E}^2$. Near degeneracies do sometimes occur in asymmetric rotors, and then the Stark effects have a mixed character; the displacements being proportional to $\mathcal{E}^2$ at low fields and to $\mathcal{E}$ at high field strengths. The above treatment applies to molecules with *no quadrupolar nuclei*. When quadrupole fine structure is present in the rotational spectrum more detailed considerations are necessary; these are beyond the scope of this volume.

The most important application of the Stark effect is in the measurement of permanent dipole moments. According to (8) the first-order Stark displacements vary linearly with μ, and hence μ can be calculated from the displacements provided the field strength and the quantum numbers J, K, and M are known. Second-order Stark displacements depend upon $(\mu\mathcal{E})^2$ and B as well as on the quantum numbers. For a symmetric top molecule it is possible to determine μ from the first- or second-order displacements, whichever is more convenient; but for a linear molecule only the second-order displacements are available. Dipole moments obtained from the Stark effect are accurate to a few parts per thousand in favourable circumstances, the uncertainty being imposed by the difficulty of producing a uniform electric field within the absorption cell. Even so the advantages are very great, and no other method for the determination of dipole moments is capable of such precision. Moreover, the measurement refers to a particular vibrational state and the change of dipole moment with vibrational amplitude can be studied; thus values of 0·708 D for the zero-point vibrational state, and 0·700 D for the lowest excited vibrational state ($v_2 = 1$) have been reported for carbon oxysulphide.*

*Shulman and Townes, *Phys. Rev.*, 1950, **77**, 530.

BIBLIOGRAPHY

Gordy, Smith and Trambarulo, *Microwave Spectroscopy*, Wiley, New York, 1953.
Townes and Schawlow, *Microwave Spectroscopy*, McGraw-Hill, New York and London, 1955.
Wollrab, *Rotational Spectra and Molecular Structure*, Academic Press, New York, 1967.

Chapter 6

The vibration of molecules

In our preliminary survey of spectroscopy in Chapter 4 we saw that for a molecule of N atoms there must be $3N - 6$ vibrational degrees of freedom ($3N - 5$ for a linear molecule). We must now investigate molecular vibrations in more detail and, again, it is convenient to consider first the mechanics from a classical viewpoint, postponing until later the question of quantisation.

6.1 Classical Mechanics of Molecular Vibrations

The vibration of a polyatomic molecule is apparently disorderly; but the displacements of the nuclei from their mean positions are the sum of displacements associated with special vibrations in which the nuclei move in straight lines and *in phase*. The in-phase property signifies that all the nuclei in a molecule pass through their mean positions, and reach their turning-points simultaneously. These special vibrations are known as the *normal*, or *fundamental*, modes. The number of normal modes is equal to the number of vibrational degrees of freedom possessed by the molecule.

To bring greater reality to the concept of a normal mode let us take as a simple example the stretching vibrations of a linear system of three masses. As we shall restrict our attention to nuclear movements parallel to the inter-nuclear axis, our system has three degrees of freedom of which one corresponds to a translation. The nuclear displacements can be described by three coordinates ξ_1, ξ_2 and ξ_3 chosen so that their origins are at the equilibrium positions of the nuclei (as shown in Fig. 6.1.1). Provided the displacements are very small one can assume that the restoring force acting on each nucleus is proportional to its displacement. (This is the condition for harmonic motion.) Moreover, it would seem realistic to attribute the forces opposing displacement to the resistance of the chemical bonds to extension of compression. Thus when the actual displacements are given by ξ_1, ξ_2, and ξ_3 the change in length of the bond between the first and second masses is $\xi_1 - \xi_2$ and between the second and third masses is $\xi_2 - \xi_3$, so that the expression for the potential energy has the form

$$2V = F(\xi_1 - \xi_2)^2 + F'(\xi_2 - \xi_3)^2. \tag{1}$$

Here F and F' are *force constants* which measure the resistance to a change in length of the first and second bonds, respectively. As the bonds are identical in our chosen model, F must equal F' and we may write

$$2V = F[(\xi_1 - \xi_2)^2 + (\xi_2 - \xi_3)^2]. \tag{2}$$

This implies that the contribution to the energy from the bond extensions can be thought of as additive. It neglects the consideration that an extension of the first bond may alter the resistance to stretching of the second bond, and *vice versa*: but we shall overlook mutual influences of this kind for the present.

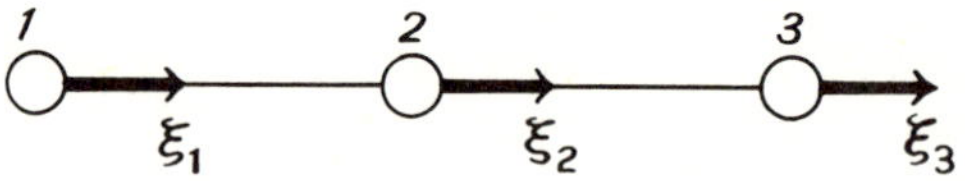

Fig. 6.1.1. Coordinates for the longitudinal vibrations of a linear triatomic molecule. The end masses are taken to be equal (*e.g.*, CO_2, CS_2).

To confirm that the potential (2) gives rise to harmonic motion, let us evaluate the restoring force acting on each of the three nuclei. The force f_i acting on the i-th nucleus is $-\partial V/\partial \xi_i$, and hence

$$f_1 = -F(\xi_1 - \xi_2) \tag{3}$$

$$f_2 = F(\xi_1 - 2\xi_2 + \xi_3) \tag{4}$$

and

$$f_3 = F(\xi_2 - \xi_3). \tag{5}$$

The restoring forces depend linearly on the displacements. If all displacements were zero except the first we would have $f_i \propto \xi_i$ that is, the ordinary expression for simple harmonic motion of a single mass.

It is convenient to replace the linear displacements ξ_i by mass-weighted displacements q_i defined by

$$q_i = \mu_i^{-\frac{1}{2}}\xi_i, \tag{6}$$

where $\mu_i = 1/m_i$ is the reciprocal of the mass of the i-th atom. If we now apply Newton's equation of motion, eqn. (3.3.1), to each of the three nuclei linear in turn (the end masses taken to be equal) we obtain a set of simultaneous linear equations,

$$\left.\begin{aligned}
\mu_1^{-\frac{1}{2}}\ddot{q}_1 + \mu_1^{\frac{1}{2}}Fq_1 - \mu_2^{\frac{1}{2}}Fq_2 &= 0 \\
\mu_2^{-\frac{1}{2}}\ddot{q}_2 - \mu_1^{\frac{1}{2}}Fq_1 + 2\mu_2^{\frac{1}{2}}Fq_2 - \mu_1^{\frac{1}{2}}Fq_3 &= 0 \\
\mu_1^{-\frac{1}{2}}\ddot{q}_3 \qquad - \mu_2^{\frac{1}{2}}Fq_2 + \mu_1^{\frac{1}{2}}Fq_3 &= 0
\end{aligned}\right\} . \tag{7}$$

A possible solution of these equations is

$$q_i = A_i \cos(\lambda^{\frac{1}{2}}t + \zeta), \qquad i = 1, 2, 3 \tag{8}$$

where A_i is the amplitude, or maximum displacement, of the i-th mass. Since $\ddot{q}_i$ then equals $-\lambda q_i$ the substitution of (8) into (7) yields a set of three simultaneous linear equations, known as the *secular equations*,

$$\left.\begin{aligned}
(\mu_1 F - \lambda)A_1 - \mu_1^{\frac{1}{2}}\mu_2^{\frac{1}{2}}FA_2 &= 0 \\
-\mu_1^{\frac{1}{2}}\mu_2^{\frac{1}{2}}FA_1 + (2\mu_2 F - \lambda)A_2 - \mu_1^{\frac{1}{2}}\mu_2^{\frac{1}{2}}FA_3 &= 0 \\
-\mu_1^{\frac{1}{2}}\mu_2^{\frac{1}{2}}FA_2 + (\mu_1 F - \lambda)A_3 &= 0
\end{aligned}\right\} . \tag{9}$$

The solution of (9) for any arbitrary value of λ is the trivial one $A_1 = A_2 = A_3 = 0$, corresponding to no displacements and thus to a non-vibrating system. The solutions of interest, for which the A's are non-vanishing, occur for special values of λ only. These special values satisfy the secular determinant,

$$\begin{vmatrix}
\mu_1 F - \lambda & -\mu_1^{\frac{1}{2}}\mu_2^{\frac{1}{2}}F & 0 \\
-\mu_1^{\frac{1}{2}}\mu_2^{\frac{1}{2}}F & 2\mu_2 F - \lambda & -\mu_1^{\frac{1}{2}}\mu_2^{\frac{1}{2}}F \\
0 & -\mu_1^{\frac{1}{2}}\mu_2^{\frac{1}{2}}F & \mu_1 F - \lambda
\end{vmatrix} = 0 . \tag{10}$$

The elements of the determinant are the coefficients of A_1, A_2, and A_3 in the secular equations (9). When a special value of λ, λ_k say, causes the determinant to vanish the displacements q_i vary with time with [according to eqn. (8)] the frequency $\nu_k = \lambda_k^{\frac{1}{2}}/2\pi$: at this frequency, therefore, the nuclei oscillate in phase and the motion characterises a normal mode of vibration. Moreover, when λ_k is substituted into the secular equation (9), the coefficients of A_1, A_2, and A_3 become fixed, so that the relative amplitudes of vibration can be determined.

By expansion of the secular determinant (10) we obtain

$$(\lambda - \mu_1 F)\lambda[\lambda - (\mu_1 + 2\mu_2)F] = 0 \tag{11}$$

which has roots $\lambda_k (k = 1, 2, 3)$

$$\lambda_1 = \mu_1 F, \quad \lambda_2 = 0, \quad \lambda_3 = (\mu_1 + 2\mu_2)F . \tag{12}$$

To find the relative amplitudes we substitute in turn the three roots λ_k into the secular equations. The amplitudes we shall designate by A_{ik} where, as usual, the subscript $i (i = 1, 2, 3)$ numbers the nuclei and the subscript $k (= 1, 2, 3)$ numbers the roots. After performing the substitution we have,

λ_k	A_{ik}
λ_1	$A_{11} = -A_{31}, A_{21} = 0$
λ_2	$A_{12} = A_{32}, A_{22} = (\mu_1/\mu_2)^{\frac{1}{2}} A_{12}$
λ_3	$A_{13} = A_{33}, A_{23} = -2(\mu_2/\mu_1)^{\frac{1}{2}} A_{13}.$

Next we normalise the amplitudes so that the total energy in each mode is the same, as required by the equipartition principle,

$$\sum_{i=1}^{3} (A_{ik})^2 = 1, \qquad k = 1, 2, 3. \tag{13}$$

This enables us to write down a unique mathematical solution for the amplitudes, which we shall distinguish by the lower case letter a_{ik}. The a_{ik} for the longitudinal motion of our three-mass system are given in Table 6.1.1. Fig. 6.1.2 is a diagram of the normalised amplitudes for carbon dioxide, chosen as a representative molecule; here the arrows are drawn so that their lengths are proportional to the a_{ik} listed in the table. The root $\lambda_2 = 0$, corresponding to a vibration of zero frequency, represents the translation of the molecule along its axis with no relative displacement of the nuclei. The two remaining roots of the secular determinant correspond to genuine normal modes of vibration. In studying Fig. 6.1.2 it should be remembered that the normalised amplitudes, a_{ik}, are expressed in terms of the mass-weighted co-ordinates q_i; therefore to visualise the actual displacements (ξ_i) the length of the arrows must be divided by the square root of the appropriate mass.

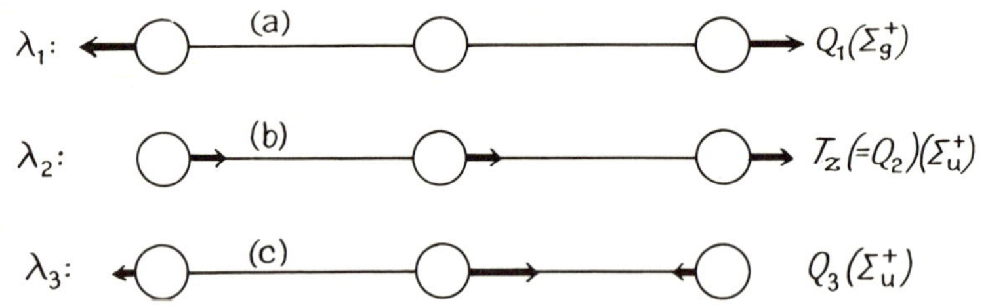

Fig. 6.1.2. Normal coordinates for the translation (*b*) and longitudinal vibrations (*a*, *c*) of CO_2.

Normal Coordinates. The normal coordinate Q_k corresponding to a root λ_k of the secular determinant is defined by the equation

Table 6.1.1. Normalised Amplitudes, a_{ik}, of a Symmetrical Linear Triatomic Molecule

$i =$	1	2	3
$k = 1$	$2^{-\frac{1}{2}}$	0	$-2^{-\frac{1}{2}}$
2	$\left(\frac{\mu_2}{\mu_1 + 2\mu_2}\right)^{\frac{1}{2}}$	$\left(\frac{\mu_1}{\mu_1 + 2\mu_2}\right)^{\frac{1}{2}}$	$\left(\frac{\mu_2}{\mu_1 + 2\mu_2}\right)^{\frac{1}{2}}$
3	$\left(\frac{\mu_1}{2(\mu_1 + 2\mu_2)}\right)^{\frac{1}{2}}$	$-2\left(\frac{\mu_2}{2(\mu_1 + 2\mu_2)}\right)^{\frac{1}{2}}$	$\left(\frac{\mu_2}{2(\mu_1 + 2\mu_2)}\right)^{\frac{1}{2}}$

$$Q_k = \sum_i a_{ik} q_i = \sum_i [A_{ik}/(\sum_i A_{ik}{}^2)^{\frac{1}{2}}] q_i. \tag{14}$$

The normal coordinates for the translation and the two genuine vibrations of the triatomic model system are, therefore,

$$Q_1 = 2^{-\frac{1}{2}}(q_1 - q_3), \tag{15}$$

$$Q_2 = [\mu_2{}^{\frac{1}{2}} q_1 + \mu_1{}^{\frac{1}{2}} q_2 + \mu_2{}^{\frac{1}{2}} q_3](\mu_1 + 2\mu_2)^{-\frac{1}{2}}, \tag{16}$$

and

$$Q_3 = [\mu_1{}^{\frac{1}{2}} q_1 - 2\mu_2{}^{\frac{1}{2}} q_2 + \mu_1{}^{\frac{1}{2}} q_3]\{2(\mu_1 + 2\mu_2)\}^{-\frac{1}{2}}. \tag{17}$$

It is evident that a diagram such as Fig. 6.1.2, which represents the actual displacements of the atoms–not in ordinary units, but in the mass-weighted scale of coordinates q_i–can also be used to represent the normal coordinates; for the length of an arrow is proportional to a_{ik}, and consequently gives not only the relative amplitudes of the atoms in the k-th normal mode of motion but also the coefficients in the expression (14) defining the normal coordinate Q_k.

To illustrate the versatility of the normal coordinates let us evaluate the expression

$$\sum_k \lambda_k Q_k{}^2. \tag{18}$$

Clearly only the non-zero roots λ_1 and λ_3 contribute to the sum. Introducing the values of λ_1 and λ_3 (eqn. 12) and those of Q_1 and Q_3 (eqns. 15 and 17) we find,

$$\lambda_1 Q_1{}^2 + \lambda_3 Q_3{}^2 = \tfrac{1}{2}\mu_1 F(q_1 - q_3)^2 + \tfrac{1}{2}F(\mu_1{}^{\frac{1}{2}} q_1 - 2\mu_2{}^{\frac{1}{2}} q_2 + \mu_1{}^{\frac{1}{2}} q_3)^2. \tag{19}$$

Reverting to displacement coordinates ξ_i, after some simplification we obtain,

$$\lambda_1 Q_1{}^2 + \lambda_3 Q_3{}^2 = \tfrac{1}{2}F[(\xi_1 - \xi_3)^2 + (\xi_1 - 2\xi_2 + \xi_3)^2]$$

$$= F[(\xi_1 - \xi_2)^2 + (\xi_2 - \xi_3)^2]. \tag{20}$$

The expression on the right-hand side of (20), according to eqn. (1), is simply $2V$; therefore the potential energy in terms of the normal coordinates is given by

$$2V = \sum_k \lambda_k Q_k{}^2. \tag{21}$$

Secondly, let us evaluate the sum

$$\sum_k Q_k{}^2. \tag{22}$$

The term contributed by Q_2 is not zero, and so we have

$$Q_1{}^2 + Q_2{}^2 + Q_3{}^2 = \tfrac{1}{2}(q_1 - q_3)^2 + [(\mu_2{}^{\frac{1}{2}}q_1 + \mu_1{}^{\frac{1}{2}}q_2 + \mu_2{}^{\frac{1}{2}}q_3)^2$$
$$+\tfrac{1}{2}(\mu_1{}^{\frac{1}{2}}q_1 - 2\mu_2{}^{\frac{1}{2}}q_2 + \mu_1{}^{\frac{1}{2}}q_3)^2](\mu_1 + 2\mu_2)^{-1}. \tag{23}$$

As the reader should check for himself, all the terms cancel except the square terms, yielding

$$Q_1{}^2 + Q_2{}^2 + Q_3{}^2 = q_1{}^2 + q_2{}^2 + q_3{}^2$$

or

$$\sum_k Q_k{}^2 = \sum_k q_i{}^2. \tag{24}$$

The general expression of (24) follows from the definition (14) of the normal coordinates, if one recalls that $\sum_i a_{ik}{}^2 = 1$; therefore (24) is to be considered part of the definition of the normal coordinates.

A similar expression holds for the time-derivatives of Q_k and q_i, namely

$$\sum_k \dot{Q}_k{}^2 = \sum_i \dot{q}_i{}^2, \tag{25}$$

where the dot, as usual, stands for $\mathrm{d}/\mathrm{d}t$. Now, for the model in Fig. 6.1.1 the kinetic energy (parallel to the internuclear axis) is given by

$$2T = \mu_1{}^{-1}\dot{\xi}_1{}^2 + \mu_2{}^{-1}\dot{\xi}_2{}^2 + \mu_3{}^{-1}\dot{\xi}_3{}^2$$
$$= \dot{q}_1{}^2 + \dot{q}_2{}^2 + \dot{q}_3{}^2 = \sum_i \dot{q}_i{}^2. \tag{26}$$

Evidently the kinetic energy in terms of the normal coordinates can be written

$$2T = \sum_i \dot{q}_i{}^2 = \sum_k \dot{Q}_k{}^2. \tag{27}$$

The Hamiltonian function, $H = T + V$ (section 3.3), of the motion is then [by eqns. (27) and (21)],

$$H = \tfrac{1}{2}(\sum_k \dot{Q}_k{}^2 + \sum_k \lambda_k Q_k{}^2)$$

$$= \tfrac{1}{2}(\dot{Q}_1{}^2 + \lambda_1 Q_1{}^2) + \tfrac{1}{2}(\dot{Q}_2{}^2 + \lambda_2 Q_2{}^2) + \ldots \quad (28)$$

The simple oscillator described in section 3.3, consisting of a particle of mass m vibrating along a linear coordinate x, has $H = \frac{1}{2}m(\dot{x}^2 + \lambda x^2)$. Thus the total energy of a molecule can be considered as the sum of mutually independent terms each of which has the form of the total energy of a simple harmonic oscillator with unit effective mass. That is to say, the motion of a bound system of atoms can be regarded as a *superposition of independent simple harmonic motions for which the coordinates are the normal coordinates* Q_k.

Summary. The normal modes of vibration of a molecule can conveniently be described by mass-weighted *displacement coordinates* q_i (where i is a representative atom) and by *the normal coordinates* Q_k, where

$$q_i = A_{ik}\cos(2\pi\nu_k t + \zeta) \quad (8)$$

and

$$Q_k = \sum_i a_{ik} q_i \quad (14)$$

with $a_{ik} = A_{ik}/(\sum_i A_{ik}{}^2)^{\frac{1}{2}}$. The amplitudes a_{ik} of the atoms in a fundamental mode of vibration (including non-genuine modes, corresponding to rotation and translation) are normalised so that

$$\sum_i a_{ik}{}^2 = \sum_k a_{ik}{}^2 = 1. \quad (29)$$

In terms of the normal coordinates the potential and kinetic energies are given by

$$2V = \sum_k \lambda_k Q_k{}^2 \,, \qquad 2T = \sum_k \dot{Q}_k{}^2 \quad (30)$$

and hence the classical Hamiltonian function $H = T + V$ can be written

$$H = \tfrac{1}{2}\sum_k (\dot{Q}_k{}^2 + \lambda_k Q_k{}^2). \quad (28)$$

For the case of the longitudinal motion of a linear triatomic molecule, eqns. (20) and (30) can be checked using the coefficients a_{ik} in Table 6.1.1. The general proofs are given in texts on spectroscopy.

6.2 Wave Mechanics of Molecular Vibration

The normal coordinates for molecular motion are obtained from classical mechanics.

To discover the vibrational wavefunctions ψ_v and energy eigenvalues E_v it is necessary to solve the Schroedinger amplitude equation

$$H\psi_v = E_v\psi_v \tag{3.4.5}$$

for the case when the classical kinetic and potential energies are given by

$$2T = \sum_k \dot{Q}_k^2 \qquad 2V = \sum_k \lambda_k Q_k^2. \tag{6.1.30}$$

The wave-mechanical treatment of a simple linear oscillator has been described in section 3.5. For the vibration of a particle of mass m along a coordinate x the classical potential and kinetic energies are

$$V = \tfrac{1}{2}Fx^2 = \tfrac{1}{2}\lambda m x^2 \quad \text{and} \quad T = \tfrac{1}{2}m\dot{x}^2,$$

when the Hamiltonian operator is found to be (eqn. 3.5.1),

$$H = -\frac{h^2}{8\pi^2 m}\frac{\mathrm{d}^2}{\mathrm{d}x^2} + \frac{1}{2}\lambda x^2. \tag{1}$$

Thus the Hamiltonian operator for the motion of the atoms in a molecule has the form

$$H = -\frac{h^2}{8\pi^2}\sum_k \frac{\partial^2}{\partial Q_k{}^2} + \frac{1}{2}\sum_k \lambda_k Q_k{}^2. \tag{2}$$

Eqn. (2) is everywhere analogous to (1), except in that the definition of the normal coordinates adjusts the effective mass to unity, so that m drops out.

The vibrational wave equation then has the form

$$\left(-\frac{h^2}{8\pi^2}\sum_k \frac{\partial^2}{\partial Q_k{}^2} + \frac{1}{2}\sum_k \lambda_k Q_k{}^2\right)\psi_V = E_V\psi_V. \tag{3}$$

The advantages of normal coordinates now emerge. Let us write ψ_V as a product,

$$\psi_V = \psi_v(Q_1)\cdot\psi_v(Q_2)\ldots\psi_v(Q_k)\ldots\psi_v(Q_{3N-6}) \tag{4}$$

and E_V as a sum,

$$E_V = E_v(1) + E_v(2) + \ldots + E_v(k) + \ldots + E_v(3N-6), \tag{5}$$

then the wave equation (3) factorises into a set of $3N-6$ equations of the type

$$\left(-\frac{h^2}{8\pi^2}\frac{d^2}{dQ_k{}^2}+\frac{1}{2}\lambda_k Q_k{}^2\right)\psi_v(Q_k) = E_v(k)\psi_v(Q_k)$$

$$k = 1, 2, \ldots (3N-6) \tag{6}$$

together with a further set of six equations that correspond to zero roots of the secular determinant and lead to the wave equations for translation and rotation. These six equations need not concern us here.

Eqn. (6) is simply the Schroedinger amplitude equation of an harmonic oscillator expressed in terms of the normal coordinate Q_k instead of the linear coordinate x. The solution has been obtained in section 3.5 and has the form of eqn. (3.5.12). Allowing for the change of variable, the wave-functions are:

$$\psi_v(Q_k) = N_{v_k}H_{v_k}(Q_k/\alpha_k{}^{\frac{1}{2}})e^{-\frac{1}{2}Q_k{}^2/\alpha_k} \tag{7}$$

with

$$\alpha_k = h/2\pi\lambda_k{}^{\frac{1}{2}} = h/4\pi^2\nu_k \tag{8}$$

and

$$N_{v_k} = (2^{v_k}v_k!\alpha_k{}^{\frac{1}{2}}\pi^{\frac{1}{2}})^{-\frac{1}{2}} \qquad v = 0, 1, 2, \ldots \tag{9}$$

Here $\nu_k = \lambda_k{}^{\frac{1}{2}}/2\pi$ is the classical frequency of the normal vibration k, and $H_{v_k}(Q_k/\alpha_k{}^{\frac{1}{2}})$ is the Hermite polynomial of degree v_k in Q_k. The first six Hermite polynomials are given by eqn. (3.2.10), with $x = Q_k/\alpha_k{}^{\frac{1}{2}}$. Thus, for the wavefunctions of the zeroth ($v_k = 0$) and first ($v_k = 1$) states of the k-th normal mode we find

$$\psi_0(Q_k) = (1/\alpha_k\pi)^{\frac{1}{4}}e^{-\frac{1}{2}Q_k{}^2/\alpha_k} \tag{10}$$

$$\psi_1(Q_k) = (4/\alpha_k{}^3\pi)^{\frac{1}{4}}Q_k e^{-\frac{1}{2}Q_k{}^2/\alpha_k}. \tag{11}$$

The functions (10) and (11) have the shape of the two lowest curves in Fig. 3.5.1.

The energy levels of the k-th normal mode of vibration are the eigenvalues belonging to the wavefunction $\psi_v(Q_k)$, namely (eqn. 3.5.10),

$$E_v(k) = (v_k + \tfrac{1}{2})h\nu_k. \tag{12}$$

Therefore the vibrational energy levels of each normal mode are uniformly spaced, the stack of levels being precisely like that of an harmonic oscillator. For the whole molecule, the manifold of vibrational energy levels is the sum over $3N-6$ normal modes,

$$E_V = (v_1 + \tfrac{1}{2})h\nu_1 + \ldots + (v_k + \tfrac{1}{2})h\nu_k + \ldots + (v_{3N-6} + \tfrac{1}{2})h\nu_{3N-6}$$

$$= \sum_{k=1}^{3N-6} (v_k + \tfrac{1}{2})h\nu$$

and the total vibrational wavefunction is the product of $3N-6$ factors each of the form (7),

$$\psi_V = \prod_{k=1}^{3N-6} \psi_v(Q_k). \tag{14}$$

The minimum energy possessed by a molecule by virtue of its internal vibrations is the *zero-point energy*,

$$E_0(1) + \ldots + E_0(k) + \ldots + E_0(3N-6) = \prod_{k=1}^{3N-6} \tfrac{1}{2} h\nu_k \tag{15}$$

—a quantity which may be of considerable magnitude in polyatomic molecules.

6.3 Symmetry of the Normal Coordinates

As we have seen in section 2.5, a molecule whose atoms are arbitrarily thought of as at rest may possess a number of *symmetry elements*, and hence the potentiality of a corresponding number of *symmetry operations* which leave the outward aspect of the molecule unchanged. The symmetry operations are those characteristic of the point group to which the molecule belongs. In a real molecule, because the atoms are never truly at rest, the symmetry elements considered are those that exist when the atoms are in their mean, or equilibrium, positions. It is evident that at least some of the normal vibrations of a molecule distort its equilibrium symmetry, and consequently that the symmetry properties form a basis for a possible classification of the normal modes. Symmetry properties were first applied to the problem of molecular vibration by Brester (1913), and the more formal group-theoretical methods by Wigner (1930).

Let us investigate the possible effects of a symmetry operation upon the normal coordinates. Fig. 6.3.1 illustrates the normal modes of water, a representative non-linear triatomic molecule, calculated by the method described in section 6.1. (A detailed calculation will be found in section 6.13).

The first and second normal modes, ν_1 and ν_2, have as much symmetry as the undistorted molecule; that is the displacements of the nuclei are left unchanged by all four operations I, C_2, σ_v, and σ_v' of the point group C_{2v} (section 2.9) to which the molecule belongs. The mode ν_3 is left unchanged by the operations I and σ_v' but the direction of the arrows, and hence the sign of the displacements, is reversed by the operations C_2 and σ_v. In other words, the effect of the symmetry operations I and σ_v' upon Q_3 is to leave Q_3 unchanged (multiplication by $+1$), whereas C_2 and σ_v change the sign of Q_3 (multiplication by -1). Symbolically we write

$$\begin{aligned} I: Q_3' &= (+1)Q_3 \qquad & \sigma_v : Q_3' &= (-1)Q_3 \\ C_2 : Q_3' &= (-1)Q_3 \qquad & \sigma_v' : Q_3' &= (+1)Q_3 \end{aligned} \tag{1}$$

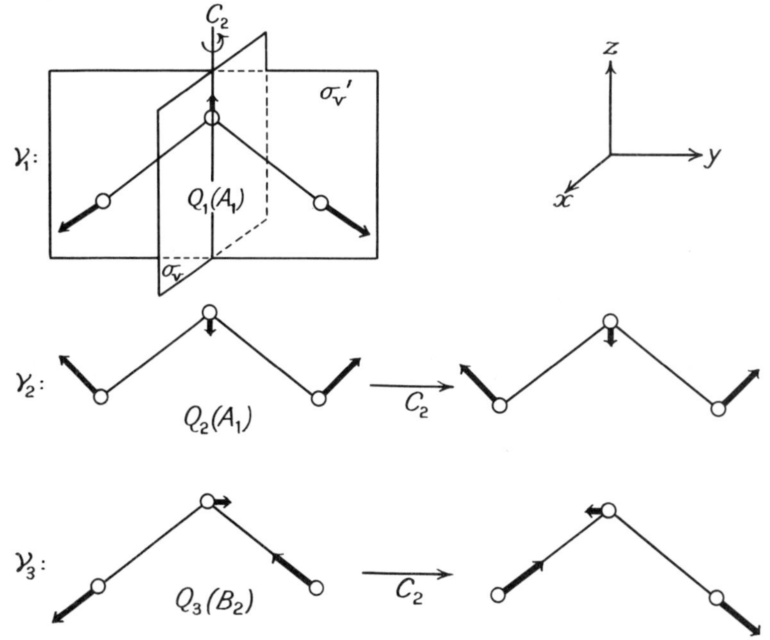

Fig. 6.3.1. Normal vibrations of H_2O. The operation C_2 leaves Q_2 unchanged, but changes the sign of Q_3.

where, in each equation of the set, the $'$ distinguishes the transformed coordinate, that is, the coordinate obtained *after the symmetry operation has been carried out.* For Q_1 and Q_2, the four operations all multiply the normal coordinate by + 1, *i.e.*,

$$\begin{aligned} I &: Q_1' = (+1)Q_1 \qquad & \sigma_v &: Q_1' = (+1)Q_1 \\ C_2 &: Q_1' = (+1)Q_1 \qquad & \sigma_v' &: Q_1' = (+1)Q_1 \end{aligned} \tag{2}$$

with a similar set of equations for Q_2.

We are now able to see that the normal coordinates of H_2O can be classified according to one or other of the four *irreducible representations*, or *symmetry species*, A_1, A_2, B_1, and B_2 of the group C_{2v}. The character table of the group has been discussed in section 2.9 and is reproduced in Table 6.3.1. The characters (χ) of the totally symmetrical representation A_1 are:

$$\chi(E) = \chi(C_2) = \chi(\sigma_v) = \chi(\sigma_v') = +1.$$

Comparison with eqn. (2) then shows that the factors which multiply Q_1 (and Q_2) when the possible symmetry operations are carried out are simply the characters of the representation A_1. Put another way, we see that Q_1 and Q_2 transform under

Table 6.3.1.

C_{2v}	I	C_2	σ_v	σ_v'
A_1	1	1	1	1
A_2	1	1	−1	−1
B_1	1	−1	1	−1
B_2	1	−1	−1	1

the symmetry operations of the group in the same manner as the symmetry representation A_1; in consequence they are said to *belong* to the representation A_1. Likewise eqn. (1) informs us that the representation to which Q_3 belongs has the characters

$$\chi(E) = \chi(\sigma_v') = +1\,, \qquad \chi(C_2) = \chi(\sigma_v) = -1,$$

which comprise the representation B_2 of the group. As the general symbol for a representation is the Greek letter Γ, we can summarise the discussion up to this point by writing the representation, Γ_V, of the normal vibrations of H_2O in the form

$$\Gamma_V = 2A_1 + B_2. \tag{3}$$

The factor 2 preceeding the species symbol A_1 indicates that two normal modes belong to this symmetry representation.

The present treatment of the normal modes of H_2O can be applied to any molecule once the normal coordinates are known. *The normal vibrations of a molecule can always be classed according to the symmetry representations of the point-group to which the molecule belongs.* But the calculation of the normal coordinates is tedious–and sometimes very difficult–for all except the simplest molecules, and the main value of symmetry considerations is that they enable us to determine the symmetry representations of the normal coordinates directly, without a detailed calculation of any kind. This information can in turn be used to simplify the determination of the normal coordinates.

Let us consider the vibrations of H_2O on the assumption that we lack any information in regard to the normal coordinates. The problem now is to determine the symmetry representation of the normal coordinates. At the mean position of each nucleus let us fix the origin of a set of cartesian displacement coordinates x_1, $y_1, z_1, x_2, \ldots z_3$, as shown in Fig. 6.3.2. (These coordinates are in all respects analgous to the displacement coordinates $\xi_1, \xi_2, \ldots$ introduced in section 6.1.) The total number of coordinates required is $3N = 9$, equal to the total number of degrees of freedom possessed by the molecule. The way in which a symmetry operation affects a displacement coordinate may be represented by an appropriate matrix. Consider the y coordinates in Fig. 6.3.2. Under the identity operation, I, the coordinates are unaltered and the operation may be represented by

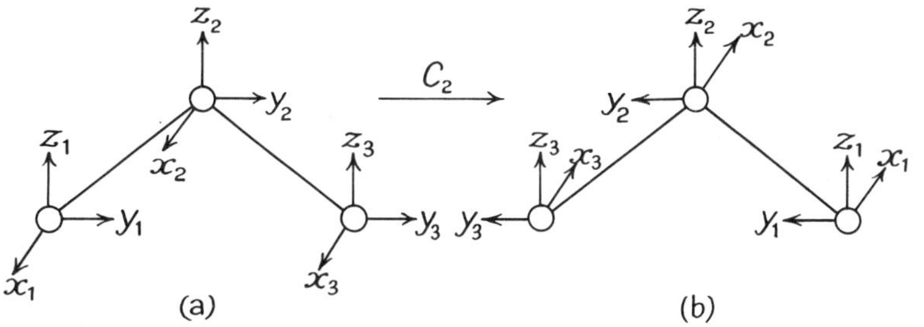

Fig. 6.3.2. Displacement coordinates for H_2O. The effect of the operation C_2 applied to the molecule in (*a*) is shown in (*b*). [The x coordinates are $\perp$ to the plane of the molecule (σ_v').]

$$I = \begin{pmatrix} 1 & 0 & 0 \\ 0 & 1 & 0 \\ 0 & 0 & 1 \end{pmatrix}, \tag{4}$$

since under this operation: $y_1' = y_1, y_2' = y_2$ and $y_3' = y_3$. The character for the y displacement coordinates for operation I is therefore 3. All three atoms make a contribution to this value. The rotation operation, C_2, may be represented by

$$C_2 = \begin{pmatrix} 0 & 0 & -1 \\ 0 & -1 & 0 \\ -1 & 0 & 0 \end{pmatrix}, \tag{5}$$

since under this operation: $y_1' = -y_3, y_2' = -y_2$ and $y_3' = -y_1$. The character of the y displacement coordinates is in this case -1, with only the oxygen atom contributing. The simple example illustrates an important general point: *Only those atoms which do not change position under the application of a symmetry operation contribute to the character of the displacement coordinates.*

Let there be U atoms with positions unaltered by a symmetry operation. Each of these atoms will have three displacement coordinates which will be transformed under a general symmetry operation by a matrix of the type:

$$\begin{pmatrix} \cos\phi & -\sin\phi & 0 \\ \sin\phi & \cos\phi & 0 \\ 0 & 0 & \pm 1 \end{pmatrix}.$$

The value of ϕ depends on the particular operation; $+1$ is taken for a proper rotation and -1 for an improper rotation. The general form for the character of the $3N$ displacement coordinates is therefore,

$$\chi(3N) = U(2\cos\phi \pm 1), \tag{6}$$

which is the sum of the characters for the translations, rotations and vibrations:

$$\chi(3N) = \chi(T) + \chi(R) + \chi(V). \tag{7}$$

In order to evaluate the character for the internal vibrations of a molecule, $\chi(V)$, we must obtain the values for $\chi(T)$ and $\chi(R)$. Translation simply corresponds to displacements along a single set of space fixed axes, x, y and z, and for a general symmetry operation the character for the translations is therefore given by,

$$\chi(T) = 2\cos\phi \pm 1. \tag{8}$$

The characters for the rotations may be obtained by imagining the molecule to be rotated slightly about each of the space fixed axes, x, y and z, in turn, and each time observing the effect of the symmetry operations on the displacements. After a little consideration it will be seen that, quite generally,

$$\chi(R) = \pm\chi(T), \tag{9}$$

with the positive sign applying for a proper rotation and the negative sign for an improper rotation. We may now combine eqns. (6), (7), (8) and (9) to give:

$$\chi(V) = (U-2)(2\cos\phi + 1) \tag{10a}$$

for a proper rotation, and

$$\chi(V) = U(2\cos\phi - 1) \tag{10b}$$

for an improper rotation. These expressions are applied to the water molecule in Table 6.3.2. with the result

$$\Gamma_V = 2A_1 + B_2, \tag{11}$$

in agreement with the direct analysis based on foreknowledge of the normal coordinates.

Table 6.3.2

C_{2v}	I	C_2	σ_v	σ_v'
U	3	1	1	3
ϕ	0°	180°	0°	0°
$\chi(V)$	3	1	1	3
		$\Gamma_V = 2A_1 + B_2$		

Eqns. (10*a*) and (10*b*) are generally applicable to any non-linear molecule, but cannot be used directly for linear molecules as there are then only two rotational degrees of freedom. In this case the characters for the $3N$ displacements may be written down from eqn. (6) and the characters for the three translations and *two* rotations (R_x and R_y) subtracted to give the $\chi(V)$ value of Γ_V. The representations of the components of translation (x, y and z) and rotation (R_x, R_y and R_z) are given in the collected tables in Appendix I.

Internal Coordinates

The judicious selection of $3N-6$ suitable coordinates for the displacement of nuclei would give the structure of the representation of the normal vibrations directly, without any confusion with translation and rotation. As vibration entails the extension of bonds and the deformation of interbond angles, an appropriate set of coordinates comprises *changes* in the length of chemical bonds and *changes* in the interbond angles. Coordinates of this type are known as *internal coordinates*. Unless the point group to which a molecule belongs contains a three-fold (or higher) axis of symmetry these are just $3N-6$ internal coordinates to choose from, so that no ambiguity attaches to their selection.

Let us first apply the method to H_2O, to see how the technique compares with the solution in cartesian coordinates. $3N-6=3$ internal coordinates are required: changes in length of the OH bond supply two of them, and a change of the $\angle$ HOH interbond angle the third. The OH bond extensions can be thought of as forming one class of coordinate (they are permuted by two of the possible operations of the group, namely C_2 and σ_v) while the HOH angular deformation falls in a second class. Each class can be considered separately. As to the OH extensions, since there are two of them and both are left unchanged by the operation I, $\chi(I)=2$. C_2 and σ_v cause both extensions to change position, so that $\chi(C_2)=\chi(\sigma_v)=0$, whereas σ_v' leaves both unchanged, $\chi(\sigma_v')=2$. The characters of the representation, listed under $\chi(r)$ in Table 6.3.3, are the sum of the characters of the irreducible representations A_1 and B_2, and so we can write

$$\Gamma(r) = A_1 + B_2 \tag{12}$$

as the structure of the representation of the bond extensions. The HOH angle increment is treated similarly. As there is only one coordinate in this class, $\chi(I)=1$.

Table 6.3.3 Characters of the Representation of Internal Coordinates of H_2O

	I	C_2	σ_v	σ_v'
$\chi(r)$	2	0	0	2
$\chi(\theta)$	1	1	1	1

C_2, σ_v, and σ_v' all leave the angle increment unchanged, hence $\chi(C_2) = \chi(\sigma_v) = \chi(\sigma_v') = 1$, as shown beside the heading $\chi(\theta)$ in the second row of Table 6.3.3. The representation of the angular deformation is then the totally symmetrical representation, *i.e.*,

$$\Gamma(\theta) = A_1. \tag{13}$$

The structure of the vibrational representation is the sum of the structures for both classes of coordinate,

$$\Gamma_V = \Gamma(r) + \Gamma(\theta) = 2A_1 + B_2, \tag{14}$$

which of course is identical with the result obtained earlier [eqns. (3) and (11)].

The advantage of using internal coordinates is now apparent. Evidently the non-totally symmetrical vibration of H_2O, ν_3 (B_2), involves OH bond extensions only; that is, it is a pure OH stretching mode. The same conclusion can be deduced from consideration of the cartesian displacement coordinates (see section 6.13) but is by no means immediately obvious. A close examination of the normal coordinate for the mode ν_3 (Fig. 6.3.1) will confirm that, for infinitesimal displacements–the length of the arrows greatly exaggerates the actual amplitude of the nuclei–the vibration involves no change of the interbond angle. The two A_1 vibrations of H_2O on the other hand may implicate both bond stretching and bond angle deformation simultaneously. Although this is as far as one can go in a rigorous analysis, it is natural to assume (by analogy with the behaviour of coupled pendulums) that the interaction between the vibratory motion associated with different internal coordinates will be weak unless the restoring forces are of comparable magnitude. A quantitative treatment shows that the restoring forces associated with the OH bond extensions are much greater than those resulting from the $\angle$ HOH deformations. Consequently we can expect that one of the A_1 normal vibrations of H_2O will be essentially a bond-stretching vibration (with a small element of angle bending), while the other will be largely a deformation vibration (with some admixture of bond stretching). How well this is borne out may again be seen from Fig. 6.3.1. Q_1 is primarily a bond-stretching coordinate, whereas Q_2 is essentially a deformation. It is an ill-defined but useful rule that a normal coordinate permitted by symmetry to involve two or more types of internal coordinates usually depends on one type to a much higher degree than on any other. Thus a combination of symmetry arguments and intuition gives a good qualitative picture of the normal vibrations of a molecule. This is often of the utmost value, for the quantitative form of the normal coordinates can only be determined by detailed and sometimes lengthy calculations.

Degenerate Vibrations

Up to this point we have considered normal vibrations which are completely defined once their amplitudes are known. However, molecules that belong to point groups of high symmetry possess a type of vibration in which the direction of the displacements relative to the framework of the nuclei is indefinite: such vibrations may take place with the same frequency in two (sometimes three) mutually perpendicular directions, and are said to be *doubly* (or *triply*) *degenerate*.

Carbon dioxide affords a good example. A vibration perpendicular to the internuclear axis can involve simple harmonic oscillation of the nuclei in any plane containing the equilibrium positions. The vibrations in different planes have, of course, the same frequency, for they can be transformed into one another by rotation about the internuclear axis. But, although there is an infinite number of planes in which the molecule can vibrate, the motion counts as only two normal modes: for the most general vibration can always be obtained from a superposition of vibrations in two mutually perpendicular directions, parallel and perpendicular, say to the plane of the paper in Fig. 6.3.3. Thus the perpendicular, or *bending*, mode of CO_2 is doubly degenerate and corresponds to two degrees of freedom. The situation resembles the familiar one of molecular translation. Translation is possible in any direction in space and yet corresponds to only three degrees of freedom, for a translation can always be expressed as a superposition of its components along three perpendicular axes.

For degenerate vibrations we need to qualify an earlier statement (section 6.1) to the effect that in a normal vibration the nuclei move in straight lines. If the components of the degenerate vibration of CO_2 are superposed with different phases, the path of the nuclei around the internuclear axis is elliptical, becoming circular when the phase difference is 90°. It is only in the separate components of the motion that the nuclear displacements are rectilinear (see Fig. 6.3.3). Evidently the execution of this vibration causes the molecule to spin about its internuclear axis; but the axial rotation does not count as a separate degree of freedom because, being inseparable from the vibration, it is already budgeted for in the two degrees of vibrational freedom.

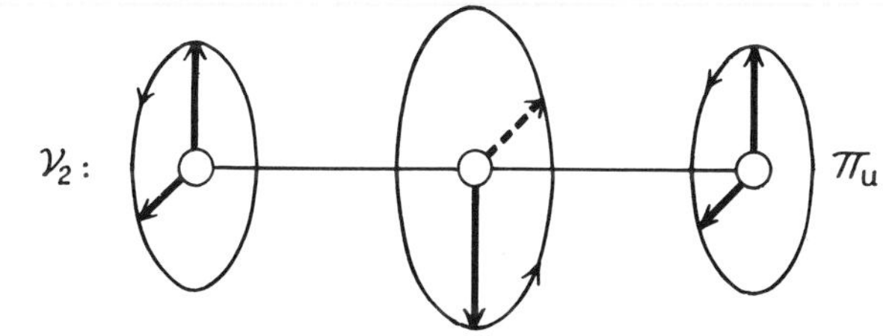

Fig. 6.3.3. Degenerate bending mode of CO_2. The circles, shown in oblique projection, represent the track of the nuclei when the components are combined with a phase difference of 90°.

Molecules with one or more three-fold (or higher) axis of symmetry possess at least one degenerate vibration, and such vibrations necessarily belong to the degenerate symmetry species of the point group. Let us determine the structure of the representation for the vibrations of CO_2. The character table of the point group, $D_{\infty h}$, is reproduced in Table 6.3.4. The group includes the operations I, C_ϕ and $C_{-\phi}$

(rotation counter-clockwise and clockwise, respectively, by an arbitrary angle ϕ about the C_∞ axis; written in the table as $2C_\phi$), the reflections σ_v in any one of the infinite number of possible planes of symmetry containing the axis C_∞ (written ∞C_2), the improper rotations S_ϕ and $S_{-\phi}(2S_\phi)$, and the inversion i. As internal coordinates we choose the two C—O bond extensions and the ∠ OCO bond angle deformation. The steps in the argument are illustrated in the lower portion of Table 6.3.4. $\chi(r)$ gives the characters of the representation of the bond extensions; the characters are 0 or 2, depending on whether an operation exchanges the extensions or leaves them unmoved. As the molecule can bend in two, mutually perpendicular, directions we must consider that there are two angular deformation coordinates. These transform in exactly the same way as the components of translation x and y; they remain unaltered by I, are reversed in sign by i, and are 'mixed' by the other operations. The characters $\chi(\theta)$ comprise the irreducible representation Π_u and therefore the degenerate mode of CO_2 involves a pure ∠ OCO deformation and belongs to the symmetry species Π_u.

As to the bond extensions, the characters $\chi(r)$ are evidently the sum of the characters of the species Σ_g^+ and Σ_u^+, and the structure $\Gamma(r)$ is therefore

$$\Gamma(r) = \Sigma_g^+ + \Sigma_u^+. \tag{15}$$

The structure of the representation formed by the internal coordinates of CO_2 then is

$$\Gamma_V = \Gamma(r) + \Gamma(\theta) = \Sigma_g^+ + \Sigma_u^+ + \Pi_u. \tag{16}$$

The three normal vibrations each belong to a different symmetry species. Two modes, Σ_g^+ and Σ_u^+ (compare Fig. 6.1.2) involve only the C—O bond extensions and thus are pure stretching modes. The third, degenerate mode, Π_u, is a pure angular deformation vibration, involving no first-order change in the internuclear distance.

Table 6.3.4. Symmetry Species for the Normal Modes of Carbon Dioxide

$D_{\infty h}$	E	$2C_\phi$	...	$\infty\sigma_v$	i	$2S_\phi$	...	∞C_2		
Σ_g^+	1	1	...	1	1	1	...	1		
Σ_u^+	1	1	...	1	−1	−1	...	−1	z	
Σ_g^-	1	1	...	−1	−1	1	...	−1		R_z
Σ_u^-	1	1	...	−1	−1	−1	...	1		
Π_g	2	2 cosϕ	...	0	2	−2 cosϕ	...	0		(R_x, R_y)
Π_u	2	2 cosϕ	...	0	−2	2 cosϕ	...	0	(x, y)	
...	...	...	...	...	...	...	...	...		
$\chi(r)$	2	2	...	2	0	0	...	0		
$\chi(\theta)$	2	2 cosϕ	...	0	−2	2 cosϕ	...	0		

The reader should satisfy himself that the representation given in eqn. (16) results when cartesian displacement coordinates are used rather than the internal coordinates r and θ. This is easily done with the aid of eqns. (6) and (7) and the representations for (x, y), z and (R_x, R_y) in Table 6.3.4.

Notation

Normal modes are indexed by the same symbol, ν_k ($k = 1, 2 \ldots 3N - 6$), used to denote the classical frequency of the vibration. The assignment of the subscript numbers $1, 2, \ldots 3N - 6$ to the different modes is made on the following basis. First, symmetry arguments are used to determine the structure of the representation of the normal coordinates: this classifies the normal coordinates into sets, members of each set having the same symmetry species. Second, the sets are ordered so that the symmetry species appear in the sequence given in the group character table. Within each set the vibrations are then numbered in order of decreasing frequency (this may require knowledge of the force constants or of the normal frequencies). Take as an example the normal vibrations of H_2O (Fig. 6.3.1), where $\Gamma_V = 2A_1 + B_2$ (eqn. 11). A_1 occurs before B_2 in the group character table, consequently the A_1 modes are considered first. The A_1 stretching mode, which has the higher frequency, is labelled ν_1 so that the bending mode becomes ν_2: the B_2 mode is then denoted ν_3.

There is one exception to this system, and it concerns triatomic linear molecules. For historical reasons the degenerate mode of these molecules is always denoted ν_2. Why this is anomalous is easily seen. The structure of the normal coordinates of CO_2, for example, is $\Gamma_V = \Sigma_g^+ + \Sigma_u^+ + \Pi_u$ (eqn. 16), this being the order of precedence in the group table. The symbols used, however, are $\nu_1(\Sigma_g^+)$, $\nu_2(\Pi_u)$, and $\nu_3(\Sigma_u^+)$.

The normal coordinate is given the same numerical subscript as the vibration. Thus Q_1 corresponds to ν_1, Q_2, to ν_2, and so on.

The Teller-Redlich Product Rule

Many molecules contain *sets of equivalent atoms*, by which we mean atoms that at most exchange positions when the various symmetry operations are carried out. Thus in acetylene, C_2H_2, the carbon pair forms one set and the hydrogen pair another. The representation for the $3N$ displacement coordinates, Γ_{3N}, is just the sum of the representations for the displacement coordinates of the individual sets.

Take as an example the water molecule. The two hydrogen atoms form one set and the oxygen atom by itself forms a second set. The characters for the displacement coordinates of each of these sets may be determined by the same method as that used for determining χ_{3N}(eqn. (6)) with the results shown in Table 6.3.5.

Table 6.3.5.

C_{2v}	I	C_2	σ_v	σ_v'
χ_H	6	0	0	2
χ_O	3	-1	1	1

The displacement coordinates for the hydrogen set therefore form the representation

$$\Gamma_H = 2A_1 + A_2 + B_1 + 2B_2, \tag{17}$$

and the displacement coordinates for the oxygen set form the representation

$$\Gamma_O = A_1 + B_1 + B_2. \tag{18}$$

Suppose that the frequency, ν_k, of the k-th normal mode in a molecule becomes $\nu_k^{(i)}$ when a set of atoms is isotopically replaced. Teller and Redlich have shown that the product of the ratios $\nu_k^{(i)}/\nu_k$ for all the vibrations of *a given symmetry species* is given by

$$\prod_k \frac{\nu_k^{(i)}}{\nu_k} = \left[\left(\frac{M^{(i)}}{M}\right)^t \left(\frac{I_x^{(i)}}{I_x}\right)^{r_x} \left(\frac{I_y^{(i)}}{I_y}\right)^{r_y} \left(\frac{I_z^{(i)}}{I_z}\right)^{r_z} \prod_j \left(\frac{m_j}{m_j^{(i)}}\right)^{n_j}\right]^{\frac{1}{2}} \tag{19}$$

in which the superscript (i) distinguishes quantities relating to the isotopic molecule. Here, M is the mass of the whole molecule: the letter j is used to denote the set of masses, m_j, in which the isotopic replacement occurs, and n_j is the number of times the particular symmetry species appears in the representation for the displacement coordinates of the set: t is the number of translational degrees of freedom in the symmetry species: r_x, r_y, r_z are the number of rotational degrees of freedom in the species from rotation about the x, y, z axes, and I_x, I_y, I_z are the corresponding principal moments of inertia. On each side of the expression degeneracies are counted once only. Eqn. (19) is a statement of the *Teller-Redlich product rule*; it is strictly valid only for harmonic vibrations. Its usefulness stems from the fact that the product of the frequency ratios is independent of any force constants and depends only on the masses and relative positions of the atoms in a molecule.

Let us return to H_2O. If the set of two hydrogen atoms is isotopically substituted (giving, say, D_2O), we have in the species A_1, $n_j = 2$, $t = 1$, $r_x = r_y = r_z = 0$: therefore

$$\frac{\nu_1^{(i)}\nu_2^{(i)}}{\nu_1\nu_2} = \left[\frac{M^{(i)}}{M}\left(\frac{m_H}{m_D}\right)^2\right]^{\frac{1}{2}}. \tag{20}$$

For the B_2 species, $n_j = 2$, $t = 1$, $r_x = 1$, $r_y = r_z = 0$. The x axis coincides with the axis of greatest inertia, c, and so

$$\frac{\nu_3^{(i)}}{\nu_3} = \left[\frac{M^{(i)}}{M}\frac{I_c^{(i)}}{I_c}\left(\frac{m_H}{m_D}\right)^2\right]^{\frac{1}{2}}. \tag{21}$$

Note that if an oxygen isotope is introduced the mass ratio enters only to the power one; for then $n_j = 1$ (see eqn. (18)). Hence for the pair of molecules H_2O and $H_2{}^{18}O$

we find

$$\frac{\nu_1{}^{(i)}\nu_2{}^{(i)}}{\nu_1\nu_2} = \left(\frac{M^{(i)}}{M}\,\frac{m_{\mathrm{O}}}{m_{^{18}\mathrm{O}}}\right)^{\frac{1}{2}}; \quad \frac{\nu_3{}^{(i)}}{\nu_3} = \left[\frac{M^{(i)}}{M}\,\frac{I_c{}^{(i)}}{I_c}\,\frac{m_{\mathrm{O}}}{m_{^{18}\mathrm{O}}}\right]^{\frac{1}{2}}. \tag{22}$$

We shall apply the Teller-Redlich product rule to acetylene in the vibrational analysis of this molecule in section 7.6.

6.4 Symmetry of the Vibrational Wavefunction

It was shown in section 6.2 that a normal vibration is associated with a quantum number v_k and a normal frequency ν_k. Eqn. (6.2.13) then gives the energy eigenvalues in terms of the ν_k: these energy values comprise the *vibrational manifold* of molecular energy levels, the *ground* (or *zero-point*) level being that for which all the quantum numbers are zero.

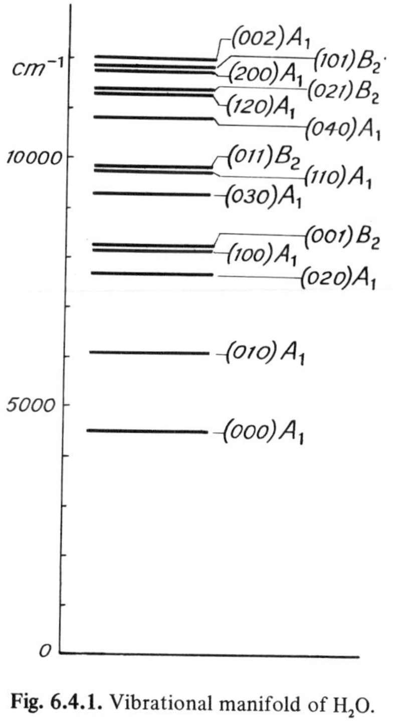

Fig. 6.4.1. Vibrational manifold of H_2O.

Fig. 6.4.1 represents the lower portion of the vibrational manifold of H_2O. The vibrational levels are indexed by the quantum numbers, (v_1, v_2, v_3), of the three normal modes. (100), (010), and (001) are termed the *fundamental* levels, that is, they are levels in which one quantum of one vibration only is excited. Levels of the type (020), (003), *etc*., in which several quanta of one vibration are excited while all other quantum numbers are zero, are known as *overtone* levels; while levels in which more than one quantum number differs from zero are called *combination* levels. Fig. 6.4.1 is drawn on the assumption that the vibrations are harmonic, and therefore does not show the convergence observed in the actual molecule where the harmonic oscillation formulae are not strictly valid.

Just as the normal coordinates may be classified by their symmetry properties, so may the vibrational wavefunctions. For H_2O, Q_1, Q_2, and Q_3 belong to the respective species A_1, A_2, and B_2 of the group C_{2v}. The vibrational wavefunction is then a product of three harmonic oscillator wavefunctions, one for each normal mode. The vibrational wavefunction of H_2O in its ground state is, therefore

$$\psi_V(000) = (1/\alpha_1\alpha_2\alpha_3\pi^3)^{\frac{1}{4}} \exp[-\tfrac{1}{2}(Q_1^2/\alpha_1 + Q_2^2/\alpha_2 + Q_3^2/\alpha_3)]. \qquad (1)$$

Now, in the group C_{2v} the application of a symmetry operation multiplies the normal coordinates by ± 1: therefore the wavefunction ψ_V (000), which contains quadratic terms only, must be left unchanged by every operation of the group. In other words, the characters of the representation of $\psi_V(000)$, are all $+1$, and hence the wavefunction belongs to the totally symmetrical representation, A_1, of the group.

The wavefunction of the fundamental level (001) is

$$\psi_V(001) = (4/\alpha_1\alpha_2\alpha_3{}^3\pi^3)^{\frac{1}{4}}\, Q_3 \exp[-\tfrac{1}{2}(Q_1^2/\alpha_1 + Q_2^2/\alpha_2 + Q_3^2/\alpha_3)]. \qquad (2)$$

As before, every operation of the group must leave the exponential factor unchanged. However,the pre-exponential factor, which contains Q_3, transforms like Q_3 itself, and so $\psi_V(001)$ must belong to the same representation, B_2, as the normal coordinate. This conclusion is a perfectly general one: *the fundamental levels of the manifold have the same symmetry species as the normal coordinate*. Likewise the wavefunctions $\psi_V(002)$, $\psi_V(003)$, . . . of the overtone levels of ν_3 transform like the polynomials $H_{v_3}(Q_3/\alpha^{\frac{1}{2}})$ of degree 2, 3, . . . , *viz*.,

$$v_3 = 2, \qquad H_{v_3}(Q_3/\alpha_3{}^{\frac{1}{2}}) = 4Q_3{}^2/\alpha_3 - 2$$

$$v_3 = 3, \qquad H_{v_3}(Q_3/\alpha_3{}^{\frac{1}{2}}) = 8Q_3{}^3/\alpha_3{}^{\frac{3}{2}} - 12Q_3/\alpha_3{}^{\frac{1}{2}}$$

. . .   .

This is sufficient to show the pattern of events. The wavefunction of an even-numbered overtone contains only even powers of the normal coordinate and is totally symmetrical. The odd overtones involve odd powers of the normal coordinate and have the same representation as the coordinate itself. When it happens that the normal coordinate is totally symmetrical—as are Q_1 and Q_2 of H_2O—the wavefunctions of the fundamental and of all overtone levels are totally symmetrical.

The species of a combination level is deduced in the same way. Take for illustration the (011) level of H_2O: we have

$$\psi_V(011) = \psi_0(Q_1) \cdot \psi_1(Q_2) \cdot \psi_1(Q_3). \tag{3}$$

The representation of $\psi_0(Q_1)$ and of $\psi_1(Q_2)$ is A_1, while that of $\psi_1(Q_3)$ is B_2. Therefore the species of $\psi_V(011)$ is the species of the direct product (see section 2.9) $A_1 \times A_1 \times B_2 = B_2$. The reader is advised to check for himself the species of other combination levels appearing in Fig. 6.4.1.

These considerations apply directly to any molecule which has non-degenerate vibrations only. Thus it is worth while to summarise the conclusions, and this is done in the set of rules given below.

(1) The wavefunction of the vibrational ground state is always totally symmetrical.

(2) The wavefunction of the k-th fundamental level has the same symmetry representation, $\Gamma(Q_k)$, as the normal coordinate Q_k.

(3) Odd overtones have the representation $\Gamma(Q_k)$. Even overtones are totally symmetrical.

(4) A combination level has the species of the direct product of the fundamental or overtone levels of which it is composed.

The rules (1), (2), and (4) are valid also for molecules with degenerate modes of vibration. Rule (3) requires reconsideration for the overtones of degenerate vibrations, but we shall not discuss this case in detail.

6.5 Thermal Population of Vibrational Levels

In wave mechanics the energy a molecule may possess by virtue of its vibrational motion is limited to one of the discrete values given by eqn. (6.2.13). We consider first any molecule that possesses non-degenerate vibrations only. In an assembly of a large number of identical molecules, the population of a given vibrational level is proportional to its Boltzmann factor, $e^{-E_V/kT}$, and thus is proportional to

$$e^{-(v_1+\frac{1}{2})h\nu_1/kT} \cdot e^{-(v_2+\frac{1}{2})h\nu_2/kT} \cdots$$
$$= e^{-\frac{1}{2}(\nu_1+\nu_2+\ldots)h/kT}(e^{-v_1\nu_1 h/kT} \cdot e^{-v_2\nu_2 h/kT} \ldots). \tag{1}$$

The number of factors on the left-hand side of (1) equals $3N - 6$, the number of vibrational degrees of freedom. On the right-hand side the zero-point energy, possessed by every molecule in the assembly, has been withdrawn as a separate factor.

Taking the ground state as the energy reference point, when the successive energies are $v_k h\nu_k$, the *vibrational partition function* is

$$f_V = \left(\sum_{v_1=0}^{\infty} e^{-v_1\nu_1 h/kT}\right) \times \left(\sum_{v_2=0}^{\infty} e^{-v_2\nu_2 h/kT}\right) \times \ldots \tag{2}$$

Each of the $3N-6$ factors in (2) is a geometrical series the sum of which is given by

$$\sum_{v_k=0}^{\infty} e^{-v_k \nu_k h/kT} = (1-e^{-h\nu_k/kT})^{-1}. \tag{3}$$

Thus the vibrational partition function of a molecule with non-degenerate vibrations is

$$f_V = (1-e^{-h\nu_1/kT})^{-1}(1-e^{-h\nu_2/kT})^{-1}\ldots \tag{4}$$

To cope with a degenerate vibration, let two of the ν_k coincide: then the appropriate factor in (4) must be taken as many times as the degeneracy implies, so that the expression (4) becomes

$$f_V = (1-e^{-h\nu_1/kT})^{-s_1}(1-e^{-h\nu_2/kT})^{-s_2}\ldots \tag{5}$$

in which s_k is the degree of degeneracy of the vibration ν_k.

For actual calculations it is necessary to recall that vibration frequencies are usually recorded in wavenumber units (σ) rather than absolute frequency (ν), with $\sigma = \nu/c$. Inserting the numerical values of the constants, h, c, and k, one obtains

$$h\nu_k/kT = hc\sigma_k/kT = 1{\cdot}439\,\sigma_k/T. \tag{6}$$

Values of f_V at different temperatures are given for H_2O and CO_2 in Table 6.5.1.

The fractional number of molecules in the ground state,

$$(v_1 = v_2 = \ldots = 0),$$

is

$$N_{0,0,\ldots}/N = 1/f_V. \tag{7}$$

Table 6.5.1. Vibrational Partition Function of some Gaseous Molecules

Molecule		$T = 300\,K$	$T = 1000\,K$
H_2O	$(1-\exp(-hc\sigma_1/kT))^{-1}$	1·000 00	1·005 2
	$(1-\exp(-hc\sigma_2/kT))^{-1}$	1·000 47	1·111 9
	$(1-\exp(-hc\sigma_3/kT))^{-1}$	1·000 00	1·004 5
	f_V	1·000 47	1·122 7
CO_2	$(1-\exp(-hc\sigma_1/kT))^{-1}$	1·001 58	1·168 8
	$(1-\exp(-hc\sigma_2/kT))^{-2}$	1·008 90	2·626 7
	$(1-\exp(-hc\sigma_3/kT))^{-1}$	1·000 01	1·035 2
	f_V	1·010 50	3·178 2

Thus we see from the table that, at ordinary temperatures, the proportion of molecules in the ground state is virtually unity: by comparison all excited levels are thinly populated. This has the important consequence, when we consider the excitation of vibrations by the absorption of radiation, that by far the greatest contribution to intensity is made up from transitions emanating from the vibrational ground state.

6.6 The Vibrational Spectrum

The absorption of radiation may cause a molecule to undergo a transition to an excited level of the vibrational manifold. The frequency absorbed, according to the *Bohr condition*, is proportional to the energy jump,

$$E_V' - E_V'' = h\nu_{v',\,v''}. \tag{1}$$

The energy required to excite vibrational transitions falls in the infrared region of the spectrum.

Absorption frequencies in a vibration spectrum are generally recorded in *wave-numbers* (σ). Wavenumber is defined as the reciprocal of the wavelength *in vacuo* expressed in cm, $\sigma(\text{cm}^{-1}) = 1/\lambda$. As the absolute frequency $\nu = c/\lambda$ we have the relation $\nu = c\sigma$, where c is the velocity of light. The wavenumber of a vibrational transition is then given by

$$E_V' - E_V'' = hc\sigma_{v',\,v''}. \tag{2}$$

Transitions in which the vibrational energy changes by one quantum only often fall in the range 200–4000 cm^{-1}, the corresponding wavelength limits being 50–2·5 microns (μ) ($1\mu = 10^{-4}$ cm). This region can be explored with prism spectrometers, or alternatively a grating may be used. Alkali halide crystals, for example sodium chloride (to 15μ) and potassium bromide (to 25μ), are used as prism materials. Grating spectrometers as well as instruments based on the interferometer principle have been designed to bridge the gap between the infrared and microwave region.

The infrared spectrum of a substance consists of a number of irregularly placed *bands* which, relative to one another, vary considerably (and seemingly unpredictably) in their intensity. The absorption of benzene, Fig. 6.6.1, suffices to illustrate this point. As the force between atoms is considerably more powerful than the force between molecules the vibration spectrum in the liquid and solid state is often very similar to that of the vapour. Although for special purposes (section 6.9) it may be necessary to record the vapour spectrum, this is not generally necessary and hence low vapour pressure is not an obstacle to the measurement of the spectrum. The irregular placing of the bands leads to the important application of infrared spectra as an aid to the identification of compounds, for the spectrum is characteristic

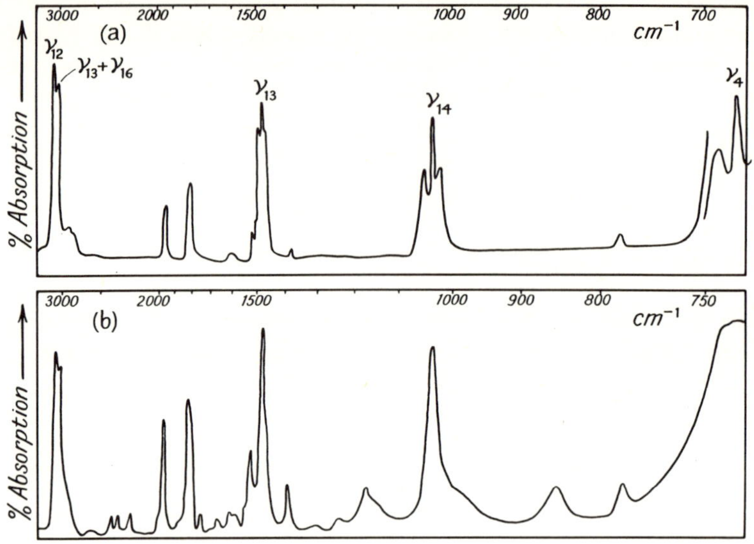

Fig. 6.6.1. Infrared bands of benzene under low resolution (NaCl prism). (*a*) is a spectrum of the vapour, (*b*) of liquid benzene.

of the molecule that gives rise to it and is not *exactly* reproduced by any other molecule, even by one of similar constitution. The analogy with human fingerprints has often been remarked. Yet it is true that a specific group of atoms in similar environments gives rise to a very similar set of normal frequencies, irrespective of the composition of more distant parts of the molecule. Thus infrared frequencies, upon comparison with the frequencies recorded for substances of known molecular structure, are often of the utmost value in establishing the constitution of complex molecules.*

6.7 Intensity of Infrared Transitions

Absorption intensity in the infrared region is subject to the same considerations which govern pure rotational transitions. These considerations have been outlined already; but as the wave mechanical treatment of vibration has been developed in some detail in Chapter 3, it is appropriate that we should seek a more quantitative understanding of intensity in vibrational spectra. This will be the goal of the present section.

* Bellamy, *The Infrared Spectra of Complex Molecules*, Methuen, London, 1958.

Fig. 6.7.1. Normal mode of an unsymmetrical diatomic molecule. In mass-adjusted displacement coordinates the normal coordinate

$$Q = (\mu_1^{\frac{1}{2}}q_1 - \mu_2^{\frac{1}{2}}q_2)(\mu_1 + \mu_2)^{-\frac{1}{2}},$$

(see section 6.10) in which the factor $(\mu_1^{\frac{1}{2}}q_1 - \mu_2^{\frac{1}{2}}q_2)$ equals the bond extension r.

Harmonic Oscillator Selection Rules (Diatomic Molecules)

In order to avoid inessential complications we shall first develop the theory for a simple example which nevertheless gives the important elements in the problem. The simplest vibrating system is a diatomic molecule, for then there is just one vibrational normal mode: its vibrational wavefunction is the simple harmonic oscillator wavefunction $\psi_v(Q)$, and the energy eigenvalues form the evenly-spaced stack,

$$E_V = (v + \tfrac{1}{2})h\nu, \tag{1}$$

ν being the classical frequency of the vibration.

In section 3.11 it was shown that the probability that a charge e in harmonic motion along a single coordinate will undergo a transition from one energy level to another with absorption of radiation is proportional to the square of the integral,

$$M_z = \int \psi_{v'}{}^{*} ez\psi_{v''}\mathrm{d}z. \tag{3.11.15}$$

Here z is the displacement of the particle, and the prime and double-prime are used in place of m and n, eqn. (3.11.15), to designate final and initial states. Instead of a single charge, a molecule consists of a group of charges representing the nuclei and electrons. In a normal vibration all charges oscillate with the same frequency; and moreover, in a *diatomic molecule* the displacements are parallel to the internuclear axis which, if we ignore rotation, we can also take as the space-fixed axis (z). Therefore to obtain the transition moment, M_z, for the molecule we replace the electric dipole moment, ez, of the particle, by the electric moment, μ_z, of the molecule, the wave-function $\psi_v(z)$ by the vibrational wavefunction $\psi_v(Q)$, and dz by the element dQ. We then have

$$M_z = \int \psi_{v'}(Q)^{*}\mu_z\psi_{v''}(Q)\mathrm{d}Q. \tag{2}$$

Symmetry requires the electric moment of the molecule to coincide at all times with the internuclear axis. But we must consider that μ_z is a function of Q, for it is probable that the actual value of the moment will change with the bond extension. Not much is known as to how μ_z varies with the extension, but it can be presumed that the moment is small both for very large and very small internuclear distances,

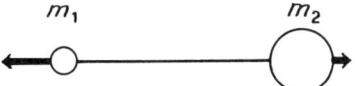

and that the behaviour in between is as represented in Fig. 6.7.2. The atom displacements in vibration are extremely small, essentially differential quantities; therefore it is *assumed* that the variation between turning-points is linear, *i.e.*, that the moment at any instant is given approximately by the expression,

$$\mu_z = \mu_z^0 + (\mathrm{d}\mu_z/\mathrm{d}Q)_0 Q. \tag{3}$$

μ_z^0, the moment when the nuclei are in their equilibrium positions, is virtually the permanent electric dipole moment of the molecule. $(\mathrm{d}\mu_z/\mathrm{d}Q)_0$ is a constant term which , for very small displacements, gives the moment change per unit displacement of the nuclei from their mean positions.

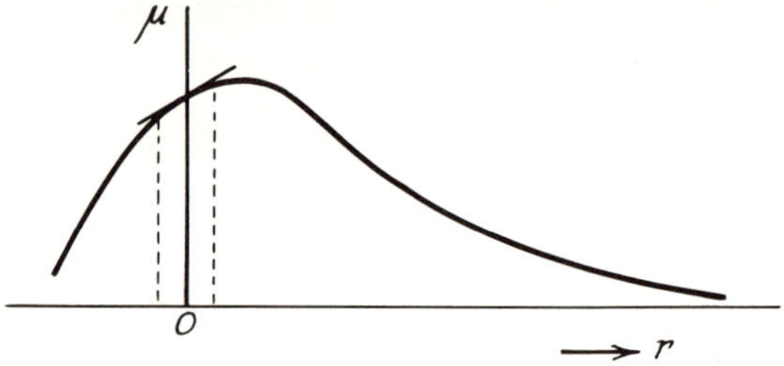

Fig. 6.7.2. Relation between the dipole moment of a diatomic molecule and the bond extension, r. See footnote to Fig. 6.7.1.

Substituting (3) into (2) we have

$$M_z = \int \psi_{v'}(Q)^* \left(\mu_z^0 + \left(\frac{\mathrm{d}\mu_z}{\mathrm{d}Q}\right)_0 Q \right) \psi_{v''}(Q)\mathrm{d}Q. \tag{4}$$

The first term, $\mu_z^0 \int \psi_{v'}(Q)^* \psi_{v''}(Q)\mathrm{d}Q$, is zero unless $v'' = v'$ owing to the orthogonality of the wavefunctions (Appendix II.1). Evidently the permanent moment does not enter into the expression for M_z [though it does control intensity in the pure rotation spectrum (section 5.4)] . The second term yields

$$M_z = \left(\frac{\mathrm{d}\mu_z}{\mathrm{d}Q}\right)_0 \int \psi_{v'}(Q)^* Q \psi_{v''}(Q)\mathrm{d}Q. \tag{5}$$

Thus the transition moment is given by the product of the moment change,$(\mathrm{d}\mu_z/\mathrm{d}Q)_0$, times the integral $\int \psi_{v'}(Q)^* Q \psi_{v''}(Q)\mathrm{d}Q$. Should either of these quantities vanish, the transition probability must vanish also. The moment change is not expected to vanish unless some symmetry property renders it identically zero. As to the integral, we recall (section 3.12) that its value is zero unless v' and v'' differ by one unit only. Consequently transitions between vibrational levels can occur if, and only if

$$\Delta v = v' - v'' = \pm 1. \tag{6}$$

The rule $\Delta v = +1$ applies to absorption of radiation. The absorption frequency, $\nu_{v', v''}$, which can now be written down using the Bohr frequency condition, $E_v' - E_v'' = h\nu_{v', v''}$, and eqn. (1), is given by

$$h\nu_{v', v''} = E_V(v+1) - E_V(v) = h\nu. \tag{7}$$

Hence the absorption frequency equals the classical frequency ν of the normal vibration. This band for $\Delta v = 1$ is known as a *fundamental* band, and its intensity depends on the value of the dipole moment change.

To sum up, let us review the premises on which the *selection rule* (6) depends. They are: (*i*) that the $\psi_v(Q)$ are harmonic oscillator wavefunctions, given by eqn. (6.7.2); and (*ii*) that the electric moment changes linearly with the normal coordinate, as assumed in eqn. (3). Each is an approximation merely, and the rule will fail to the extent that the approximations are not strictly valid.

Symmetry Selection Rules (Diatomic Molecules)

To this point we have not considered the symmetry properties of the molecule, which may impose further restrictions on the number of allowed transitions.

Before entering into a formal treatment let us discuss the influence of symmetry from a qualitative point of view. The moment, M_z, of a transition allowed by the harmonic oscillator selection rule (6) is proportional to the dipole moment change. If it should happen that $(\mathrm{d}\mu_z/\mathrm{d}Q)_0$ vanishes the intensity will be zero even for the fundamental band. In order that $(\mathrm{d}\mu_z/\mathrm{d}Q)_0$ shall be identically zero some symmetry property must be involved. This is the way in which symmetry restricts the occurrence of transitions, over and above the restriction encompassed by the rule (6).

Diatomic molecules are of two kinds: either the nuclei are the same (homonuclear) or they are different (heteronuclear). Homonuclear diatomic molecules have zero dipole moment, essentially because the element of a centre of symmetry is present: further , the vibrational displacements are equal and opposite, so that the symmetry centre is preserved in every phase of the motion. This means that the electric moment is constant (equal to zero) throughout the period of vibration, and hence that its derivative $(\mathrm{d}\mu_z/\mathrm{d}Q)_0 = 0$. Therefore *all* vibrational transitions of a homonuclear diatomic molecule are forbidden to occur by absorption of radiation, including the fundamental transitions allowed by the harmonic oscillator selection rules. For a heteronuclear molecule, on the other hand, there is no symmetry reason requiring μ_z or its derivative to be zero, and so symmetry considerations add nothing to the restriction imposed by (6).

The infrared spectrum of a heteronuclear diatomic molecule therefore consists of the fundamental band. Homonuclear molecules show no absorption in the infra-red region.

Selection Rules for Polyatomic Molecules

The selection rules are again of two sorts, the harmonic oscillator rules and the symmetry rules, and we shall consider them in this order.

In a polyatomic molecule the dipole moment μ and the transition moment M are vectors with components μ_x and M_x, μ_y and M_y, . . . on xyz cartesian axes fixed in the molecule; and we shall ignore rotation so that the axes are space-fixed as well. The dipole moment is in principle a function of the normal coordinates and thus, in first approximation,

$$\mu_x = \mu_x^0 + \sum_k \left(\frac{\partial \mu_x}{\partial Q_k}\right)_0 Q_k, \tag{8}$$

with similar equations for μ_y and μ_z. The expression for the x-component of the transition moment then becomes

$$M_x = \int \psi_{V'}{}^* \mu_x \psi_{V''} \mathrm{d}\tau \tag{9a}$$

$$= \mu_x^0 \int \psi_{V'}{}^* \psi_{V''} \mathrm{d}\tau + \sum \left(\frac{\partial \mu_x}{\partial Q_k}\right)_0 \int \psi_{V'}{}^* Q_k \psi_{V''} \mathrm{d}\tau, \tag{9b}$$

in which $\mathrm{d}\tau$ is the volume element $\mathrm{d}Q_1, \mathrm{d}Q_2 \ldots \mathrm{d}Q_k \ldots \mathrm{d}Q_{3N-6}$. If ψ_V is supposed to have strictly the form (6.2.7), that is, a product of harmonic oscillator wavefunctions, the first term in (9*b*) gives

$$\mu_x^0 \{ \int \psi_{v_1'}(Q_1)^* \psi_{v_1''}(Q_1) \mathrm{d}Q_1 \ldots \int \psi_{v_k'}(Q_k)^* \psi_{v_k''}(Q_k) \mathrm{d}Q_k \ldots \}. \tag{10}$$

The expression (10) vanishes unless $v_1' = v_1''$, . . . , $v_k' = v_k''$, . . . and so has the value zero for any change of the vibrational quantum numbers. Likewise, each term in the summation in (9*b*) splits into a product

$$\left(\frac{\partial \mu_x}{\partial Q_k}\right)_0 \int \psi_{V'}{}^* Q_k \psi_{V''} \mathrm{d}\tau = \left(\frac{\partial \mu_x}{\partial Q_k}\right)_0 \left\{ \int \psi_{v_1'}(Q_1)^* \psi_{v_1''}(Q_1) \mathrm{d}Q_1 \right.$$

$$\left. \times \int \psi_{v_2'}(Q_2)^* \psi_{v_2''}(Q_2) \mathrm{d}Q_2 \ldots \int \psi_{v_k'}(Q_k)^* Q_k \psi_{v_k''}(Q_k) \mathrm{d}Q_k \ldots \right\}. \tag{11}$$

If the expression (11) is not to vanish we must have $v_1' = v_1''$, $v_2' = v_2''$, *etc.*, except for the k-th normal vibration where $v_k' = v_k'' \pm 1$. Therefore, for polyatomic molecules, the harmonic oscillator selection rule restricts transitions occurring with absorption or emission of radiation to those in which *one vibrational quantum number changes by one unit only*. The frequencies absorbed (or emitted) are then the fundamental frequencies of the normal modes; and the intensity of the *fundamental bands* is determined by the value of the dipole moment change associated with the normal coordinates. But it must be kept in mind that the harmonic oscillator selection rule depends on the form of the expression (11): hence it is subject to the assumption that ψ_V is strictly a product of harmonic oscillator wavefunctions, $\psi_{v_k}(Q_k)$, and that the dipole moment at any instant is given by expressions of the type (8). Transitions will appear in violation of the selection rule should either of these assumptions fail.

As with diatomic molecules, when symmetry requires a dipole moment derivative to vanish, the corresponding fundamental band is forbidden in the infrared spectrum. Though it is easy to discover when the moment derivative is zero for a diatomic system the decision is apt to be less simple for polyatomic molecules. What is needed, then, is a formula which will tell us automatically that a moment derivative is identically zero. This involves group-theoretical considerations along the lines of section 3.12. The x-component of the transition moment is given by the integral (9*a*). If the value of the integral is not to vanish the integrand, $\psi_{V'}{}^*\mu_x\psi_{V''}$, must be totally symmetrical: otherwise there will be at least one symmetry operation which changes its sign, and this is only possible if the integral has the precise value zero. Should the integrals for M_x, M_y, and M_z vanish simultaneously the transition is a forbidden one.

Consider a transition emanating from the vibrational ground state of a molecule. (The thermal distribution of molecules is such that transitions from the ground state give rise to all the important bands in an infrared absorption spectrum.) $\psi_{V''}$ is then the ground state wavefunction and thus is totally symmetrical. If $\psi_{V'}{}^*\mu_x\psi_{V''}$ is to be totally symmetrical $\psi_{V'}$ must belong to the same symmetry species as μ_x; for the direct product of a symmetry species multiplied by itself is either totally symmetrical (if the species is non-degenerate), or contains the totally symmetrical species (if the species is a degenerate one). Then unless $\psi_{V'}$ and μ_x have the same symmetry species the value of M_x will be identically zero. Likewise M_y and M_z will vanish when $\psi_{V'}$ belongs to a different symmetry species from μ_y and μ_z, respectively. Therefore to get a direct answer to the question of whether a transition is forbidden we have only to discover the symmetry species of μ_x, μ_y, and μ_z.

Now the components of the dipole moment are given by the expressions

$$\mu_x = \sum_{i=1}^{N} e_i x_i \qquad \mu_y = \sum_{i=1}^{N} e_i y_i \qquad \mu_z = \sum_{i=1}^{N} e_i z_i \tag{12}$$

in which x_i, y_i, z_i are the coordinates and e_i the effective charge of the i-th atom. In a group containing non-degenerate species only, a symmetry operation at most permutes equivalent atoms and thus, inasmuch as equivalent atoms have the same effective charge, the dipole moment components must have the same symmetry properties as Σx_i, Σy_i, and Σz_i, the sums being taken over equivalent atoms only. An operation cannot do more than change a subscript i for some other subscript corresponding to another atom equivalent to i, which merely changes the order of the terms in the summation, so that Σx_i, Σy_i, and Σz_i behave under a symmetry operation in the same manner as x, y, and z. The conclusion is the same for groups that possess degenerate species.

To sum up, the components μ_x, μ_y, and μ_z have the same symmetry species as the coordinate axes x, y, z themselves. Consequently the *symmetry selection rule* allows transitions in absorption from the ground state to any excited vibrational state whose wavefunction has *the same symmetry species as at least one of the axes, x, y, z*. This rule places no restriction on the quantum change of a given vibration, or on the number of normal vibrations which simultaneously may change their quantum numbers. The harmonic oscillator selection rule, however, restricts changes to one quantum of one vibration only; therefore the only transitions from the ground state permitted by both sets of rules are to *the fundamental levels of the normal vibrations whose normal coordinates have the same symmetry species as the coordinates x, y, or z.*

Qualitative Intensities in Infrared Spectra: the Effect of Anharmonicity

On the basis of the selection rules, only fundamental frequencies should appear in an infrared spectrum. Not all fundamentals will be active, but merely those associated with a change of the electric dipole moment. These conclusions are of the utmost value in seeking to decipher the spectra of actual molecules.

Let us consider the spectrum of water vapour. The character table of the group, C_{2v}, appears in Table 6.3.1. The coordinates x, y, z, have the species B_1, B_2, and the totally-symmetrical A_1 species, respectively, whilst the vibrational states are either A_1 or B_2 (Fig. 6.4.1). The symmetry selection rule therefore allows any pair of states to combine in a transition. But, considering the thermal population of levels (Table 6.5.1) and the harmonic oscillator selection rule, the most important transitions will be from the ground state, $(000)(A_1)$, to the fundamental levels $(100)(A_1)$, $(010)(A_1)$, and $(001)(B_2)$. The first two, $(000) \rightarrow (100)$ and $(000) \rightarrow (010)$, have $M_x = M_y = 0, M_z \neq 0$, and so involve a moment change parallel to the z-axis; whereas the last, $(000) \rightarrow (001)$ has only $M_y \neq 0$, corresponding to a moment change in the molecular plane perpendicular to the z-axis. (This is confirmed by the form of the normal mode ν_3, Fig. 6.3.1. Because the bond moment change in the first O—H bond is equal and opposite to that in the second, the value of the dipole moment is unaltered by the vibration; instead it pivots to and fro about the x-axis, the rocking motion being equivalent to a moment change parallel to y.)

Table 6.7.1. Infrared Bands of H_2O Vapour

			Selection rule according to:	
σ(obs.) cm^{-1}	Intensity*	Interpretation	Harmonic Oscillator approximation	Symmetry
1595·0	*vs*	$(000)(A_1) \rightarrow (010)(A_1)$	Allowed	Allowed
3151·4	*m*	$(000)(A_1) \rightarrow (020)(A_1)$	Forbidden	Allowed
3651·7	*s*	$(000)(A_1) \rightarrow (100)(A_1)$	Allowed	Allowed
3755·8	*vs*	$(000)(A_1) \rightarrow (001)(B_2)$	Allowed	Allowed
5332·0	*m*	$(000)(A_1) \rightarrow (011)(B_2)$	Forbidden	Allowed
6874	*w*	$(000)(A_1) \rightarrow (021)(B_2)$	Forbidden	Allowed

* *s* = strong, *m* = medium, *w* = weak, etc.

Frequencies of the infrared bands of H_2O vapour are given in Table 6.7.1. The spectrum has three outstandingly strong bands at 1595, 3652, and 3756 cm^{-1}, which are assigned to the fundamental frequencies ν_2, ν_1, and ν_3 respectively. There remain several weaker bands which must represent transitions occurring in violation of the harmonic oscillator rule. 3154 cm^{-1} obviously marks the transition, (000) → (020), to the first overtone level of the vibration ν_2, while the other transitions involve quantum changes of two vibrations simultaneously. Thus the harmonic oscillator rule holds to a fair approximation (the fundamentals appear with much the greater intensity) but not rigorously. Part of the reason for its failure is that the overtone $2\nu_2$ (at 3151 cm^{-1}) occurs at less than twice the frequency of the fundamental ($2 \times 1595 = 3190$ cm^{-1}). A second possibility is that the dipole moment is not the linear function of the displacements assumed in writing eqn. (8). Both effects are believed to be of importance.

Take as a second example the fundamental transitions of the molecule CO_2. The species of the coordinates x, y, z appear in Table 6.3.4. No coordinate has the same species as $Q_1(\Sigma_g^+)$ and thus the fundamental frequency ν_1 is forbidden in the infrared spectrum. $Q_2(\Pi_u)$* and $Q_3(\Sigma_u^+)$ correspond to the species of the coordinates

Table 6.7.2. Fundamental Transitions of CO_2

Normal mode	Symmetry species: Ground level (v_1, v_2, v_3)	Symmetry species: Fundamental level (v_1, v_2, v_3)	Symmetry selection rule	Infrared absorption intensity
ν_1	(000) (Σ_g^+)	(100) (Σ_g^+)	Forbidden	Absent
ν_2	(000) (Σ_g^+)	(010) (Πu)	Allowed	Strong
ν_3	(000) (Σ_g^+)	(001) (Σ_u^+)	Allowed	Strong

(x, y) and z, respectively, so that both ν_2 and ν_3 are allowed by the symmetry selection rule. Table 6.7.2 summarises the experimental information. As expected, both ν_2 and ν_3 are observed as strong bands, but no absorption due to the forbidden fundamental ν_1 has ever been detected. Thus *the symmetry selection rule holds rigorously*. The reason is undoubtedly that the derivation of the symmetry rule does not take the vibrations as strictly harmonic, nor does it make any explicit assumption as to the form by which the dipole moment varies with displacements. The missing fundamental ν_1 can be observed in the Raman effect, where it is allowed by the appropriate selection rule (see section 7.3).

These two examples establish the pattern of normal behaviour. The harmonic oscillator selection rule holds in first approximation only, in consequence of the *mechanical anharmonicity* of the displacements and the *electrical anharmonicity* of

* We must not confuse the normal coordinate Q_2 associated with the genuine vibration ν_2 (Fig. 6.3.3) with the coordinate for the translation T_z, shown in Fig. 6.1.2 (*b*).

the dipole moment change. By contrast, the symmetry selection rules in general hold rigorously for gases at low pressure. In the liquid state, and in gases at high pressure, molecules influence one another appreciably and occasional transitions do appear in violation of the symmetry rules. Benzene, whose spectrum is reproduced in Fig. 6.6.1, affords a good illustration, Owing to the high symmetry (point group D_{6h}) only four fundamentals are allowed by the symmetry selection rules. In the vapour spectrum the active fundamentals appear at 671, 1037, 1485, and 3099 cm^{-1}, the remaining bands, including that at 3045 cm^{-1}, being combination bands. The liquid spectrum displays all the vapour bands but in addition shows weak absorption at 1174 and 854 cm^{-1}. That the 1174 and 854 cm^{-1} bands are actually infrared-forbidden fundamentals can be deduced from a study of the Raman effect where both frequencies are allowed and appear strongly. Partial breakdown of selection rules in the liquid state is one explanation of the small differences that are often observed as between liquid and vapour spectra.

There is no hard and fast rule as to the intensity of overtone and combination bands. Normally the overtone is between one and two powers of ten weaker than the fundamental band; but the fundamental frequencies are distributed with a wide range of intensity, so that it can happen that the overtone of a strong fundamental is more intense than another fundamental for which the dipole moment change, and hence the intensity, is small. In special circumstances an overtone or combination band becomes abnormally intense. This occurs when the energy of, say, a combination level chances to coincide with the fundamental level of a different vibration. A resonance phenomenon can then arise. Classically one thinks of the molecule as gradually transferring its vibration from the fundamental mode to the combination tone, and back again. In quantum mechanics the resonance pushes the levels further apart and mixes their character, that is to say, each actual level becomes in part combination tone and in part fundamental. If a transition to the fundamental level is allowed, the resonance gives rise to *a pair of transitions of similar intensity*, as each of the resonating levels partakes of the character of the fundamentals.

A good example can be seen in the infrared spectrum of benzene (Fig. 6.6.1). The active C—H stretching fundamental of benzene, species E_{1u}, happens almost to coincide with the combination level formed from one quantum of each of two other fundamentals, 1485 cm^{-1} (E_{1u}) and 1585 cm^{-1} (E_{2g}). Neglecting anharmonicity, the two last fundamentals give a combination level at 3070 cm^{-1}: its species is $E_{1u} \times E_{2g} = B_{1u} + B_{2u} + E_{1u}$ (see section 6.3, and Table I.10) and so the level consists of three sub-levels of the same energy, their species being B_{1u}, B_{2u}, and E_{1u}. It can be shown that resonance occurs between levels of the same symmetry species only; otherwise the interaction integral is zero. In this example, the E_{1u} sub-level of the combination interacts with the nearby E_{1u} fundamental. The mixing that results shows in the selection rules; both levels combine with the ground state to give infrared bands of about equal intensity, whereas in the absence of mixing one expects one strong band (the fundamental) and one weak band. With roughly equal intensities observed, we conclude that each level has about one-half the character of the fundamental and one-half that of the combination. As the actual bands fall at

3099 and 3045 cm^{-1} it would seem that, in the absence of resonance, both the fundamental and the combination must occur roughly midway between, *i.e.*, near 3070 cm^{-1}; and this is in agreement with the known frequencies of the components of the combination.

The phenomenon is known generally as *Fermi resonance*. Its occurrence is widespread, especially in complex molecules where indications of this type of resonance are to be found in nearly every example.

Quantitative Infrared Intensities

In the harmonic oscillator approximation the value of the dipole moment change associated with a normal vibration governs the intensity of the fundamental band in the infrared spectrum. The moment change is a quantity which cannot be predicted from the merely mechanical properties of the vibration, and in this section our aim must be to calculate the change from the experimentally observed intensity of infrared bands. This makes it possible to draw conclusions about the charge distribution within molecules.

It is first necessary to define a quantity which measures the intensity of an infrared transition. For monochromatic radiation of frequency ν the fundamental law governing absorption by molecules in the gas phase is

$$I = I_0 e^{-\kappa p l}. \tag{13}$$

Hence

$$\kappa = (1/pl) \ln I_0/I. \tag{14}$$

Here p is the pressure in atm. (reduced to 0°C), l is the optical path in cm, and I_0 and I are the incident and transmitted light intensities. This definition of the *absorption coefficient*, κ, makes κ a function of frequency only. Since every absorption band has a definite width a more fundamentally important quantity, known as the *integrated absorption intensity*, A, is defined by the expression,

$$A = \int_{\text{band}} \kappa(\nu)\mathrm{d}\nu = (1/pl)\int_{\text{band}} \ln(I_0/I)\mathrm{d}\nu\,, \tag{15}$$

or, when frequency is stated in wavenumber units, by the equivalent expression

$$A = c\int_{\text{band}} \kappa(\sigma)\mathrm{d}\sigma. \tag{16}$$

The actual measurement of κ requires careful experimentation for reasons connected with the performance of infrared spectrometers. However, once κ is known as a function of frequency it is a simple matter to calculate A by numerical or graphical integration.

The intensity of an infrared band is a measure of the *transition probability*. For the linear oscillator discussed in section 3.11 it was found that the probability of a transition occurring with absorption of radiation is proportional to the square of the transition moment (eqn. 3.11.14). This result is valid for molecules to the extent that the normal vibrations are perfectly harmonic. Theory gives $8\pi^3 n_0 \nu_k/3ch$ for the overall proportionality constant, where n_0 is the number of molecules per cm^3 in the gas at 0°C at 1 atm. pressure, and thus we have

$$A_k = \frac{8\pi^3 n_0 \nu_k}{3ch}[M_x(k)^2 + M_y(k)^2 + M_z(k)^2] \tag{17}$$

as the expression relating the integrated intensity, A_k, of the k-th *fundamental band* ($v_k = 0 \rightarrow 1$) to the square of the corresponding transition moment. Here $M_x(k)$ represents the integral

$$M_x(k) = \left(\frac{\partial \mu_x}{\partial Q_k}\right)_0 \int \psi_1(Q_k) Q_k \psi_0(Q_k) \mathrm{d}Q_k, \tag{18}$$

with similar definitions for the y- and z- components.

As we are assuming that the vibration is strictly harmonic we may take $\psi_1(Q_k)$ and $\psi_0(Q_k)$ to be the wavefunctions (6.2.11) and (6.2.10). Then

$$\int \psi_1(Q_k) Q_k \psi_0(Q_k) \mathrm{d}Q_k = (2/\alpha_k{}^2\pi)^{\frac{1}{2}} \int Q_k{}^2 e^{-Q_k{}^2/\alpha_k} \mathrm{d}Q_k. \tag{19}$$

The limits of integration are effectively $\pm\infty$. Using the standard integral

$$\int_{-\infty}^{\infty} x^2 e^{-x^2/a} \mathrm{d}x = 2\int_0^{\infty} x^2 e^{-x^2/a} \mathrm{d}x = (\pi a^3/4)^{\frac{1}{2}}$$

and recalling the substitution

$$\alpha_k = h/4\pi^2 \nu_k, \tag{6.2.8}$$

we find

$$(2/\alpha_k{}^2\pi)^{\frac{1}{2}} \int Q_k{}^2 e^{-Q_k{}^2/\alpha_k} \mathrm{d}Q_k = (\alpha_k/2)^{\frac{1}{2}} = (h/8\pi^2 \nu_k)^{\frac{1}{2}}. \tag{20}$$

Introducing this result into (18) we obtain

$$M_x(k) = \left(\frac{\partial \mu_x}{\partial Q_k}\right)_0 \left(\frac{h}{8\pi^2 \nu_k}\right)^{\frac{1}{2}}. \tag{21}$$

Likewise,

$$M_y(k) = \left(\frac{\partial \mu_y}{\partial Q_k}\right)_0 \left(\frac{h}{8\pi^2 \nu_k}\right)^{\frac{1}{2}}, \qquad M_z = \left(\frac{\partial \mu_z}{\partial Q_k}\right)_0 \left(\frac{h}{8\pi^2 \nu_k}\right)^{\frac{1}{2}}. \tag{22}$$

Substitution of (21) and (22) into (17) gives

$$A_k = \frac{\pi n_0}{3c}\left[\left(\frac{\partial \mu_x}{\partial Q_k}\right)_0^2 + \left(\frac{\partial \mu_y}{\partial Q_k}\right)_0^2 + \left(\frac{\partial \mu_z}{\partial Q_k}\right)_0^2\right] \tag{23}$$

as the expression connecting the experimentally observable intensity of the k-th fundamental band with the components of the dipole moment change. This is the equation we are looking for.

When a molecule possesses sufficient symmetry only one of the quantities $(\partial \mu_x/\partial Q_k)_0, \ldots, \ldots, \ldots$ is different from zero; it can then be calculated directly from the integrated intensity of the fundamental, although there will be an ambiguity in the sign. This is done for the active infrared fundamentals of CO_2 and CS_2 in Table 6.7.3. In this case of the doubly degenerate mode, ν_2, symmetry requires that $(\partial \mu_x/\partial Q_k)_0 = (\partial \mu_y/\partial Q_k)_0$, so that the total band intensity is contributed equally by the components of the vibration along the x and y directions. The theory assumes that the wavefunctions are strictly harmonic, and that the electric dipole moment change obeys eqn. (8), and so must fail to the extent that anharmonicities enter: but the errors introduced in this way are probably no greater than the uncertainty, often of ± 10 %, associated with the experimental determination of band areas.

Table 6.7.3. Dipole Moment Derivatives for CO_2 and CS_2

	k	σ_k cm^{-1}	A_k* $c\ sec^{-1}\ cm^{-1}$	Dipole moment derivative $\lvert(\partial\mu_x/\partial Q_k)_0\rvert$	$\lvert(\partial\mu_y/\partial Q_k)_0\rvert$	$\lvert(\partial\mu_z/\partial Q_k)_0\rvert$
CO_2	2	667	720×10^{10}	62	62	0
	3	2349	8080	0	0	239
CS_2	3	1523	7700	0	0	236

* CO_2: Kaplan and Eggers, *J. Chem. Phys.*, 1956, **25**, 876 (ν_2); Eggers and Crawford, *ibid.*, 1951, **19**, 1556 (ν_3).
CS_2: McKean, Callomon, and Thompson, *J. Chem. Phys.*, 1952, **20**, 520.

The dipole moment derivatives can be transformed into more familiar quantities by introducing the concept of *bond moments* (see section 11.3). Carbon dioxide, for instance, has no dipole moment. Yet we can consider that each C═O bond has a moment μ' associated with it, as symmetry will cause the two moments to cancel in the undistorted molecule. In the vibration ν_3, Fig. 6.1.2(c), each bond moment changes slightly (in opposite directions) with the distortion so that the cancellation is no longer perfect. The C—O internuclear distance being denoted by r, the bond moment change, $d\mu'/dr$, should be calculable from the derivative $(\partial \mu_z/\partial Q_3)_0$.

Let r_{12} and r_{23} represent the bond extensions in the first and second C═O bonds.

Then , recalling the definition (section 6.1) of the mass-adjusted displacement coordinates q_i, we have

$$r_{12} - r_{23} = \xi_1 - 2\xi_2 + \xi_3 = \mu_1^{\frac{1}{2}} q_1 - 2\mu_2^{\frac{1}{2}} q_2 + \mu_1^{\frac{1}{2}} q_3. \tag{24}$$

Eqn. (6.1.17) gives Q_3 as a function of the q_i. Substituting (24) into this eqn. we find

$$Q_3 = (r_{12} - r_{23})[2(\mu_1 + 2\mu_2)]^{-\frac{1}{2}}. \tag{25}$$

The value of bond moment μ' must change with the bond extensions. Let μ_{12}' and μ_{23}' be the instantaneous moments in the first and second bonds, then

$$\mu_z = \mu_{12}' - \mu_{23}' \tag{26}$$

and hence

$$\frac{\partial \mu_z}{\partial r_{12}} = \left(\frac{d\mu'}{dr}\right)_0, \tag{27}$$

where $(d\mu'/dr)_0$ represents the equilibrium value of the derivative of bond moment with respect to bond length. Now $\partial\mu_z/\partial r_{12}$ and $\partial\mu_z/\partial Q_3$ are related by the general equation between partial derivatives

$$\frac{\partial \mu_z}{\partial r_{12}} = \frac{\partial \mu_z}{\partial Q_3}\frac{\partial Q_3}{\partial r_{12}}. \tag{28}$$

Eqn. (25) gives $\partial Q_3/\partial r_{12} = [2(\mu_1 + 2\mu_2)]^{-\frac{1}{2}}$, and hence

$$\left(\frac{d\mu'}{dr}\right)_0 = \left(\frac{\partial \mu_z}{\partial Q_3}\right)_0 [2(\mu_1 + 2\mu_2)]^{-\frac{1}{2}}. \tag{29}$$

$[2(\mu_1 + 2\mu_2)]^{-\frac{1}{2}}$ has the value $1{\cdot}904 \times 10^{-12}$ g for CO_2, and $1{\cdot}926 \times 10^{-12}$ g for CS_2. Therefore, from the values of the moment derivative given in Table 6.7.3, we obtain

for CO_2,

$$(d\mu'/dr)_0 = \pm 5{\cdot}59 \times 10^{-10} \text{ esu}$$

and for CS_2,

$$(d\mu'/dr)_0 = \pm 5{\cdot}52 \times 10^{-10} \text{ esu}$$

Values of this order are expected, for two unit charges, $+e$, and $-e$, give rise to an electric moment er, and hence to a moment derivative of

$$d(er)/dr = e = 4{\cdot}8 \times 10^{-10} \text{ esu.}$$

The magnitude of the bond moment derivatives in CO_2 (and CS_2) suggests the following explanation. Mesomerism in carbon dioxide involves the structures

$$\overset{+}{\mathrm{O}}\!\equiv\!\mathrm{C}\!-\!\overset{-}{\mathrm{O}} \qquad \mathrm{O}\!=\!\mathrm{C}\!=\!\mathrm{O} \qquad \overset{-}{\mathrm{O}}\!-\!\mathrm{C}\!\equiv\!\overset{+}{\mathrm{O}}\ .$$

In the undistorted molecule the structure O═C═O is the most important perhaps overwhelmingly so. But distortion in the mode ν_3 shortens one bond and lengthens the others so that at the turning-points the contribution of the polar structures is enhanced. Consequently the polarity changes abruptly from its mean value, zero, as the bond distances are simultaneously extended and compressed.

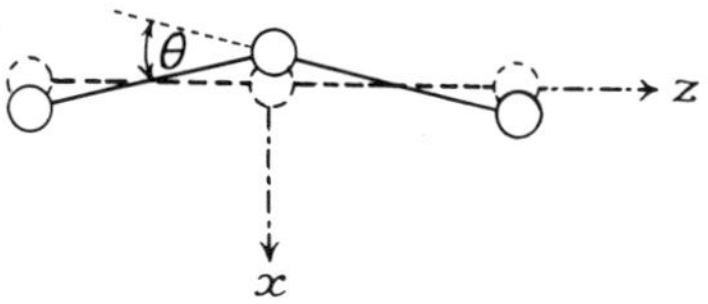

Fig. 6.7.3. Each bond moment contributes a component $\mu' \sin\frac{1}{2}\theta$ to the dipole moment in the x direction: thus $\mu_x = 2\mu' \sin\frac{1}{2}\theta \approx \mu'\theta$ for small displacements.

The bending vibration, ν_2, of CO_2 gives information as to the value of the bond moment itself. The internal coordinate θ measures the deviation of the angle between the first and second C═O bonds from 180°. The normal coordinate is $Q_2 = \theta r[2(\mu_1 + 2\mu_2]^{-\frac{1}{2}}$ (see section 6.11), and thus $\partial Q_2/\partial\theta = r[2(\mu_1 + 2\mu_2)]^{-\frac{1}{2}}$. Take the component of the vibration in the xz plane: for infinitesimal displacements we have $\mu_x = \mu'\theta$ (Fig. 6.7.3), hence $(\partial\mu_x/\partial\theta)_0 = \mu'$. But $(\partial\mu_x/\partial\theta)_0 = (\partial\mu_x/\partial Q_2)_0(\partial Q_2/\partial\theta)$, and so we find

$$\mu' = \left(\frac{\partial\mu_x}{\partial\theta}\right)_0 = r\,[2(\mu_1 + 2\mu_2)]^{-\frac{1}{2}}\left(\frac{\partial\mu_x}{\partial Q_2}\right)_0 . \tag{30}$$

Introducing the atomic masses, the equilibrium C═O bond distance $r = 1{\cdot}162$ Å, and the numerical value of the moment derivative $(\partial\mu_x/\partial Q_2)_0$ (Table 6.7.3) then gives

$$\mu' = \pm\, 1{\cdot}37 \times 10^{-18}\ \text{esu}\cdot\text{cm}\ (1{\cdot}37\ \text{D}).$$

The interesting point is that the C═O bond moment obtained in this way is obviously of the right order of magnitude. It is not, however, as large as the value 2·3 D* assigned to the C═O bond in aldehydes and ketones. The difference can be traced back to shortcomings in the bond moment concept, but it would take us too far afield to discuss this in any detail.

*Smyth, *Dielectric Behaviour and Structure*, McGraw-Hill, New York and London, 1955, chap. VIII.

Infrared intensity is also related to the molar atomic polarisability, P_A, of molecules. A molecule exposed to electromagnetic radiation is polarised by the electric field of the radiation (section 11.1); that is to say, the field induces a small dipole moment whose magnitude $\mu_i = \alpha\mathcal{E}$ is proportional to the electric field intensity $\mathcal{E}$. The constant of proportionality α is the *polarisability* of the molecule. When the frequency of the radiation coincides with an infrared-active fundamental, transitions occur and the associated dipole moment change is then large in comparison with the moment induced at frequencies where there is no absorption. The resulting transitory, increase in the polarisability can be seen in Fig. 11.5.1. At frequencies substantially above and below that of an active vibration the polarisability is found to differ by an amount that represents the contribution of the vibration. It turns out that the contribution to the polarisability is proportional to the square of the dipole moment change connected with the normal vibration. Therefore forbidden infrared fundamentals, for which the moment change is zero, do not influence the polarisability.

We shall not give the theory, merely the results. It can be shown* that the contribution, $\alpha(k)$, of the k-th normal vibration is given by the equation,

$$\alpha(k) = 2\,[M_x(k)^2 + M_y(k)^2 + M_z(k)^2\,]/3h\nu_k. \tag{31}$$

With the definition, $P_A = (4\pi N_A/3)\alpha$ (section 11.5), of the molar atomic polarisability, and that Avogadro's number $N_A = n_0 M/\mathrm{d}$, we can use eqn. (17) to eliminate the transition moment. We obtain

$$P_A(k) = (Mc/3\pi^2\,\mathrm{d})A_k/\nu_k^{\,2} \tag{32}$$

in which $P_A(k)$ is the contribution of the k-th vibration to P_A. P_A is then given by

$$P_A = \sum_k P_A(k) = (Mc/3\pi^2\,\mathrm{d}) \sum_k A_k/\nu_k^{\,2}\,. \tag{33}$$

For an ideal gas the molar volume, M/d, has the approximate value 22 400 cm^3 at 0°C and 1 atm. pressure.

Table 6.7.4 gives the steps in the calculation of the molar atomic polarisability of carbon dioxide from intensity data. The contribution of the fundamental bands appears in the upper part of the table. The effect of overtone and combination bands† is small in comparison, probably less important in the final result than the experimental errors in the intensity measurements on the fundamentals. The sum of the $P_A(k)$ gives $P_A = 0{\cdot}78\ \mathrm{cm}^3$, in good agreement with the value of $0{\cdot}81\ \mathrm{cm}^3$ obtained from dielectric constant measurements. Similar results for a number of molecules are summarised and discussed in section 11.5.

Thus the interpretation of infrared intensity is securely established in both theory and experiment. The measured intensities yield, on one hand, dipole moment

* Pitzer, *Quantum Chemistry*, Constable, London, 1953, appendix 20.
† Eggers and Crawford, *loc. cit.*

Table 6.7.4. Molar Atomic Polarisability of CO_2

Infrared band		A_k		
Assignment	cm^{-1}	$c\ sec^{-1}\ cm^{-1}$ $\times 10^{-10}$	A_k/ν_k^2 $\times 10^{14}$	$P_A(k)$ cm^3
ν_1	1344	0	0	0
ν_2	667	720	1·80	0·408
ν_3	2349	8080	1·62	0·368
$3\nu_2$	1932	0·01	0·000 003	0·000
$\nu_1 + \nu_2$	2076	0·43	0·000 1	0·000
$2\nu_2 + \nu_3$	3614	82	0·007	0·002
$\nu_1 + \nu_3$	3716	117	0·009	0·002
				0·78

derivatives which can be used to get information about bond moments; while, on the other, they give polarisability data that can be checked independently. It was not always so. For many years, polarisabilities calculated from intensity measurements were in disagreement with the values found from dielectric theory. The difficulty was resolved by Wilson and his collaborators* who showed that the early intensity measurements had been in error. As we shall see later (section 6.9) the infrared bands of a vapour are made up of a large number of individual lines which involve simultaneous changes of vibrational and rotational energy quanta. These lines are normally so sharp that spectrometers have insufficient resolving power to measure the true absorption coefficient. Correct absorption coefficients are obtained only after a sufficient pressure of a transparent gas has been added to broaden the individual lines.

6.8 Torsional Vibrations and Allied Topics

If a bond stretching mode is excited to progressively higher energy levels, the vibration becomes so violent that at some point the bond fractures and dissociation ensues. Likewise, the unrestricted excitation of a deformation vibration leads eventually to dissociation, though the two sorts of vibration do not necessarily yield the same products. (For instance, the vibration ν_3 of H_2O (Fig. 6.3.1) can be expected to give rise to H + OH, whereas ν_2 may yield H_2 + O.) There exists, however, a class of vibration which on excitation, often by a modest amount, leads to a rotation of one part of the molecule with respect to another, that is, to the interconversion of different conformations of the molecule. This class comprises the *torsional modes* of vibration, the restoring forces being provided by the resistance to torsion of chemical single and double bonds.

* Wilson and Wells, *J. Chem. Phys.*, 1946, **14**, 578. Thorndyke, Wells, and Wilson, *ibid.*, 1947, **15**, 157.

Ethane and ethylene suitably illustrate the nature of torsional vibrations. The stable configuration of ethane is a staggered on (Fig. 2.5.1); and there are three symmetrically-equivalent staggered configurations, separated by an internal rotation of 120°. To pass from one such configuration to another involves crossing an energy barrier. At low energies the torsional mode represents the oscillation of one CH_3 group relative to the other in a potential trough whose energy minima correspond to the staggered configurations; but, on increasing the amplitude, the vibration can become sufficiently vigorous for the atoms to traverse the barrier separating one configuration from another. A normal vibration of this kind, that involves the *torsion* (or *twisting*) of a chemically single bond, is descriptively called an 'hindered internal rotation'. In *ethylene* the equivalent configurations are separated by an internal rotation through 180°; also, because twisting a double bond involves uncoupling the π-electrons, the resistance to torsion is much higher than in ethane-like compounds. In ethylene derivatives the torsional oscillation provides a possible mechanism for the thermal *cis* ⇌ *trans* interconversion of isomers.

In order to make any progress in understanding torsional vibrations it is necessary to make some assumption as to the shape of the energy barrier restricting internal rotation. Ethane has its maxima and minima 60° apart (Fig. 6.8.1). The simplest function which has this property is a cosine function,

$$V = \tfrac{1}{2}V_0(1 + \cos 3\phi), \tag{1}$$

in which ϕ is the azimuthal angle between symmetry planes of the methyl groups, measured from a configuration of maximum energy. The energy eigenvalues of a normal vibration subject to the potential (1) have been worked out and are shown qualitatively on the right-hand side of the figure. Naturally they are somewhat different from the eigenvalues of an harmonic vibration. The potential (1) can be taken to apply not merely to ethane, but to other molecules (*e.g.*, CH_3OH, CH_3CH_2Cl) in which three equivalent configurations are attained during a complete internal rotation. When the configurations are not equivalent, as with 1 : 2-dichloroethane (section 11.5), further terms must be added to the potential.

Much work has been directed to the determination of the barrier height, V_0, opposing internal rotation. Entropy and heat capacity measurements, the most common source of information, are discussed in texts on thermodynamics;* and dipole moments are sometimes useful (see section 11.5). The purely spectroscopic methods may involve difficult calculations but give a more accurate value for V_0 when they can be worked out. Essentially there are two such methods. First, the fundamental frequency of the torsional vibration, if it can be observed in the infra-red (or Raman) spectrum, may be used to calculate V_0, for the energy levels of the hindered rotator depend upon the barrier height. But torsional fundamentals are always of low frequency and therefore difficult to measure; moreover, in some key cases they are forbidden (this is true of the torsional mode in ethane, for example,

* Fowler and Guggenheim, *Statistical Thermodynamics*, Cambridge, 1939, chap. III.

which has the species A_{1u} in the group D_{3d}). Sometimes the fundamental frequency can be obtained indirectly from microwave spectra. Thus the microwave spectrum may contain a pair of lines, one marking a pure rotational transition of molecules in the zeroth level, and the other of molecules in the fundamental level of the torsional vibration. The relative intensity of the lines is then proportional to the Boltzmann factor of the upper level, so that the energy separation may be calculated. The second spectroscopic method also involves the use of microwaves. Resonance between equivalent configurations causes each energy level of the torsional vibration to split into a number of sub-levels, not all with the same energy. The splitting is shown on an enlarged scale in Fig. 6.8.1. It may be possible to observe transitions between different sub-levels in the microwave region. The barrier height controls the extent of the splitting and can then be calculated from the transition frequencies. The last is the most sensitive method available at the present time.

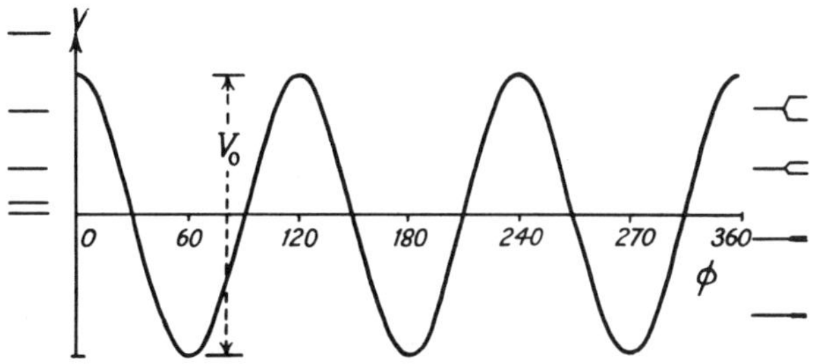

Fig. 6.8.1. Barrier hindering the rotation of a methyl group. The energy levels for free internal rotation and torsional oscillation are given to the left and right of the diagram, respectively.

Some potential barriers hindering free internal rotation of methyl groups are listed in Table 6.8.1. The scope of the phenomenon is, of course, broader than this: similar barriers confer a certain resistance to torsion upon every single bond, although in some instances the barriers are small. The origin of the potential is a repulsive force between electron pairs in the bonds emanating from the atoms at either end of the link about which internal rotation may take place. Thus in *ethane* it is considered that the C—H bonding electron-pairs mutually repel one another; this leads directly to the conclusion that the configuration of minimum energy is the staggered one in which the C—H bonds attached to *different* carbon atoms are as far apart as possible. Rotation into an eclipsed configuration brings three pairs of bonds into closer proximity and involves the expenditure of energy. The character of the repulsive force is mysterious at present; but it is evidently of short range for in dimethylacetylene, where the interacting methyl groups are held apart by an acetylenic group, the potential barrier is at most a fraction of a kilocalorie and may even be zero.

It should be realised that the quantitative theory of hindered rotation depends on the assumption that the potential barrier has the shape of a simple cosine function. If this is an approximation only, then the barrier heights will be in error by as much as the approximation fails. It is important, therefore, that some experiment should be devised which will throw light on the actual shape. This was first

Table 6.8.1 Potential Barriers Opposing Free Rotation of Methyl Groups.

Molecule	V_0 k cal mole^{-1}	Method*	Authors
CH_3CH_3	2·88	T	Pitzer (1951)
CH_3CCl_3	2·70	T	Rubin, Levedahl and Yost (1944)
	3·00	M	Luft (1955)
C_2H_5Cl	3·69	M	Schwendeman and Jacobs (1962)
C_2H_5Br	3·57	M	Lide (1959)
CH_3NH_2	~ 1·9	T	Aston and Gittler (1955)
	1·96	M	Lide (1957)
$CH_3CH{=}CH_2$	1·98	M	Lide and Mann (1957)
CH_3SH	1·46	T	Russel, Osborne and Yost (1942)
	1·26	M	Kojima and Nishikawa (1957)
CH_3OH	1·07	M	Ivash and Dennison (1953)
$(CH_3)_2CO$	0·78	M	Swalen and Costain (1959)
CH_3CHO	1·16	M	Kilb, Lin and Wilson (1957)
CH_3COOH	0·48	M	Tabor (1957)
$CH_3C{\equiv}CCH_3$	~ 0	T	Yost, Osborne and Garner (1941)

* T : Thermodynamic, M : Microwave

done in 1956.* If the internal rotation has three-fold symmetry the potential can be represented accurately, apart from a constant term, by the Fourier expansion,

$$V = a\cos 3\phi + b\cos 6\phi + c\cos 9\phi + d\cos 12\phi + \dots, \tag{2}$$

of which the simple expression (1) takes care merely of the first term. Consider the molecule nitromethane, CH_3NO_2. For the first N—O bond, taking the initial three terms in the expansion (2), the potential can be written

$$V(1) = a\cos 3\phi + b\cos 6\phi + c\cos 9\phi. \tag{3}$$

The azimuthal angle between the first and second N—O bonds is 180°; therefore, for the second bond,

$$V(2) = a\cos 3(\phi + 180^\circ) + b\cos 6(\phi + 180^\circ) + c\cos 9(\phi + 180^\circ). \tag{4}$$

As $\cos n(\phi + 180^\circ) = (-1)^n \cos n\phi$ when n is an integer, the potential hindering internal rotation in nitromethane is $V(1) + V(2) = 2b\cos 6\phi$. The limits of $2b\cos 6\phi$ are $\pm 2b$, so that the barrier height $V_0 = 4b$. The experimentally observed frequencies in the microwave region give $V_0 = 0{\cdot}0060$ kcal mole^{-1}, and therefore $b = 0{\cdot}0015$ kcal mole^{-1}. As the barrier, $2a$, opposing the rotation of a *single* N—O group is probably 2 ± 1 kcal mole^{-1}, the first term in the Fourier series (3) must be of the order one thousand times more important than the second. Evidently

* Tannerbaum, Myers, and Gwinn, *J. Chem. Phys.*, 1956, **25**, 42.

the simple assumption represented by (1) is an excellent approximation to the potential opposing rotation of the individual N—O bonds in nitromethane.

A related phenomenon is that of *inversion*. The outstanding example is ammonia. If the coordinates of all atoms in NH_3 are reflected in the *centre of mass* the resulting configuration is an enantiomer of the original one; it cannot be realised by simple rotations of the original configuration, inasmuch as the rotations fail to bring the hydrogen atoms into the same relative order that they occupy after inversion. The mirror-image configurations must of course have the same energy, and so a plot of potential energy against the pyramid height (the vertical distance of the nitrogen atom from the plane of the hydrogen atoms) has a double minimum. Its general aspect is shown in Fig. 6.8.2. Resonance between the configurations splits the vibrational energy levels into the pairs of sub-levels indicated in the figure. Transitions between the lowest pair of sub-levels are readily observed in the microwave spectrum; and the separation of sub-levels can then be used to calculate the height of the barrier resisting inversion. This microwave transition of ammonia is of special interest historically, for it was not only the first known instance of microwave absorption* but also the first for which the hyperfine structure due to nuclear quadrupole coupling was resolved.†

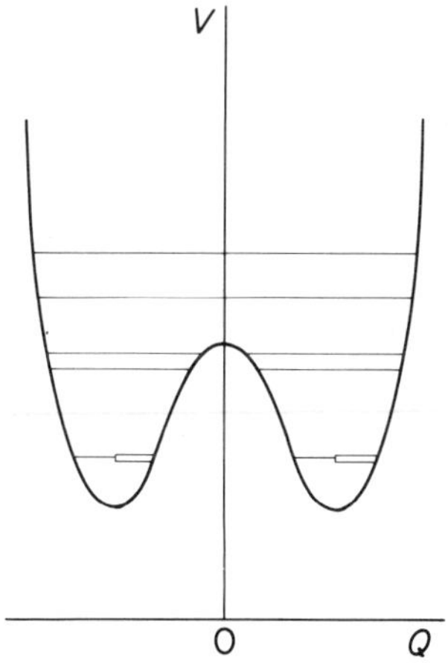

Fig. 6.8.2. Potential energy and energy levels for the ∠ HNH deformation vibration of ammonia. The height of the central barrier is 5·93 kcal $mole^{-1}$.

In methylamine, CH_3NH_2, we have an instance of hindered internal rotation (about the C—N bond) and inversion (of the nitrogen pyramid) in the same molecule. The resulting microwave spectrum is complicated but has been worked out.‡ The barrier to inversion is somewhat lower than that for NH_3.

* Cleeton and Williams, *Phys. Rev.*, 1934, **45**, 234.
† Good, *Phys. Rev.*, 1946, **70**, 213.
‡ Lide and Mann, *J. Chem. Phys.*, 1957, **27**, 343, 353, Shimoda and Itoh, *ibid.*, 1954, **22**, 1456.

6.9 Details of Vibrational Transitions

To this point in the chapter we have described vibrational spectra in a way which would suggest that the vibrational energy alone changes during a transition. This is merely a convenience. Actually, lesser quanta of energy add to or subtract from the pure vibrational jump, the effect being to broaden the absorption region into a band of appreciable width. In the *liquid state* the thermal energy of a molecule may be different before and after the absorption of radiation; that is to say, the transition may add to or draw upon the energy of the 'Brownian motion' of a molecule. The probability that energy will be transformed in this manner falls off sharply with the quantity of energy involved in the transfer. Consequently, instead of the exceedingly sharp line expected from a pure vibrational jump, the absorption bands of liquids and solutions appear as bell-shaped curves whose width at half the peak height is often in the range 10–30 cm^{-1}. The spectrum of liquid benzene, Fig. 6.6.1 (*b*), is typical in this respect. Band widths in the spectra of liquids tend to be characteristic of the vibration involved, and so are useful for identification, but no quantitative information can be derived from them. This is in contrast to the fine structure of the bands of gaseous molecules, which are next in usefulness to pure rotation spectra as a source of structural data.

In the *vapour state*, where rotation is quantised, a pure vibrational energy level is merely the lowest member ($J = 0$) of a stack of rotational energy levels each distinguished by a different set of rotational quantum numbers. For *linear molecules*, when we have only the quantum number J to reckon with, the arrangement of the rotational levels attached to two vibrational energy levels is as shown in Fig. 6.9.1 (*a*). Let the lower of the two levels in the figure represent the zero-point vibrational level and let the upper be some fundamental level. Then the distribution of molecules over the lower state rotational levels will be given by the expression (5.3.9) and hence will have qualitatively the form shown in Fig. 5.3.1. A majority of molecules in the lower state therefore possess rotational energy (that is, they are in levels for which $J \neq 0$). Should the quantum number J change in course of a transition, we have a situation in which both vibrational and rotational energies alter simultaneously. The allowed changes in J accompanying a vibrational transition are determined by very similar considerations to those used in obtaining the pure rotational selection rules (section 5.4). The essential difference is that in vibration-rotation spectra we must focus our attention on the change of dipole moment accompanying the vibration, rather than the permanent dipole moment itself. The vibrations in a linear molecule may cause a change of dipole moment either parallel or perpendicular to the molecular axis. The former situation is exemplified by the ν_3 vibration of CO_2 which we may term a *parallel* vibration, and the rotational selection rule for its fundamental ($\Sigma_g^+ \rightarrow \Sigma_u^+$) is

$$\Delta J = \pm 1. \qquad (1)$$

The ν_2 vibration of CO_2 produces a dipole moment change perpendicular to the molecular axis and we term this a *perpendicular* vibration. The selection rule for the fundamental in this case ($\Sigma_g^+ \rightarrow \Pi_u$) is

$$\Delta J = 0, \pm 1. \tag{2}$$

In what follows we shall consider linear molecules only, for they illustrate all the essential qualities of vibration-rotation spectra. The rotational energy of a linear molecule (assumed rigid) is given by

$$E_R = J(J+1)h^2/8\pi^2 I_b \tag{5.2.16}$$

and the vibrational energy by the expression,

$$E_V = \sum_k (v_k + \tfrac{1}{2})h\nu_k = \sum_k (v_k + \tfrac{1}{2})hc\sigma_k. \tag{6.2.13}$$

For the total energy we then have

$$E_{V,R} = \sum_k (v_k + \tfrac{1}{2})hc\sigma_k + BchJ(J+1), \tag{3}$$

in which

$$B = h/8\pi^2 cI_b. \tag{4}$$

The expression (4) differs from the earlier definition of B (eqn. 5.2.14) in that the velocity of light, c, is present in the denominator. The effect is that B is now given in wavenumbers (cm^{-1}) instead of absolute frequency units (MHz).

For the moment let us consider that the value of B does not change between the upper and lower states. From (3), the wavenumbers of the lines forming an absorption band are then

$$\sigma = (E_{V,R'} - E_{V,R''}) = \sigma_k + B\{J'(J'+1) - J''(J''+1)\}. \tag{5}$$

The transitions are in three sets corresponding to the permitted values of $\Delta J = -1$, 0, or $+1$. The set for which $\Delta J = -1$ forms the *P*-branch of the band. The wavenumbers, $P(J)$, of lines in this set are found by substituting $J' = J'' - 1$ into (5), when we have

$$P(J) = \sigma_k - 2BJ \quad : \quad J = 1, 2, 3, \ldots \tag{6}$$

The double-prime has been dropped because the quantum number J in (6) refers to one state only—the lower state—so that a distinguishing mark is unnecessary. The first transition of the *P*-branch has $J = 1$ and this occurs at the wavenumber $P(1) = \sigma_k - 2B$. The next number has the wavenumber $P(2) = \sigma_k - 4B$, and the branch as a whole runs to low frequency from the *origin*, σ_k, the line spacing being uniform and equal to $2B$. Likewise, the set with $\Delta J = +1 (J' = J'' + 1)$, known as the R-branch, is easily found to correspond to the wavenumbers $R(J)$ given by

$$R(J) = \sigma_k + 2B + 2BJ \quad ; \quad J = 0, 1, 2, \ldots \tag{7}$$

This branch runs from the origin towards higher frequencies with a uniform spacing $2B$, the first member being at $R(0) = 2B$. Note that the band origin σ_k is not present in either the P- or the R-branch. The idealised structure of a band comprising a P- and R-branch only is shown in Fig. 6.9.1 (*b*).

The third set of lines forms the Q-branch. In this set $J' = J''$ and hence

$$Q(J) = \sigma_k \quad ; \quad J = 0, 1, 2, \ldots \tag{8}$$

Therefore all lines in the Q-branch coincide, their frequency being that of the pure vibrational jump. As mentioned above, $\Delta J = 0$ is forbidden for $\Sigma - \Sigma$ transitions of linear molecules; consequently transitions of this type will be marked by bands composed of P- and R-branches only.

As in pure rotation spectra the relative intensity of lines in a vibration-rotation band is governed by the relative thermal population of the ground levels. Therefore the intensity distribution in the P- and R- branches has the qualitative appearance of Fig. 5.3.1, that is, intensity is low near the origin and passes through a maximum in both directions from the centre. The maxima mark P and R transitions from the ground-state level which has the greatest population of molecules. Let the quantum number of this level be J_{max}. Then the separation of maxima in the R- and P-branches is $R(J_{max}) - P(J_{max})$ which, according to eqns. (6) and (7), is given by

$$R(J_{max}) - P(J_{max}) = 4B(J_{max} + \tfrac{1}{2}). \tag{9}$$

But we have the relation

$$J_{max} + \tfrac{1}{2} = (kT/2Bch)^{\frac{1}{2}} \tag{5.3.10}$$

(in which B is now expressed in cm^{-1}), and hence

$$R(J_{max}) - P(J_{max}) = (8BkT/ch)^{\frac{1}{2}} = 2{\cdot}36(BT)^{\frac{1}{2}}. \tag{10}$$

The application of this equation we shall discuss presently.

The appearance of the infrared fundamentals of carbon dioxide under medium resolution is shown in Fig. 6.9.2 (*a*) and (*b*). Even though individual lines are not resolved the different branches are easily seen. The Q-branch is very prominent in the fundamental $\nu_2(\Sigma_g^+ \rightarrow \Pi_u)$. $\nu_3(\Sigma_g^+ \rightarrow \Sigma_u^+)$ lacks a Q-branch, as expected, but the P and R structure is clearly marked. Evidently the structure of a vibration-rotation band, even under conditions of incomplete resolution, may suffice to establish the assignment to a particular mode of vibration. Further, the envelope of an incompletely resolved band can be used to calculate a very rough value of the rotational constant, from the separation of the R and P maxima. Thus in the ν_3 fundamental band of CO_2, Fig. 6.9.2 (*a*), the peaks are separated by 22 cm^{-1} at 303 K. Substitution in eqn. (10) then gives $B \approx 0{\cdot}3$ cm^{-1} for carbon dioxide, the precise value being

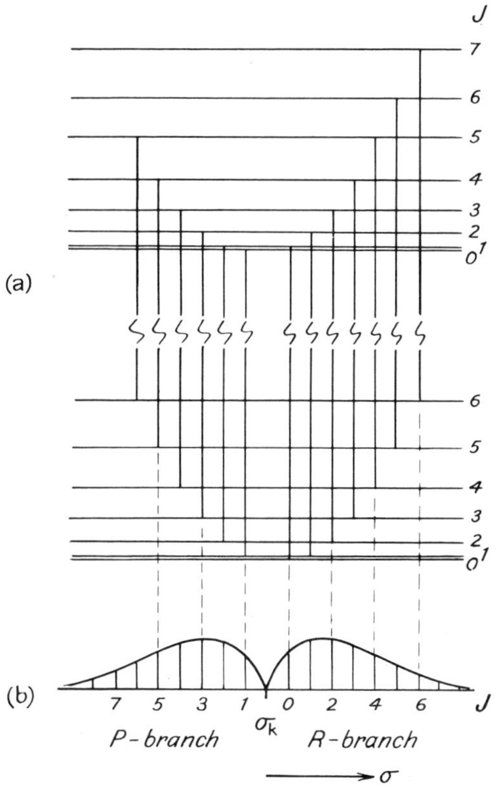

Fig. 6.9.1. Energy levels and transitions in a simple vibration-rotation band showing *P*- and *R*-branches only (schematic).

0·389 cm^{-1}. It is of interest that the first attempts to determine inertial constants from spectra were made by this method (Burmeister, 1913).

In a less approximate treatment we must drop the assumption that the rotational constant has identically the same value in the two vibrational states. A small difference is to be expected because the nuclear displacements, and therefore the mean inertial constants, change with the vibrational quantum number. As in the approximate treatment, the wavenumber of the individual lines is found by combining the selection rule (1) with the expression (3) for the energy levels. One obtains

$$P(J) = \sigma_k - (B' + B'')J + (B' - B'')J^2; \quad J = 1, 2, 3, \ldots \tag{11}$$

$$Q(J) = \sigma_k + (B' - B'')J + (B' - B'')J^2; \quad J = 0, 1, 2, \ldots \tag{12}$$

and

$$R(J) = \sigma_k + 2B' + (3B' - B'')J + (B' - B'')J^2; \quad J = 0, 1, 2, \ldots \tag{13}$$

in which B' and B'' are the respective rotational constants of the upper and lower

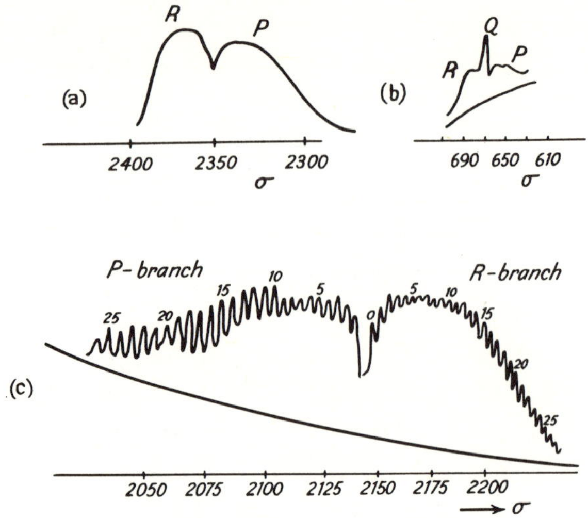

Fig. 6.9.2. (*a*) and (*b*) Fundamental bands ν_3 and ν_2 of CO_2 under low resolution. (*c*) Fundamental band of CO under good resolution. The wavenumber scale increases to the left in (*a*) and (*b*), but to the right in (*c*).

states, and $J(\equiv J'')$ is the quantum number of the lower state. These equations do not predict uniform spacing in the *P*- and *R*-branches, or that the lines in the *Q*-branch (if a *Q*-branch is present) will coincide. The uneven spacing explains why the envelope of the *P*- and *R*-branches in the partly-resolved spectra of carbon dioxide is not symmetrical. The non-uniform spacing can also be seen in Fig. 6.9.2 (*c*) which represents the fundamental band of carbon monoxide recorded under conditions which resolve individual lines.

The sense of the non-uniformity is determined by the sign of $(B'-B'')$, the coefficient of J^2 in eqns. (11)–(13). In the stretching vibration of a diatomic molecule such as carbon monoxide the rotational constant is invariably smaller in the excited state than in the ground state. The coefficient $(B'-B'')$ is then negative, so that the term $(B'-B'')J^2$ causes the line spacing progressively to increase in the *P*-branch, and progressively to decrease in the *R*-branch. In other words, the lines tend to converge as one scans through the band in the direction of increasing frequency. The tendency to converge towards high frequency is typical of bond stretching modes in general: it can be seen, for instance, in the envelope of the ν_3 fundamental of carbon dioxide in Fig. 6.9.2 (*a*). In bending modes of vibration $B'-B''$ is often positive and then the band convergence appears in the opposite direction.

To determine accurate rotational constants from a vibration-rotation band one looks for a pair of lines whose separation is governed by the rotational constant of the ground state, or that of the excited state, but not by both. Various combinations are possible but two are especially useful, namely,

$$R(J-1)-P(J+1) = 2B''(2J+1) \tag{14}$$

and

$$R(J)-P(J) = 2B'(2J+1). \tag{15}$$

These expressions are obtained without difficulty from eqns. (11) and (13). Then, to determine B'', all that is necessary is to plot the wavenumber given by the left-hand side of (14) against $2J + 1$, when one obtains a straight line through the origin whose slope is $2B''$. A similar treatment based on the expression (15) yields $2B'$. In Table 6.9.1 we give the observed wavenumbers and wavenumber differences of a representative number of lines in the fundamental band of carbon monoxide. The last two columns give the quotient of , respectively, $R(J-1)-P(J+1)$ and $R(J)-P(J)$ divided by $2J + 1$. The rotational constants obtained in this way show a definite trend owing to the fact that we have neglected to take centrifugal stretching into account. In an exact treatment centrifugal stretching is allowed for by introducing an extra parameter into the formula for the wavenumbers of the rotational lines. The method of least squares then gives the final values,

$$\text{CO}: \begin{array}{l} B_0(\equiv B'') = 1{\cdot}922\,6_5 \text{ cm}^{-1} \\ B_1(\equiv B') = 1{\cdot}905\,1_7 \text{ cm}^{-1}. \end{array} \quad (16)$$

Table 6.9.1 Wavenumbers of Lines in the Fundamental Band of CO*

J	$R(J)$	$P(J)$	$R(J-1)$ $-P(J+1)$	$R(J)$ $-P(J)$	$2B''$	$2B'$
0	2147·19					
1	50·97	2139·53	11·54	11·44	3·847	3·813
2	54·71	35·65	19·23	19·06	3·846	3·812
3	58·41	31·74	26·92	26·67	3·846	3·810
4	62·08	27·79	34·60	34·29	3·844	3·810
5	65·71	23·81	42·29	41·90	3·844	3·809
		19·79				
...	...	...	...	...	...	...
18	2209·61					
19	12·73	2064·51	149·57	148·22	3·835	3·801
20		60·03				

* after Rao, *J. Chem. Phys.*, 1950, **18**, 213.

Here the subscript numeral attached to B indicates explicitly that the value belongs to either to the zeroth or to the first quantum state of the vibrational manifold. Given the rotational constant in the two lowest vibrational states it should be possible, by extrapolation through an interval equivalent to one-half the vibrational quantum, to discover the value of the constant B_e appropriate to the virtual state in which the atoms are motionless (corresponding to the minimum of potential energy). Experience shows that the extrapolation is linear and hence the extrapolation formula is

$$B_v = B_e - \alpha(v + \tfrac{1}{2}). \quad (17).$$

Introducing the numerical values given in (16) one obtains

$$\text{CO}: \quad \begin{aligned} \alpha &= 0{\cdot}017\,48 \\ B_e &= 1{\cdot}931\,3_9 \text{ cm}^{-1}. \end{aligned}$$

The internuclear distance r_e calculated from B_e is given in Table 6.9.2.

Table 6.9.2 Structure of Some Simple Molecules as determined from Infrared and Microwave Spectra

Molecule	Vibration-rotation bands* B_e cm^{-1}	Vibration-rotation bands* B_e MHz	Pure-rotation bands† B_e MHz	Bond distances (Å) and angles					
CO	1·931 39	57902·0	57897·5	CO	1·1282				
OCS	0·203 56	6102·6	6099·35	CO	1·161	CS	1·561		
	B_0		B_0						
N_2O	0·419 2	1256_7	12561·7	NN	1·126	NO	1·191		
HCN	1·478·2	4431_7	44316·0	CH	1·068	CN	1·156		
DCN	1·207 5	3620_0	36207·5						
CH_3F	$0{\cdot}851_5$	255_{27}	25536·1	CH	1·1	CF	1·39	HCH	110°
$CH_3{}^{35}Cl$	$0{\cdot}443_2$	132_{66}	13292·8	CH	1·113	CCl	1·781	HCH	110·5°
$CH_3{}^{37}Cl$	$0{\cdot}436_2$	130_{56}	13088·1						
$CH_3C{\equiv}CH$	0·285 0	8544	8545·8	CH (in CH_3)	1·10	C—C	1·46		
				C≡C	1·21	C—H	1·06	HCH	108°

* With the exception of CO, from the work of Thompson and collaborators (1953–56).
† Townes and Schawlow, *Microwave Spectroscopy*, McGraw-Hill, New York and London, 1955. Apart from CO and HCN, inertial constants for more isotopic species than are listed in the table are required to calculate the structural parameters. When the isotopic mass is not given the common isotope (^{12}C, ^{14}N, ^{16}O, . . .) is implied.

In connection with the spectra of carbon dioxide discussed here a further point arises. Under conditions of high resolution the approximate spacing between successive lines in P- and R-branches is $4B$ rather than $2B$. This occurs since the presence of two equivalent oxygen atoms with spin $I = 0$ requires only levels of even J be populated in the ground state (section 5.6). Nuclear spin statistical weight effects in high resolution spectra can be very informative. The resolved spectra of molecules such as acetylene, water, formaldehyde and diazomethane, for example, show beautiful 3:1 intensity alternations due to the presence of equivalent proton pairs in accordance with their accepted symmetries. On the other hand, the absence of such effects in the infrared spectrum of nitrous oxide at once showed that this linear molecule was not a symmetrical structure with equivalent nitrogen atoms.

Finally, it is worthwhile to consider how vibration-rotation spectra stand in relation to pure rotation spectra as a source of structural information. When both methods can be applied to the same molecule the results are in exceedingly good agreement. A few examples are given in Table 6.9.2. In general the advantage is with pure rotation spectra inasmuch as the resolving power of a microwave spectrometer, and therefore the accuracy of the observations, is superior to that of an infrared instrument. This is often the most important single consideration. But polyatomic molecules that lack a permanent dipole moment do not absorb in the microwave region, yet they give rise to vibration-rotation bands which can be analysed. Moreover, vibration-rotation bands yield values of the rotational constants in two vibrational states, so that the correction for the zero-point amplitude in the vibration concerned is automatically available.

There remains one more spectral method for the determination of structural data. This is the pure rotational Raman spectrum, discussed in section 7.5.

6.10 The Calculation of Force Constants

The method described in section 6.1, based on Newton's equations of motion, grows unwieldy in all except the simplest ealculations. Much more powerful techniques are now available, largely owing to the work of E. Bright Wilson and his collaborators; and for small, symmetrical, molecules the calculation of force constants by these methods reduces to a simple mathematical exercise. Here our purpose is to give an elementary introduction to the method, without seeking mathematical proofs. The reader who desires to learn more of the subject should consult the monograph by Wilson, Decius, and Cross.* We shall use the notation of these authors, in order to assist comparison.

The object is two-fold. First, coordinates are chosen for the nuclear displacements which eliminate the zero roots (corresponding to translation and rotation) from the secular determinant, thus reducing its rank and order from $3N$ to $3N-6$.This is accomplished by introducing *internal coordinates* (section 6.3) instead of the cartesian displacement coordinates used in section 6.1. Secondly, the secular determinant is set up in such a manner that it has the form of a product of factors, one corresponding to each symmetry species in the structure of the normal coordinates. To achieve this the internal coordinates are combined to give *symmetry coordinates* which transform according to one or other of the symmetry species of the molecular point group.

We shall discuss the application of the method to a few simple polyatomic molecules, for it is then easier to follow than in the abstract form.

Definitions

Internal coordinates are normally the bond extensions (r) and bond angle deformations (θ) discussed in section 6.3. Let the general symbol for an internal coordinate be S. The total number of internal coordinates needed is equal to the number of normal modes; and we index the internal coordinates by the subscript letter t, so

* Wilson, Decius and Cross, *Molecular Vibrations*, McGraw-Hill, New York and London, 1955.

that S_t indicates the t-th member of the set

$$S_1, S_2, S_3, \ldots S_t, \ldots S_{3N-6}, \tag{1}$$

which describes the deformation of the molecule by vibration.

Cartesian displacement coordinates, $\xi_1, \xi_2, \ldots \xi_i, \ldots$ were introduced in section 6.1 to describe the displacement of the atoms from their mean positions. In general $3N$ such coordinates are necessary (cf. Fig. 6.11.1), and thus i runs from 1 to $3N$. We use the same symbol, ξ_i or S_t, for the coordinate as well as the actual displacement of the atom measured along the coordinate. The *displacements* S_t can then be expressed in terms of the *displacements* ξ_i by an equation of the form

$$S_t = \sum_{i=1}^{3N} B_{ti}\xi_i \tag{2}$$

in which coefficients B_{ti} are constants fixed by the geometry of the molecule. The expression

$$G_{tt'} = \sum_{i=1}^{3N} \mu_i B_{ti} B_{t'i} \tag{3}$$

(where μ_i is the reciprocal mass of the atom to which the coordinate ξ_i is attached) can be thought of as defining the elements $G_{tt'}$ of a matrix $\boldsymbol{G}$ having the form

$$\boldsymbol{G} = \begin{pmatrix} G_{11} & G_{12} & \cdots & G_{1t} & \cdots \\ G_{21} & G_{22} & \cdots & G_{2t} & \cdots \\ \cdots & \cdots & \cdots & \cdots & \cdots \\ G_{t1} & G_{t2} & \cdots & G_{tt} & \cdots \\ \cdots & \cdots & \cdots & \cdots & \cdots \end{pmatrix}. \tag{4}$$

Thus $\boldsymbol{G}$ is a square matrix of rank $3N-6$($3N-5$ if the molecule is linear). It turns out that the elements $G_{tt'}$ and $G_{t't}$ are equal, hence $\boldsymbol{G}$ is symmetrical about its leading diagonal.

The potential energy of the molecule in terms of the internal coordinates is given by the expression

$$2V = \sum_{t,t'} F_{tt'} S_t S_{t'}, \tag{5}$$

in which the coefficients $F_{tt'}$ are force constants. The F_{tt} can be considered as elements of a second matrix $\boldsymbol{F}$

$$F = \begin{pmatrix} F_{11} & F_{12} & \cdots & F_{1t} & \cdots \\ F_{21} & F_{22} & \cdots & F_{2t} & \cdots \\ \cdots & \cdots & \cdots & \cdots & \cdots \\ F_{t1} & F_{t2} & \cdots & F_{tt} & \cdots \\ \cdots & \cdots & & \cdots & \end{pmatrix}. \quad (6)$$

The $\boldsymbol{F}$ matrix, like the $\boldsymbol{G}$ matrix, has rank and order $3N-6$ and is symmetrical about its leading diagonal.

It can be shown that the *secular determinant* is given by the expression

$$|\boldsymbol{F}-\boldsymbol{G}^{-1}\lambda| = 0, \quad (7)$$

in which $\boldsymbol{G}^{-1}$ is the inverse of the matrix $\boldsymbol{G}$.

We have now established definitions of an apparently formidable array of quantities which, however, are surprisingly easy to use.

A Simple Example

Consider first the normal vibration of a diatomic molecule. As the displacements are parallel to the internuclear axis we require merely two displacement coordinates, ξ_1 and ξ_2, as shown in Fig. 6.10.1. The internal coordinate, S_1, is the bond extension, and thus

$$S_1 = \xi_1 - \xi_2. \quad (8)$$

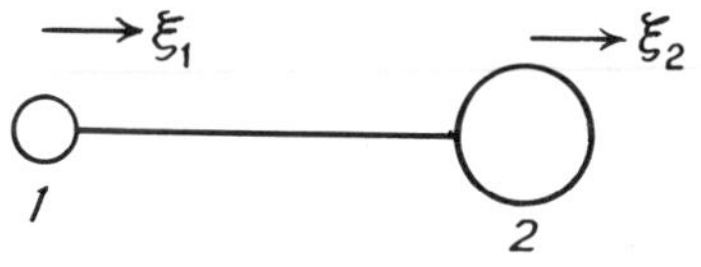

Fig. 6.10.1. Displacement coordinates for a diatomic molecule.

Comparing (8) with the expression (2) we find ($t = 1, i = 1$ and 2)

$$B_{11} = 1, \qquad B_{12} = -1. \quad (9)$$

As there is only one normal mode the matrix $\boldsymbol{G}$ is one-dimensional. Its single element ($t = t' = 1$) is, according to the definition (3),

$$G_{11} = \mu_1 B_{11}{}^2 + \mu_2 B_{12}{}^2 = \mu_1 + \mu_2 \quad (10)$$

and thus $\boldsymbol{G} = ((\mu_1 + \mu_2))$. The inverse of a one-dimensional matrix is simply the inverse of the element. Therefore

$$G^{-1} = ((\mu_1 + \mu_2)^{-1}).$$

The potential function is

$$2V = F_r(\xi_1 - \xi_2)^2 = F_r S_1{}^2 \tag{11}$$

in which F_r is the bond stretching force constant. Comparison with (5) shows that $F_r = F_{11}$, so that $\boldsymbol{F} = (F_r)$. Therefore the secular determinant is

$$|\boldsymbol{F} - \boldsymbol{G}^{-1}\lambda| = F_r - (\mu_1 + \mu_2)^{-1}\lambda = 0. \tag{12}$$

This is the equation we are looking for: it relates the force constant to the atomic masses and the normal frequency $\nu = \lambda^{\frac{1}{2}}/2\pi$ of the vibration.

Further considerations are necessary to determine the *normal coordinate*. In mass-weighted displacement coordinates, $q_i = \xi_i \mu_i{}^{-\frac{1}{2}}$, the kinetic energy is given by

$$2T = \dot{q}_1{}^2 + \dot{q}_2{}^2. \tag{13}$$

Let the momentum conjugate to q_i be p_i. Take Hamilton's equations in the form $\partial T/\partial \dot{q}_i = p_i$, then

$$p_i = \frac{\partial T}{\partial \dot{q}_i} = \dot{q}_i, \tag{14}$$

and so we have

$$2T = p_1{}^2 + p_2{}^2. \tag{15}$$

But eqn. (14) may be written, using the fundamental equation for partial derivatives,

$$p_i = \frac{\partial T}{\partial \dot{S}_1} \frac{\partial \dot{S}_1}{\partial \dot{q}_i}. \tag{16}$$

$\partial T/\partial \dot{S}_1$ is the momentum, P_1, conjugate to S_1. Further, $\partial \dot{S}_1/\partial \dot{q}_i = \partial S_1/\partial q_i$; thus (since $S_1 = \mu_1{}^{\frac{1}{2}} q_1 - \mu_2{}^{\frac{1}{2}} q_2$), $p_1 = \mu_1{}^{\frac{1}{2}} P_1$ and $p_2 = \mu_2{}^{\frac{1}{2}} P_1$, and therefore

$$2T = (\mu_1 + \mu_2) P_1{}^2. \tag{17}$$

Using Hamilton's equation a second time we obtain

$$\dot{S}_1 = \partial T/\partial P_1 = (\mu_1 + \mu_2) P_1. \tag{18}$$

Substitution of (18) into (17) then yields

$$2T = (\mu_1 + \mu_2)^{-1}\dot{S}_1{}^2. \tag{19}$$

The normal coordinate, Q_1, is related to S_1 by an equation of the form

$$S_1 = L_{11}Q_1, \tag{20}$$

in which the coefficient L_{11} is a constant. (It will emerge that the value of L_{11} is determined by the atomic masses.) On substituting (20) into (11) and (19) we find

$$2V = F_r L_{11}{}^2 Q_1{}^2 \tag{21}$$

and

$$2T = (\mu_1 + \mu_2)^{-1} L_{11}{}^2 \dot{Q}_1{}^2. \tag{22}$$

But the potential and kinetic energies in normal coordinates are given by (see section 6.1)

$$2V = \lambda Q_1{}^2 \text{ and } 2T = \dot{Q}_1{}^2. \tag{23}$$

Therefore

$$(\mu_1 + \mu_2)^{-1} L_{11}{}^2 = 1, \qquad L_{11} = (\mu_1 + \mu_2)^{\frac{1}{2}} \tag{24}$$

and thus

$$Q_1 = (\mu_1 + \mu_2)^{-\frac{1}{2}} S_1, \tag{25}$$

which is the desired expression for the normal coordinate in terms of the bond extension.

Note that if eqn. (24) is substituted into the expression (21) and the result compared with (23) one obtains

$$(\mu_1 + \mu_2)F_r = \lambda, \tag{26}$$

which establishes the secular determinant (12) in so far as diatomic molecules are concerned.

Force constants for a number of diatomic molecules are listed in Table 6.11.1. Since $F_r r^2$ has the dimension of energy, the units of a *stretching force constant* F_r, are dynes per centimetre. The value of stretching force constants is normally of the order 10^5 dyne cm^{-1}; thus it is convenient to tabulate the constants in the units of millidynes per angstrom,

$$1 \times 10^5 \text{ dyne cm}^{-1} = 1 \text{ md Å}^{-1}.$$

6.11 Linear Y—X—Y Molecules (Point Group $D_{\infty h}$)

A suitable set of cartesian displacement coordinates is illustrated in Fig. 6.11.1 (a). As to the internal coordinates, let r_{12} and r_{23} denote the extension of the first and second X—Y bonds, and let θ_x and θ_y be the components of the ∠ YXY deformation θ in the xz and yz planes, respectively. Then

$$r_{12} = \xi_1 - \xi_2 \quad (1)$$

$$r_{23} = \xi_2 - \xi_3. \quad (2)$$

Consideration of Fig. 6.11.1 (b) will show that for very small displacements,

$$\theta_x = \rho(\xi_4 - 2\xi_5 + \xi_6) \quad (3)$$

$$\theta_y = \rho(\xi_7 - 2\xi_8 + \xi_9), \quad (4)$$

where ρ^{-1} is the equilibrium internuclear distance X—Y.

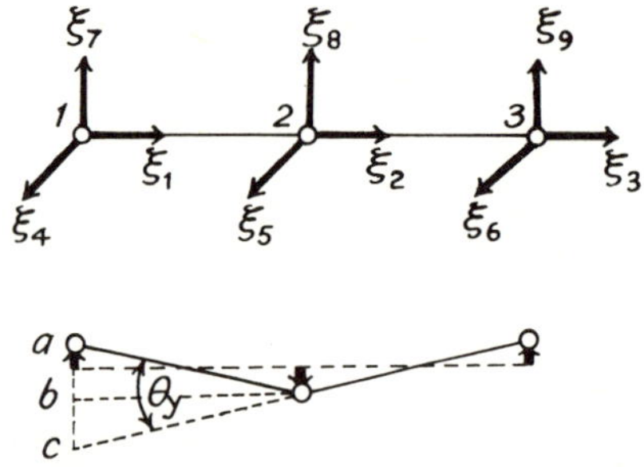

Fig. 6.11.1. In (b) the distance ab is $\xi_7 - \xi_8$, while bc equals $\xi_9 - \xi_8$: hence $\theta_y \approx \rho \times ac = \rho(\xi_7 - 2\xi_8 + \xi_9)$.

Certain symmetry operations of the group permute r_{12} and r_{23}. But the combinations $r_{12} + r_{23}$ and $r_{12} - r_{23}$ do not mix under any symmetry operation and so belong to definite symmetry species (actually $r_{12} + r_{23}$ belongs to Σ_g^+ and $r_{12} - r_{23}$ to Σ_u^+). Likewise the species of the coordinate pair (θ_x, θ_y) is Π_u. The coordinates used to describe the atom displacements are:

$$\left.\begin{aligned} S_1 &= 2^{-\frac{1}{2}}(r_{12} + r_{23}) = 2^{-\frac{1}{2}}(\xi_1 - \xi_3) \\ S_{2x} &= \theta_x = \rho(\xi_4 - 2\xi_5 + \xi_6) \\ S_{2y} &= \theta_y = \rho(\xi_7 - 2\xi_8 + \xi_9) \\ S_3 &= 2^{-\frac{1}{2}}(r_{12} - r_{23}) = 2^{-\frac{1}{2}}(\xi_1 - 2\xi_2 + \xi_3) \end{aligned}\right\}. \quad (5)$$

S_1–S_3 are internal coordinates of a special kind, chosen so as to belong to a definite symmetry species of the group $D_{\infty h}$. They are known as *internal symmetry coordinates.* The coordinates S_{2x} and S_{2y} give rise to identical elements in the $\boldsymbol{G}$ matrix, so that we retain merely one of them and use the plain subscript 2. The elements $G_{tt'}$ are then

$$\left.\begin{array}{ll} G_{11} = \mu_1 & G_{22} = 2(\mu_1 + 2\mu_2)\rho^2 \\ G_{12} = G_{21} = 0 & G_{23} = G_{32} = 0 \\ G_{13} = G_{31} = 0 & G_{33} = \mu_1 + 2\mu_2 \end{array}\right\}. \tag{6}$$

Evidently $\boldsymbol{G}$ is a diagonal matrix, for only the diagonal elements differ from zero. In this event it is easy to obtain the matrix $\boldsymbol{G}^{-1}$, the inverse of the matrix $\boldsymbol{G}$. The elements $(G^{-1})_{tt'}$, of $\boldsymbol{G}^{-1}$ are simply the reciprocals of the corresponding elements, G_{tt}, of $\boldsymbol{G}$.* Thus

$$\boldsymbol{G}^{-1} = \begin{pmatrix} \mu_1^{-1} & 0 & 0 \\ 0 & \{2(\mu_1 + 2\mu_2)\rho^2\}^{-1} & 0 \\ 0 & 0 & (\mu_1 + 2\mu_2)^{-1} \end{pmatrix}. \tag{7}$$

As to the potential, we consider that there are forces which oppose the extension or compression of chemical bonds, and similar forces which resist a change of the angle between bonds. In this approximation the potential function is

$$2V = F_r(r_{12}^2 + r_{23}^2) + F_\theta(\theta_x^2 + \theta_y^2). \tag{8}$$

(8) is known as a valence force potential. It neglects the mutual influence of the displacement of atoms not bonded to one another and so cannot be expected to be strictly valid. As $r_{12}^2 + r_{23}^2 = S_1^2 + S_3^2$ and

$$\theta_x^2 + \theta_y^2 = S_{2x}^2 + S_{2y}^2 = S_2^2,$$

the substitution of symmetry coordinates for internal coordinates gives

$$2V = F_r(S_1^2 + S_3^2) + F_\theta S_2^2. \tag{9}$$

Comparing (9) with the general expression (6.10.5) the elements of $\boldsymbol{F}$ are seen to be

$$\left.\begin{array}{ll} F_{11} = F_r & F_{22} = F_\theta \\ F_{12} = F_{21} = 0 & F_{23} = F_{32} = 0 \\ F_{13} = F_{31} = 0 & F_{33} = F_r \end{array}\right\}. \tag{10}$$

* This applies *only* to a diagonal, square matrix. It is by no means true in the general case.

Table 6.11.1 Force Constants of some Diatomic and Triatomic Molecules

	Molecule		Fundamental frequency (cm^{-1})	Force constant md $Å^{-1}$	
Diatomic	H_2		4277	F_r	5·3
	HCl		2938		4·9
	HBr		2604		4·0
	Cl^2		561		3·3
	O_2		1568		11·7
	CO		2157		18·8
	N_2		2345		22·7
Linear XY_2	CO_2	σ_1	1345	F_r	17·0
		σ_2	667	$\rho^2 F_\theta$	0·57
		σ_3	2349	F_r	14·2
	N_3^-(in KN_3)	σ_1	1344	F_r	15·1
		σ_2	645	$\rho^2 F_\theta$	0·58
		σ_3	2041	F_r	11·6
	NO_2^+(in N_2O_5)	σ_1	1400	F_r	18·5
		σ_2	538	$\rho^2 F_\theta$	0·42
		σ_3	2375	F_r	16·2
	CS_2	σ_1	656	F_r	8·1
		σ_2	397	$\rho^2 F_\theta$	0·23
		σ_3	1523	F_r	6·9
Non-linear XY_2	H_2O ($\angle$ HOH = 104·5°)	σ_1	3652	F_r	7·6
		σ_2	1595	$\rho^2 F_\theta$	0·70
		σ_3	3756	F_r	7·7
	SO_2 ($\angle$ OSO = 120°)	σ_1	1151	F_r	9·9
		σ_2	519	$\rho^2 F_\theta$	0·76
		σ_3	1361	F_r	10·0
	NO_2 ($\angle$ ONO = 134°)	σ_1	1322	F_r	11·3
		σ_2	750	$\rho^2 F_\theta$	1·1
		σ_3	1616	F_r	8·4

The secular determinant (6.10.7) is then, taking account of the usual rule for subtraction of matrices,

$$\begin{vmatrix} F_r - \mu_1{}^{-1}\lambda & 0 & \\ 0 & F_\theta - \{2(\mu_1 + 2\mu_2)\rho^2\}^{-1}\lambda & \\ 0 & 0 & F_r - (\mu_1 + 2\mu_2)^{-1}\lambda \end{vmatrix} = 0, \tag{11}$$

which has solutions

$$\lambda_1 = \mu_1 F_r, \qquad \lambda_2 = 2(\mu_1 + 2\mu_2)\rho^2 F_\theta, \qquad \lambda_3 = (\mu_1 + 2\mu_2)F_r. \tag{12}$$

As each mode of vibration has a different symmetry species the normal coordinates are given by expressions similar to that for a diatomic molecule, *viz*.,

$$Q_t = (G^{-1})_{tt}{}^{\frac{1}{2}} S_t. \tag{13}$$

Therefore

$$Q_1 = (2\mu_1)^{-\frac{1}{2}}(r_{12} + r_{23}) \qquad Q_2 = \{2(\mu_1 + 2\mu_2)\rho^2\}^{-\frac{1}{2}}\theta \qquad (14)$$
$$Q_3 = \{2(\mu_1 + 2\mu_2)\}^{-\frac{1}{2}}(r_{12} + r_{23}).$$

Force constants for several representative Y—X—Y molecules, including carbon dioxide and carbon disulphide, are listed in Table 6.11.1. The deformation constants are given as $\rho^2 F_\theta$, that is to say, they are reduced to the same units, md Å^{-1}, as the stretching constants. It is evident that the force necessary to bend a bond is of the order one-tenth of that needed for stretching or compression. An interesting feature of the stretching constants of linear XY_2 molecules is that the value calculated from the in-phase mode ν_1 is uniformly higher than that obtained from the out-of-phase motion ν_3. This means that resistance to the symmetrical displacements is higher than to the antisymmetrical. Take for illustration the molecule CO_2. Stretching the first CO bond enhances the importance of the formal resonance structure $\overset{-}{\mathrm{O}}$—C≡$\overset{+}{\mathrm{O}}$ (cf. section 6.7): therefore the multiplicity of the second bond increases, and hence, inasmuch as increased multiplicity tends to shorten a bond, it has less resistance to compression than to extension. This is one way in which the distortion of a bond may influence the stiffness of its neighbours.

6.12 Linear Y—X—X—Y Molecules

Let $\xi_1 - \xi_4$ be oriented parallel to the internuclear axis and attach one to each of the four masses $Y^{(1)}$—$X^{(2)}$—$X^{(3)}$—$Y^{(4)}$. The internal coordinates comprising the bond extensions are related to the ξ_i by the expressions

$$r_{12} = \xi_1 - \xi_2 \qquad (1)$$

$$r_{23} = \xi_2 - \xi_3 \qquad (2)$$

$$r_{34} = \xi_3 - \xi_4. \qquad (3)$$

Here r_{12} and r_{34} are the extensions of the first and second X—Y bonds, and r_{23} that of the X—X bond. To cope with the bending vibrations, let $\xi_5 - \xi_8$ be attached to the first to fourth masses, respectively, oriented perpendicular to the internuclear axis and lying in a plane containing this axis. Suppose that the molecule bends in the plane containing $\xi_5 - \xi_8$; then, if θ_{13} and θ_{24} represent the deviation from 180° of interbond angles at the second and third masses along the chain, simple geometrical considerations give the relationships

$$\theta_{13} = \rho_1\xi_5 - (\rho_1 + \rho_2)\xi_6 + \rho_2\xi_7 \qquad (4)$$

$$\theta_{24} = -\rho_2\xi_6 + (\rho_1 + \rho_2)\xi_7 - \rho_1\xi_8 \qquad (5)$$

in which $\rho_1{}^{-1}$ is the internuclear distance X—Y and $\rho_2{}^{-1}$ that between X—X.

Symmetry operations tend to permute r_{12} and r_{34} as well as θ_{13} and θ_{24}. The internal symmetry coordinates are, therefore,

$$\Sigma_g^+: \left.\begin{aligned} S_1 &= 2^{-\frac{1}{2}}(r_{12}+r_{34}) = 2^{-\frac{1}{2}}(\xi_1-\xi_2+\xi_3-\xi_4) \\ S_2 &= r_{23} = \xi_2-\xi_3 \end{aligned}\right\}, \tag{6}$$

$$\Sigma_u^+: \; S_3 = 2^{-\frac{1}{2}}(r_{12}-r_{34}) = 2^{-\frac{1}{2}}(\xi_1-\xi_2-\xi_3+\xi_4)\,, \tag{7}$$

$$\Pi_g: \; S_4 = 2^{-\frac{1}{2}}(\theta_{13}+\theta_{24}) = 2^{-\frac{1}{2}}[\rho_1\xi_5-(\rho_1+2\rho_2)\xi_6+ (\rho_1+2\rho_2)\xi_7-\rho_1\xi_8]\,, \tag{8}$$

$$\Pi_u: \; S_5 = 2^{-\frac{1}{2}}(\theta_{13}-\theta_{24}) = 2^{\frac{1}{2}}\rho_1[\xi_5-\xi_6-\xi_7+\xi_8]\,. \tag{9}$$

For the *non-zero* elements of G, we find (since $\mu_1=\mu_4$ and $\mu_2=\mu_3$)

$$\left.\begin{aligned} G_{11} &= \mu_1+\mu_2 & G_{33} &= \mu_1+\mu_2 \\ G_{12} &= G_{21} = -2^{\frac{1}{2}}\mu_2 & G_{44} &= \mu_1\rho_1{}^2+\mu_2(\rho_1+2\rho_2)^2 \\ G_{22} &= 2\mu_2 & G_{55} &= (\mu_1+\mu_2)\rho_1{}^2 \end{aligned}\right\}. \tag{10}$$

The valence-force potential function is

$$2V = F_r(r_{12}{}^2+r_{34}{}^2)+F_{r'}r_{23}{}^2+F_\theta(\theta_{13}{}^2+\theta_{24}{}^2)\,, \tag{11}$$

where F_r is the force-constant for extension of the X—Y bonds, $F_{r'}$ that for the X—X bond, and F_θ that corresponding to the change of interbond angle. As $r_{12}{}^2+r_{23}{}^2=S_1{}^2+S_3{}^2$ and $\theta_{13}{}^2+\theta_{24}{}^2=S_4{}^2+S_5{}^2$, comparison with eqn. 6.10.5 gives for the elements of $\boldsymbol{F}$

$$\left.\begin{aligned} F_{11} &= F_r & F_{33} &= F_r \\ F_{22} &= F_{r'} & F_{44} &= F_{55} = F_\theta \end{aligned}\right\}. \tag{12}$$

The elements of $\boldsymbol{F}$ not listed in (12) are zero. Thus $\boldsymbol{F}$ is a diagonal matrix (this is always true when a valence-force potential is assumed); but $\boldsymbol{G}$ has off-diagonal elements G_{12} and G_{21}, corresponding to the fact that S_1 and S_2 have the same symmetry species. Since the $\boldsymbol{G}^{-1}$ matrix cannot be found merely by inspection of $\boldsymbol{G}$,* it is more economical in time to multiply the secular determinant (6.10.7) by the determinant of G when one obtains the alternative form

$$|\boldsymbol{GF}-\boldsymbol{I}\lambda| = 0 \tag{13}$$

* $\boldsymbol{G}^{-1}$ can, however, be obtained from $\boldsymbol{G}$ by routine methods. See, for example, Aitken, *Determinants and Matrices*, Oliver and Boyd, Edinburgh and London, 1948.

in which $\boldsymbol{I}$ is the unit matrix. The determinantal equation (13) has the form

$$\begin{pmatrix} G_{11} & G_{12} & 0 & 0 & 0 \\ G_{21} & G_{22} & 0 & 0 & 0 \\ 0 & 0 & G_{33} & 0 & 0 \\ 0 & 0 & 0 & G_{44} & 0 \\ 0 & 0 & 0 & 0 & G_{55} \end{pmatrix} \times$$

$$\begin{pmatrix} F_r & 0 & 0 & 0 & 0 \\ 0 & F_{r'} & 0 & 0 & 0 \\ 0 & 0 & F_r & 0 & 0 \\ 0 & 0 & 0 & F_\theta & 0 \\ 0 & 0 & 0 & 0 & F_\theta \end{pmatrix} -$$

$$\begin{pmatrix} \lambda & 0 & 0 & 0 & 0 \\ 0 & \lambda & 0 & 0 & 0 \\ 0 & 0 & \lambda & 0 & 0 \\ 0 & 0 & 0 & \lambda & 0 \\ 0 & 0 & 0 & 0 & \lambda \end{pmatrix} = 0, \tag{14}$$

in which the elements which have the value zero are shown as zeros. Taking account of the rules for multiplication and subtraction of matrices, (14) becomes

$$\begin{vmatrix} (\mu_1+\mu_2)F_r-\lambda & -2^{\frac{1}{2}}\mu_2F_{r'} & 0 & 0 & 0 \\ -2^{\frac{1}{2}}\mu_2F_r & 2\mu_2F_{r'}-\lambda & 0 & 0 & 0 \\ 0 & 0 & (\mu_1+\mu_2)F_r-\lambda & 0 & 0 \\ 0 & 0 & 0 & [\mu_1\rho_1{}^2+\mu_2(\rho_1+2\rho_2)^2]F_\theta-\lambda & 0 \\ 0 & 0 & 0 & 0 & [(\mu_1+\mu_2)\rho_1{}^2]F_\theta-\lambda \end{vmatrix} = 0, \tag{15}$$

which factorises into one quadratic and three linear equations. The quadratic expression

$$\begin{vmatrix} (\mu_1+\mu_2)F_r-\lambda & -2^{\frac{1}{2}}\mu_2F_{r'} \\ -2^{\frac{1}{2}}\mu_2F_r & 2\mu_2F_{r'}-\lambda \end{vmatrix} = 0 \tag{16}$$

yields

$$\lambda^2 - [(\mu_1 + \mu_2)F_r + 2\mu_2 F_{r'}]\,\lambda + 2\mu_1\mu_2 F_r F_{r'} = 0, \tag{17}$$

which, according to the theory of quadratic equations, has roots given by

$$\left.\begin{aligned} \lambda_1 + \lambda_2 &= (\mu_1 + \mu_2)F_r + 2\mu_2 F_{r'} \\ \lambda_1\lambda_2 &= 2\mu_1\mu_2 F_r F_{r'} \end{aligned}\right\}. \tag{18}$$

The linear factors give

$$\lambda_3 = (\mu_1 + \mu_2)F_r \tag{19}$$

$$\lambda_4 = \{\mu_1\rho_1{}^2 + \mu_2(\rho_1 + 2\rho_2)^2\}F_\theta \tag{20}$$

and

$$\lambda_5 = (\mu_1 + \mu_2)\rho_1{}^2 F_\theta\ . \tag{21}$$

In section 7.6 the constants F_r, $F_{r'}$ and F_θ are denoted by F, F' and f, respectively.

6.13 Non-Linear XY_2 Molecules

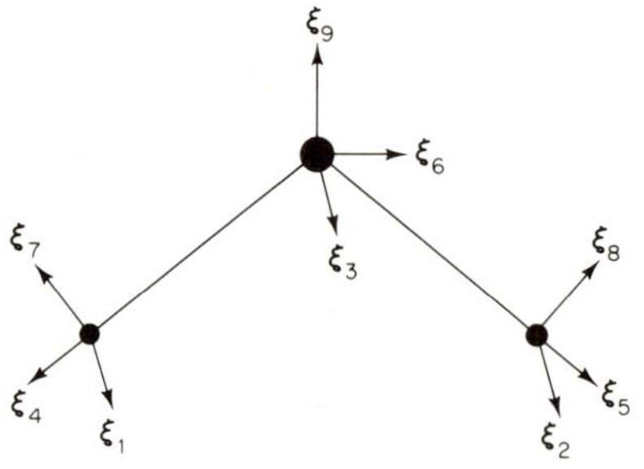

Fig. 6.13.1. Alternative displacement coordinates for H_2O.

A suitable set of cartesian displacement coordinates is shown in Fig. 6.13.1. Let r_{31} and r_{32} be the X—Y bond extensions and θ the $\angle$ YXY deformation. For very small displacements the relationships between the internal coordinates and the ξ_i are

$$r_{31} = \xi_4 + s\xi_6 + c\xi_9 \tag{1}$$

$$r_{32} = \xi_5 - s\xi_6 + c\xi_9 \tag{2}$$

$$\theta = \rho(\xi_7 + \xi_8 - 2s\xi_9), \tag{3}$$

in which $s = \sin \frac{1}{2}\alpha$ and $c = \cos \frac{1}{2}\alpha$, where α is the YXY interbond angle, and ρ^{-1} is the XY bond distance. The internal symmetry coordinates are:

$$A_1: \; S_1 = 2^{-\frac{1}{2}}(r_{31} + r_{32}) = 2^{-\frac{1}{2}}(\xi_4 + \xi_5 + 2c\xi_9) \tag{4}$$

$$A_1: \; S_2 = \theta = \rho(\xi_7 + \xi_8 - 2s\xi_9) \tag{5}$$

$$B_2: \; S_3 = 2^{-\frac{1}{2}}(r_{31} - r_{32}) = 2^{-\frac{1}{2}}(\xi_4 + \xi_5 - 2s\xi_6). \tag{6}$$

The elements of $\boldsymbol{G}$ are then (using the trigonometrical relations $\sin \alpha = 2sc$, $1 + \cos \alpha = 2c^2$, $1 - \cos \alpha = 2s^2$),

$$\left.\begin{array}{llll} G_{11} = \mu_1 + \mu_3(1 + \cos \alpha) & & G_{22} = 2\rho^2[\mu_1 + \mu_3(1 - \cos \alpha)] \\ G_{12} = G_{21} = -2^{\frac{1}{2}}\mu_3\rho \sin \alpha & & G_{23} = G_{32} = 0 \\ G_{31} = G_{13} = 0 & & G_{33} = \mu_1 + \mu_3(1 - \cos \alpha) \end{array}\right\} . \tag{7}$$

The simple valence-force potential is

$$2V = F_r({r_{31}}^2 + {r_{32}}^2) + F_\theta \theta^2, \tag{8}$$

and hence the non-zero elements of $\boldsymbol{F}$ are

$$F_{11} = F_r \qquad F_{22} = F_\theta \qquad F_{33} = F_r. \tag{9}$$

With the secular determinant in the form

$$|\boldsymbol{GF} - \boldsymbol{I}\lambda| = 0. \tag{6.12.13}$$

one obtains, after the multiplication and subtraction of matrices,

$$\begin{vmatrix} [\mu_1 + \mu_3(1 + \cos \alpha)]\, F_r - \lambda & -2^{\frac{1}{2}}\mu_3\rho \sin \alpha\, F_\theta & 0 \\ -2^{\frac{1}{2}}\mu_3\rho \sin \alpha\, F_r & 2\rho^2[\mu_1 + \mu_3(1 - \cos \alpha)]\, F_\theta - \lambda & 0 \\ 0 & 0 & [\mu_1 + \mu_3(1 - \cos \alpha)]\, F_r - \lambda \end{vmatrix} = 0 \tag{10}$$

The determinant breaks down into one quadratic and one linear factor. The quadratic factor gives on expansion

$$\lambda^2 - \lambda\{[\mu_1 + \mu_3(1 + \cos \alpha)]\, F_r + 2\rho^2[\mu_1 + \mu_3(1 - \cos \alpha)]\, F_\theta\} + 2\rho^2({\mu_1}^2 + 2\mu_1\mu_3)F_rF_\theta = 0, \tag{11}$$

which has roots

$$\lambda_1 + \lambda_2 = [\mu_1 + \mu_3(1 + \cos \alpha)]\, F_r + 2\rho^2[\mu_1 + \mu_3(1 - \cos \alpha)]\, F_\theta \tag{12}$$

and

$$\lambda_1\lambda_2 = 2\rho^2(\mu_1 + 2\mu_1\mu_3)F_rF_\theta. \tag{13}$$

Finally, the linear factor of (10) gives

$$\lambda_3 = [\mu_1 + \mu_3(1 - \cos\alpha)]\, F_r\,. \tag{14}$$

Note that, when $\alpha = 180°$, eqn. (10) has the aspect of the secular determinant (6.11.11) for a linear XY_2 molecule.

A convenient means of solving for the force constants is to substitute numerical values of μ and λ into (10). For the water molecule ($\alpha = 104{\cdot}5°$) one obtains, on substitution of the atomic masses,

$$\begin{vmatrix} 0{\cdot}625\times10^{24}F_r - \lambda & -0{\cdot}0532\times10^{24}\rho F_\theta & 0 \\ -0{\cdot}0532\times10^{24}\rho F_r & 1{\cdot}290\times10^{24}\rho^2F_\theta - \lambda & 0 \\ 0 & 0 & 0{\cdot}645\times10^{24}F_r - \lambda \end{vmatrix} = 0, \tag{15}$$

which (with $\lambda_1 = 4{\cdot}736$, $\lambda_2 = 0{\cdot}901$, and $\lambda_3 = 5{\cdot}008$, all 10^{29}) has solutions,

$$F_r = 7{\cdot}5_6 \qquad \rho^2F_\theta = 0{\cdot}70_2 \text{ md Å}^{-1}, \tag{16}$$

and

$$F_r = 7{\cdot}7_5 \text{ md Å}^{-1}. \tag{17}$$

The relative unimportance of the off-diagonal elements in (15) is noteworthy. If the off-diagonal elements were identically zero the quadratic part of (15) would reduce to two linear factors, $0{\cdot}625\times10^{24}F_r = \lambda_1$ and $1{\cdot}290\times10^{24}\rho^2F_\theta = \lambda_2$. In this event the first A_1 mode (ν_1) would be a pure O—H stretch, and the second (ν_2) a pure $\angle$ HOH deformation. Neglecting the off-diagonal elements one would calculate from the force constants in (16) that $\sigma_1 = 3649$ and $\sigma_2 = 1597$ cm^{-1}, which can be thought of as the normal frequencies that would be observed if each vibration involved just one internal coordinate. Interaction mixes somewhat the characters of the two modes and pushes their frequencies apart: thus the A_1 frequencies of the water molecule are actually 3652 and 1595 cm^{-1} (see Table 6.11.1). The interaction is extremely small in this example because the normal frequencies are well separated.

BIBLIOGRAPHY

Herzberg, *Infrared and Raman Spectra of Polyatomic Molecules,* Van Nostrand, New York, 1945.
Wilson, Decius, and Cross, *Molecular Vibrations,* McGraw-Hill, New York and London, 1955.
Allen and Cross, *Molecular Vib-Rotors,* Wiley, New York and London, 1963.

Chapter 7

The Raman effect

7.1 The Scattering of Light

When light passes through a transparent medium, a small fraction of the light is scattered by the molecules present. The phenomenon is known generally as Rayleigh scattering and is believed to account for the colour of the sea and the sky. One specific kind of scattering, the *Raman effect*, affords a method of studying molecular vibration and rotation that is comparable in importance with the analysis of infrared and microwave spectra.

The mechanism of light scattering is the following. When an atom or molecule is placed in an electric field, the electrons are repelled by the field, and an electric dipole moment develops in the molecule pointing roughly in the direction of the field. This induced dipole moment is rather small as long as the perturbing electric field is small compared with the electric field of electrons and nuclei within the molecule. Provided the external field is not too great the induced moment μ_i is proportional to the field strength $\mathcal{E}$, and thus

$$\mu_i = \alpha \mathcal{E},$$

where α, the constant of proportionality, is known as the *polarisability* and is expected to be different for different molecules.* If the field fluctuates with time, so does the induced dipole. Therefore the electric field associated with a light wave induces in an atom or molecule in its path a dipole that vibrates with the same frequency as the incident radiation. From the classical viewpoint the vibrating dipole in turn emits electromagnetic waves, and these comprise the scattered radiation. It will be seen that the normal mechanism of scattering changes the direction of the propagation of light but not its frequency. The intensity of the scattered light depends on the polarisability of the molecules, and it is possible to measure the

* When describing the induced moment and polarisability of a single molecule we shall use the plain symbols μ_i and α. The measured value of, for instance, the polarisability is however an average value, taken over a large number of randomly oriented molecules: we then use the symbol $\bar{\alpha}$.

magnitude of the polarisability by determining the intensity of the scattering. As a rule light scattering is observed for radiation in the visible, or sometimes the ultraviolet region. The polarisability is then that due to the electron displacements only, and is related to the molar electronic polarisability (or molar refractivity) by the equation $P_E = (4/3)\pi N_A \bar{\alpha}$ (Section 11.5).

Further information can be obtained from a study of the polarisation of the scattered light. Let us suppose first that the scattering particles are atoms of one of the rare gases, for which the electron density is spherically symmetrical. For such atoms the polarisability is the same in all directions. Plane polarised light† may be expected to induce a moment parallel to the plane of polarisation so that the light emitted–the scattered light–is polarised in the same plane as the incident radiation. For plane polarised incident radiation the scattered radiation should then be completely polarised in the same plane. This assumption is fulfilled for rare gas atoms and for molecules of tetrahedral or octahedral symmetry (*e.g.*, CCl_4 or SF_6), because the external electric force, the induced dipole moment, and hence the electric vector of the scattered radiation, all vibrate in one and the same plane. However it is by no means true in the general case.

Polarisability is said to be *anisotropic* when α is different for different directions in a molecule. The ordinary methods of measurement, for instance those described in Section 11.5, then give a mean value $\bar{\alpha}$, averaged over all possible orientations of the molecule with respect to the external field. In this event the magnitude of the

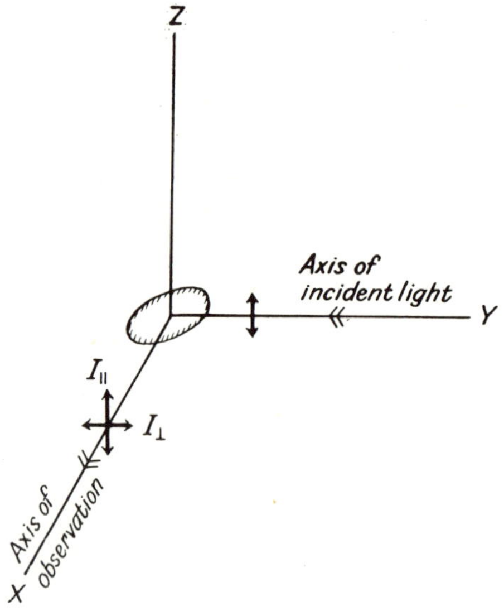

Fig. 7.1.1. Depolarisation of scattered radiation (schematic).

† The optical plane of polarisation is the plane of the electric vector $\mathcal{E}$.

induced moment must depend on the orientation of the molecule in the electric field. Further, the induced moment is not parallel to the incident electrical field but deviates from it by tending to be oriented in a direction of greater polarisability. The scattered light, which vibrates in the same plane as the induced dipole, is not in this case polarised in the same plane as the incident light. As a very large number of randomly oriented molecules contribute to the scattering, an incident beam of plane polarised light will give rise to scattered radiation which is in some degree depolarised. Imagine that the incident light travels along the Y-axis of Fig. 7.1.1, and that it is polarised in the YZ-plane; the scattered light is observed in some direction perpendicular to Y, say in the direction of the X-axis. The scattered light will have components polarised in the XY and XZ planes, as shown in the figure. Let the intensity of the components be $I_\perp$ and $I_\parallel$, respectively: then the ratio $I_\perp/I_\parallel$ is known as the *depolarisation factor* ρ, and can be measured experimentally. For atoms and molecules whose polarisability is isotropic the depolarisation factor is zero; that is, the scattered radiation is plane polarised *irrespective of whether the incident light is polarised or not*. The highest degree of anisotropy is reached when the polarisability along one particular direction in a molecule is very great compared to the polarisabilities perpendicular to this direction. In this limiting case the depolarisation factor for incident polarised light is $\frac{1}{3}$.

7.2 The Polarisability Ellipsoid

It is necessary to develop a more detailed picture of the polarisability of an anisotropic molecule. Let us choose three perpendicular directions in the molecule and label them x, y, and z. The axes would normally be taken to coincide with symmetry elements in the molecule, though for the moment this is immaterial. The orientation of the axes is, of course, fixed once the choice of axes is made; and we shall ignore rotation, so that the axes are space-fixed as well. As a beam of plane polarised light impinging on the molecule will not as a rule coincide with one of the axes, the electric vector will have non-zero components $\mathscr{E}_x$, $\mathscr{E}_y$, and $\mathscr{E}_z$ (where $\mathscr{E}^2 = \mathscr{E}_x^2 + \mathscr{E}_y^2 + \mathscr{E}_z^2$) along the axes. Likewise the induced moment, which is not necessarily parallel to $\mathscr{E}$, has components μ_x, μ_y, and μ_z on the selected axes. Let

$$\alpha_{xx} \quad \alpha_{xy} \quad \alpha_{xz}$$

be the proportionality constants (see Section 7.1) between μ_x and $\mathscr{E}_x$, μ_x and $\mathscr{E}_y$, and μ_x and $\mathscr{E}_x$, respectively. Then

$$\mu_x = \alpha_{xx}\mathscr{E}_x + \alpha_{xy}\mathscr{E}_y + \alpha_{xz}\mathscr{E}_z. \tag{1}$$

The first term on the right is the contribution to μ_x from the field component $\mathscr{E}_x$, the second from $\mathscr{E}_y$, and the third from $\mathscr{E}_z$. The second and third terms are not zero, because in the most general situation the moment induced by the component

$\mathcal{E}_y$, say, is not parallel to y and so has a component along x. Similarly for μ_y and μ_z we have

$$\mu_y = \alpha_{yx}\mathcal{E}_x + \alpha_{yy}\mathcal{E}_y + \alpha_{yz}\mathcal{E}_z \tag{2}$$

and

$$\mu_z = \alpha_{zx}\mathcal{E}_x + \alpha_{zy}\mathcal{E}_y + \alpha_{zz}\mathcal{E}_z. \tag{3}$$

Eqns. (1)–(3) can be written as the matrix equation,

$$\begin{pmatrix} \mu_x \\ \mu_y \\ \mu_z \end{pmatrix} = \begin{pmatrix} \alpha_{xx} & \alpha_{xy} & \alpha_{xz} \\ \alpha_{yx} & \alpha_{yy} & \alpha_{yz} \\ \alpha_{zx} & \alpha_{zy} & \alpha_{zz} \end{pmatrix} \times \begin{pmatrix} \mathcal{E}_x \\ \mathcal{E}_y \\ \mathcal{E}_z \end{pmatrix}. \tag{4}$$

The proper representation of α, then, is the matrix

$$\alpha = \begin{pmatrix} \alpha_{xx} & \alpha_{xy} & \alpha_{xz} \\ \alpha_{yx} & \alpha_{yy} & \alpha_{yz} \\ \alpha_{zx} & \alpha_{zy} & \alpha_{zz} \end{pmatrix}. \tag{4a}$$

The elements $\alpha_{xx}, \alpha_{xy}, \ldots$ are not all different, for it can be shown that

$$\alpha_{xy} = \alpha_{yx}, \quad \alpha_{xz} = \alpha_{zx}, \quad \alpha_{yz} = \alpha_{zy}. \tag{5}$$

In other words, the matrix (4*a*) is symmetrical about its leading diagonal.

It is a mathematical property of the symmetrical form of the matrix (4*a*) that there must be a special set of axes, x', y', and z', for which all the off-diagonal elements are zero,

$$\alpha = \begin{pmatrix} \alpha_{x'x'} & 0 & 0 \\ 0 & \alpha_{y'y'} & 0 \\ 0 & 0 & \alpha_{z'z'} \end{pmatrix}. \tag{6}$$

These special axes are directions in the molecule for which the induced moments are parallel to the perturbing field, for in this situation the eqns. (1), (2), and (3) simplify to

$$\mu_{x'} = \alpha_{x'x'}\mathcal{E}_{x'}, \qquad \mu_{y'} = \alpha_{y'y'}\mathcal{E}_{y'}, \qquad \mu_{z'} = \alpha_{z'z'}\mathcal{E}_{z'}. \tag{7}$$

The physical significance of the transformation from (4*a*) to (6) is simply this. If the set of axes x, y, and z is given any arbitrary orientation in the molecule, it cannot be expected that the components of the induced moment will be parallel to the components of the applied field; hence the off-diagonal elements of the matrix α as given by eqn. (4*a*) are different from zero. But by rotating the axes relative to the frame of the molecule it will be possible to discover the special orientation of axes (distinguished by the prime) for which the induced moments are parallel to the field components. α then has the form of the diagonal matrix (6), for all the off-diagonal elements must be zero.

The locus of points formed by plotting $\alpha^{-\frac{1}{2}}$ in any direction from the origin is the surface of a figure known as the *polarisability ellipsoid*, whose principal axes are the special axes x', y', and z'.* For a completely anisotropic molecule $\alpha_{x'x'} \neq \alpha_{y'y'} \neq \alpha_{z'z'}$ and the figure is a triaxial ellipsoid (*i.e.*, it has three axes of unequal length), similar to the momental ellipsoid of an asymmetric top. A fully isotropic molecule has $\alpha_{x'x} = \alpha_{y'y'} = \alpha_{z'z'}$, when the polarisability ellipsoid is a sphere. The value of the polarisability obtained from refractive index measurements (Section 11.5), or from the intensity of Rayleigh scattering, is the average of the values along the principal axes,

$$\bar{\alpha} = \tfrac{1}{3}(\alpha_{x'x'} + \alpha_{y'y'} + \alpha_{z'z'}). \tag{8}$$

It is easy to pick out the orientation of the principal axes of the polarisability ellipsoid in a molecule that has elements of symmetry. An axis of symmetry must coincide with one, and a plane of symmetry must contain two, of the principal axes. Take as an example the ammonia molecule, NH_3. Let z' coincide with the C_3 axis of symmetry; then x' and y' must be perpendicular to this axis and of equal length, so that $\alpha_{x'x'} = \alpha_{y'y'}$. The polarisability ellipsoid of ammonia is therefore an ellipsoid of revolution. The reader should satisfy himself that the polarisability ellipsoid is an ellipsoid of revolution for any molecule that possesses a single three-fold (or higher) symmetry axis. When a molecule has more than one three-fold proper symmetry axis, the polarisability ellipsoid is a sphere.

7.3 The Vibrational Raman Effect

The light scattered by molecules is not composed entirely of radiation of the same frequency as the incident light. Some of the scattering contains the molecular vibration frequencies added to or subtracted from the frequency of the original radiation. This effect was predicted theoretically by Smekal: it is named after Raman, who discovered it. Historically much of its importance lies in the circumstance that,

* The equation of the polarisability ellipsoid is then

$$\alpha_{x'x'}x'^2 + \alpha_{y'y'}y'^2 + \alpha_{z'z'}z'^2 = 1,$$

which is the equation of a figure whose half axes are $\alpha_{x'x'}^{-\frac{1}{2}}$, $\alpha_{y'y'}^{-\frac{1}{2}}$ and $\alpha_{z'z'}^{-\frac{1}{2}}$.

However, some authors prefer to define the polarisability ellipsoid by the equation

$$\xi^2/\alpha_{\xi\xi} + \eta^2/\alpha_{\eta\eta} + \zeta^2/\alpha_{\zeta\zeta} = 1,$$

where $\alpha_{\xi\xi}$, $\alpha_{\eta\eta}$, and $\alpha_{\zeta\zeta}$ are components of the polarisability along a cartesian set of axes ξ, η, ζ. The half-axes are then $\alpha_{\xi\xi}^{\frac{1}{2}}$, $\alpha_{\eta\eta}^{\frac{1}{2}}$, and $\alpha_{\zeta\zeta}^{\frac{1}{2}}$, and the coordinates ξ, η, ζ are inversely related to (but coincident with) the set x', y', z'.

until the development of recording infrared spectrometers, the observation of vibration frequencies in the *Raman spectrum* was technically easier than the measurement of infrared absorption.

The origin of the Raman effect is in the change of polarisability of a molecule in the course of vibration. If α varies with Q_k, the normal coordinate of a vibration of frequency ν_k, we may write as a first approximation

$$\alpha = \alpha^0 + \left(\frac{\partial\alpha}{\partial Q_k}\right)_0 Q_k, \tag{1}$$

in which α^0 is the polarisability of the molecule when the nuclei are in their equilibrium positions, and $(\partial\alpha/\partial Q_k)_0$ is the rate of change of α with Q_k for infinitesimal nuclear displacements. The expression (1) is then analogous to the equation (6.7.8) for the expansion of the permanent electric moment in terms of the normal coordinates. Since the nuclear displacements fluctuate with the frequency ν_k, so that $Q_k = Q_k^0 \cos 2\pi\nu_k t$ (Section 6.1), we have for the induced moment, $\mu_i = \alpha\mathcal{E}$, the expression

$$\mu_i = \{\alpha^0 + (\partial\alpha/\partial Q_k)_0 Q_k^0 \cos 2\pi\nu_k t\}\mathcal{E}^0 \cos 2\pi\nu t, \tag{2}$$

where $\mathcal{E} = \mathcal{E}^0 \cos 2\pi\nu t$ is the oscillating field of the incident light, taken to be monochromatic light of frequency ν. From (2), using well-known trigonometrical formulae, we find

$$\mu_i = \alpha^0\mathcal{E}^0 \cos 2\pi\nu t + (\mathcal{E}^0/2)(\partial\alpha/\partial Q_k)_0 Q_k^0 [\cos 2\pi(\nu + \nu_k)t + \cos 2\pi(\nu - \nu_k)t]. \tag{3}$$

The first term on the right corresponds to scattered radiation of unchanged frequency ν. This is the Rayleigh scattering, which constitutes the main part of the scattered light. The second term is much smaller than the first, because the polarisability varies only slightly during vibration; yet it gives rise to the scattering of frequencies $\nu \pm \nu_k$ which comprise the Raman effect.

According to this purely classical theory, the two frequencies $\nu + \nu_k$ and $\nu - \nu_k$ should occur in the scattered radiation with equal probability. This prediction is not borne out by experiment. The Raman frequency $\nu - \nu_k$ always appears with greater intensity, usually with considerably greater intensity. The quantum-theoretical explanation makes it clear why this is so. Scattering is measured for incident radiation to which the medium is transparent; in other words, the energy $h\nu$ of the incident quantum is not sufficient to raise the molecule to an electronically excited state. When a light quantum collides with a molecule it is retained only for the time necessary to establish the insufficiency of the energy, and then re-radiated. But it may happen that part of the energy of the incident quantum is retained by the molecule as vibrational energy, when the quantum emitted $h(\nu - \nu_k)$ has the frequency

$\nu - \nu_k$. Therefore molecules in the vibrational ground state can only subtract energy from the incident quanta. A quantum that collides with a molecule in a fundamental vibrational level may, however, be radiated with the vibrational quantum added to it, so that the scattered light contains frequencies $\nu + \nu_k$. Both possibilities are illustrated in Fig. 7.3.1. The intensity of scattering at $\nu + \nu_k$ is proportional to the Boltzmann factor exp $[-h\nu_k/kT]$ of the vibrational level and therefore is less, often considerably less, than that from the ground state.

The condition that the normal frequency ν_k shall appear in the Raman effect is, according to eqn. (3), that the quantity $(\partial\alpha/\partial Q_k)_0$ must be different from zero. Thus a normal vibration is active in the Raman spectrum if, and only if, *the nuclear displacements change the polarisability of the molecule*. As this criterion is completely different from that controlling infrared activity we can expect that some

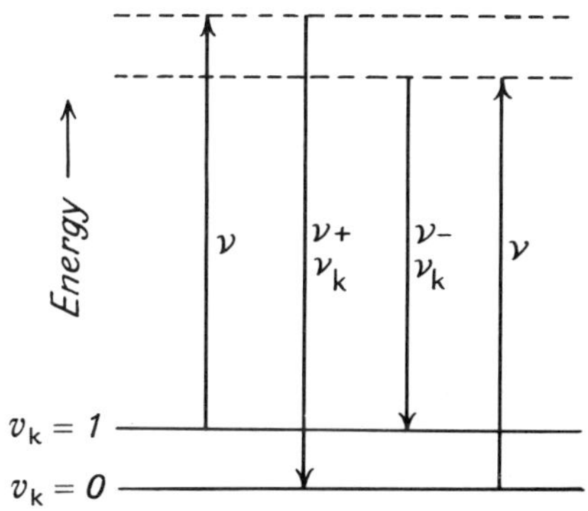

Fig. 7.3.1. Origin of the vibrational Raman effect.

vibrations will occur in the Raman effect but not in the infrared spectrum, and *vice-versa.* It will be our next task to discover which of the normal vibrations of a molecule are associated with a polarisability change, and hence to formulate the *selection rules* for the vibrational Raman effect.

Selection Rules (Classical Theory)

Let us consider the effect of the normal vibrations of a molecule upon its polarisability ellipsoid. Take as an example the molecule H_2O. For the undistorted molecule the directions of the axes of the polarisability ellipsoid are completely determined by symmetry: one of the axes of the ellipsoid must coincide with the C_2 axis, whilst the remaining two are perpendicular to this axis and lie one in each of the two symmetry planes denoted σ_v and σ_v' in Fig. 6.3.1. Let the principal axis z' coincide with C_2, let y' be perpendicular to C_2 in the plane of the molecule, and let x' be normal to the plane of the molecule. The x', y', and z' axes of the ellipsoid then coincide, *when the nuclei are in their mean positions*, with the general x, y, and z axes fixed in the molecule.

Consider the effect of the vibration ν_1. The nuclei are displaced in the directions indicated by the normal coordinate Q_1 (Fig. 6.3.1). Extension of the bonds changes their polarisability, so that the axial *lengths*, but not the orientation, of the ellipsoid change with the phase of the vibration, corresponding to the variation of $\alpha_{x'x'}$, $\alpha_{y'y'}$, and $\alpha_{z'z'}$: the polarisability ellipsoid 'breathes' with the same frequency as the normal vibration. Similar remarks apply to the second totally symmetrical vibration ν_2. The non-totally symmetrical vibration ν_3 has a different effect, however. The length of the axes now remains unchanged during the vibration because the polarisability change due to the extension of the first OH bond is exactly counter-balanced by the contraction of the second. But the axes y' and z' change *direction* owing to the unsymmetrical deformation of the molecule: the polarisability ellipsoid rocks to and fro about the x' axis, which remains fixed in direction because the nuclear displacements are entirely in the yz plane. This behaviour is shown diagrammatically in Fig. 7.3.2.

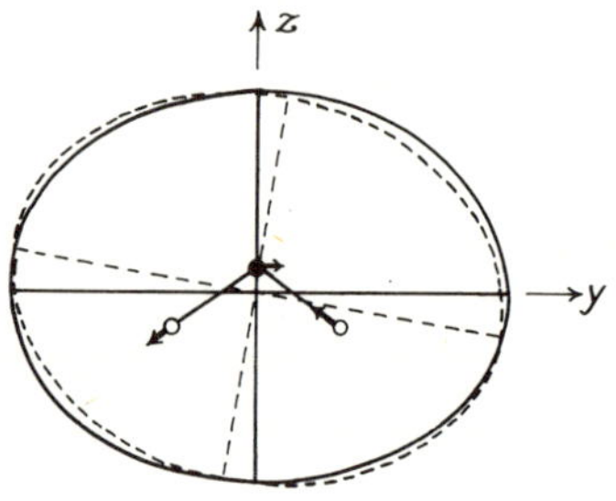

Fig. 7.3.2. Polarisability change in the vibration ν_3 (B_2) of H_2O. The full line is a principal section of the ellipsoid for the undistorted molecule: the dotted line indicates the orientation at a turning-point of the vibration. The principal axes y' and z' coincide with the y and z axes only when the nuclei are in their equilibrium positions.

Let us enumerate the polarisability changes for H_2O. We shall refer everything to the set of axes, x, y, and z fixed in the molecule; that is, to axes which do not change direction when the molecule is distorted by vibration. When the nuclei are at equilibrium, and likewise when they are displaced by the totally symmetrical vibrations ν_1 and ν_2, the axes x', y', and z' of the ellipsoid coincide with x, y, and z: hence the off-diagonal elements in the matrix (7.2.4*a*) are all equal to zero. During the vibrations ν_1 and ν_2 all axes of the ellipsoid change in length (but not in direction), hence $(\partial\alpha_{xx}/\partial Q_k)_0$, $(\partial\alpha_{yy}/\partial Q_k)_0$, and $(\partial\alpha_{zz}/\partial Q_k)_0$ are different from zero for $k = 1$ or 2. But the off-diagonal elements α_{xy}, α_{xz}, and α_{yz} are zero at all phases of the vibrations ν_1 and ν_2; therefore $(\partial\alpha_{xy}/\partial Q_k)_0$, $(\partial\alpha_{xz}/\partial Q_k)_0$, and $(\partial\alpha_{yz}/\partial Q_k)_0$ are equal to zero for the two totally symmetrical modes. The vibration ν_3 leaves the principal axes of the ellipsoid unchanged in length, and thus

$$\left(\frac{\partial \alpha_{xx}}{\partial Q_3}\right)_0 = \left(\frac{\partial \alpha_{yy}}{\partial Q_3}\right)_0 = \left(\frac{\partial \alpha_{zz}}{\partial Q_3}\right)_0 = 0.$$

Further, the x' axis of the ellipsoid always coincides with x, so that $\alpha_{xy} = \alpha_{xz} = 0$, and therefore $(\partial\alpha_{xy}/\partial Q_3)_0 = (\partial\alpha_{xz}/\partial Q_3)_0 = 0$. But the y' and z' axes of the ellipsoid fail to coincide with y and z unless the nuclei are in their equilibrium positions; hence $(\partial\alpha_{yz}/\partial Q_3)_0$ is different from zero, even though $\alpha_{yz} = 0$ for the undistorted molecule.

Take as another example the symmetrical linear triatomic molecule CO_2. Its normal vibrations are shown in Fig 6.1.2 (for the longitudinal vibrations) and Fig. 6.3.3 (bending vibration). As above, we let the x', y', and z' axes of the ellipsoid coincide with the set of translational axes, x, y, and z when the nuclei are in their equilibrium positions. The normal vibrations ν_1, ν_2, and ν_3 all leave the *orientation* of the principal axes of polarisability unchanged: therefore $\alpha_{xy} = \alpha_{xz} = \alpha_{yz} = 0$, hence $(\partial\alpha_{xy}/\partial Q_k)_0 = (\partial\alpha_{xz}/\partial Q_k)_0 = (\partial\alpha_{yz}/\partial Q_k)_0 = 0$ for $k = 1, 2$, and 3. The non-totally symmetrical vibrations ν_2 and ν_3 do not change the length of the principal axes of the ellipsoid (this is always true for non-totally symmetrical vibrations), and therefore $(\partial\alpha_{xx}/\partial Q_k)_0$, $(\partial\alpha_{yy}/\partial Q_k)_0$, and $(\partial\alpha_{zz}/\partial Q_k)_0$ are also zero for $k = 2$ or 3. For the non-totally symmetrical vibrations, therefore, no one of the six independent elements of the polarisability matrix (7.2.4*a*) changes with the nuclear displacements. In the totally symmetrical mode ν_1, the axial lengths change; thus $(\partial\alpha_{xx}/\partial Q_1)_0$, $(\partial\alpha_{yy}/\partial Q_1)_0$, and $(\partial\alpha_{zz}/\partial Q_1)_0$ are different from zero.

According to the classical theory given earlier in this section, a vibration ν_k is expected to appear in the Raman effect if $(\partial\alpha/\partial Q_k)_0 \neq 0$. In terms of the matrix representation of α, $(\partial\alpha/\partial Q_k)_0$ is different from zero if *at least one of the components,*

$$\begin{matrix} \left(\frac{\partial \alpha_{xx}}{\partial Q_k}\right)_0 & \left(\frac{\partial \alpha_{xy}}{\partial Q_k}\right)_0 & \left(\frac{\partial \alpha_{xz}}{\partial Q_k}\right)_0 \\ & \left(\frac{\partial \alpha_{yy}}{\partial Q_k}\right)_0 & \left(\frac{\partial \alpha_{yz}}{\partial Q_k}\right)_0 \\ & & \left(\frac{\partial \alpha_{zz}}{\partial Q_k}\right)_0 \end{matrix} \qquad (4)$$

differs from zero. It follows that all three normal vibrations of H_2O may appear in the Raman effect; for in the vibration ν_3 one of the elements of $(\partial\alpha/\partial Q_3)_0$ is non-zero, whilst for ν_2 and ν_1 three such elements differ from zero. For CO_2, the vibration ν_1 can (and does) appear, but ν_2 and ν_3 are forbidden. It is always possible to discover if a normal vibration is active by a detailed argument of this sort, but much easier methods are available. These we shall discuss in the remaining part of this section.

Selection Rules (Quantum Theory)

The quantum theoretical treatment of the Raman effect is too complex to consider here in any detail. It leads, however, to a symmetry selection rule, analogous to the symmetry selection rule for infrared absorption (Section 6.7), which is both simple in its outward form and easy to use. The rule states that a vibration ν_k is *active in the Raman effect if, and only if, at least one of the triple products,*

$$\begin{array}{lll} \Gamma(\psi_V').\Gamma(\alpha_{xx}).\Gamma(\psi_V'') & \Gamma(\psi_V').\Gamma(\alpha_{xy}).\Gamma(\psi_V'') & \Gamma(\psi_V').\Gamma(\alpha_{xz}).\Gamma(\psi_V'') \\ & \Gamma(\psi_V').\Gamma(\alpha_{yy}).\Gamma(\psi_V'') & \Gamma(\psi_V').\Gamma(\alpha_{yz}).\Gamma(\psi_V'') \\ & & \Gamma(\psi_V').\Gamma(\alpha_{zz}).\Gamma(\psi_V'') \end{array} \quad (5)$$

is totally symmetrical. Here the ψ_V' and ψ_V'' are the vibrational wavefunctions of the upper and lower states connected by a transition, and the $\alpha_{xx}, \alpha_{xy}, \ldots$ are the six independent matrix elements of the polarisability α (eqn. 7.2.4*a*).

To use this rule we must know the symmetry species of the various elements of the polarisability matrix. Suppose that an electric field is applied parallel to the x-coordinate of a molecule: then, since the field has no component in the y or z direction, we have for the components of the induced moment (eqns. 7.2.1–3)

$$\mu_x = \alpha_{xx}\mathscr{E}_x, \qquad \mu_y = \alpha_{xy}\mathscr{E}_x, \qquad \mu_z = \alpha_{xz}\mathscr{E}_x. \quad (6)$$

$\mathscr{E}_x$, being a vector quantity, has the same symmetry species as the coordinate x. The components of the induced dipole moment, likewise, have direction; so that μ_x has the same symmetry species as x, μ_y as y, and μ_z as z. Suppose that the coordinate x changes sign under some symmetry operation that can be performed on a molecule; then $\mathscr{E}_x$ and μ_x both change sign when the operation is applied and thus, according to (6), α_{xx} must remain unchanged. If the operation which changes the sign of x leaves y and z unchanged, by the same argument α_{xy} and α_{xz} change sign when the operation is carried out. Put another way, α_{xx} has the same species as the product $x \times x = x^2$, $\alpha_{xy}(=\alpha_{yx})$ as $x \times y$, and $\alpha_{xz}(=\alpha_{zx})$ as $x \times z$. By broadening the basis of the argument it is easily seen that the three remaining elements of the polarisability, α_{yy}, $\alpha_{yz}(=\alpha_{zy})$, and α_{zz}, have the species of products y^2, $y \times z$ and z^2, respectively. The species of the products $x^2, xy, \ldots$ are listed in the collected character tables in Appendix I.

As an example, consider a molecule of point group C_{2v} (Table 1.4). The symmetry species of the coordinate axes x, y, and z are B_1, B_2, and A_1, respectively. From the character table we see that the species of the product xy is A_2, that of yz is B_2, and that of xz is B_1; whilst x^2, y^2, and z^2 all belong to the totally symmetrical species A_1. Suppose that we wish to know if a transition from the zero-point vibrational level to any one of the fundamental levels is allowed in the Raman effect. The upper state vibrational wavefunction, ψ_V', of a fundamental level ($v_k = 1$) has the same species as the normal coordinate Q_k of the vibration excited by the transition and so may belong to any of the four possible symmetry species A_1, A_2, B_1

and B_2. The ground state wavefunction ψ_V'' is totally symmetrical (this is always true for the zero-point vibrational state) and thus has the species A_1. For a totally symmetrical normal vibration Q_k, and hence ψ_V', has the species A_1; in this case the triple products $\Gamma(\psi_V').\Gamma(\alpha_{xx}).\Gamma(\psi_V'')$, $\Gamma(\psi_V').\Gamma(\alpha_{yy}).\Gamma(\psi_V'')$, and $\Gamma(\psi_V').\Gamma(\alpha_{zz}).\Gamma(\psi_V'')$ are totally symmetrical since all have the species $A_1 \times A_1 \times A_1 = A_1$. The products involving α_{xy}, α_{xz}, and α_{yz} are not totally symmetrical however, as the reader should check for himself. When ψ_V' has the species A_2, the only triple product that is totally symmetrical is $\Gamma(\psi_V').\Gamma(\alpha_{xy}).\Gamma(\psi_V'') = A_2 \times A_2 \times A_1 = A_1$. Likewise, when ψ_V' has the species B_1 and B_2, the products

$$\Gamma(\psi_V').\Gamma(\alpha_{xz}).\Gamma(\psi_V'') \text{ and } \Gamma(\psi_V').\Gamma(\alpha_{yz}).\Gamma(\psi_V''),$$

respectively (but no others), are totally symmetrical. Hence the symmetry selection rule permits all the fundamentals of a C_{2v} molecule to appear in the Raman effect. This general argument, of course, encompasses the conclusions obtained above for the particular example of H_2O.

Take as another illustration a molecule in the group $D_{\infty h}$. CO_2 would be a representative molecule. The necessary symmetry species can be read from Table I.16. We see directly that α_{zz}, α_{xz}, α_{yz}, and α_{xy} have the species Σ_g^+, Π_g, Π_g, and Δ_g respectively. Owing to the degeneracy, only $\alpha_{xx} + \alpha_{yy}$ and $\alpha_{xx} - \alpha_{yy}$ actually have a definite species, namely Σ_g^+ and Δ_g, but for practical purposes this amounts to the same thing as letting both α_{xx} and α_{yy} belong to each of the species Σ_g^+ and Δ_g. Now consider transitions from the zero-point level, Σ_g^+, to fundamental levels of the vibrational manifold. Transitions to fundamental levels of species Σ_g^+, Π_g, and Δ_g only will be allowed, for these are the species which contain elements of the polarisability matrix. For instance, if the species of the upper state wavefunction is Π_g, then the triple products $\Gamma(\psi_V').\Gamma(\alpha_{xz}).\Gamma(\psi_V'')$ and $\Gamma(\psi_V').\Gamma(\alpha_{yz}).\Gamma(\psi_V'')$ are $\Pi_g \times \Pi_g \times \Sigma_g^+ = \Sigma_g^+ + \Sigma_g^- + \Delta_g$, which contains the totally symmetrical species, Σ_g^+. (Physically this corresponds to $(\partial\alpha_{xz}/\partial Q_k)_0$ and $(\partial\alpha_{yz}/\partial Q_k)_0$ being different from zero, where Q_k is the normal coordinate of the Π_g vibration). Conversely, if ψ_V' belongs to any one of the species Σ_u^+, Σ_g^-, Σ_u^-, Π_u, . . . transitions from the ground state to this state are forbidden in the Raman effect because no elements of the polarisability occur in these species: thus it is impossible for any of the products (5) to be totally symmetrical.

For the particular example of CO_2, the normal vibrations are $\nu_1(\Sigma_g^+)$, $\nu_2(\Pi_u)$, and $\nu_3(\Sigma_u^+)$; of them, the first is allowed in the Raman effect while the second and third are forbidden. This is precisely the conclusion already reached on the basis of classical theory.

We have here the germ of a useful general rule. If a molecule is centrosymmetrical, then x, y, and z must belong to a u species of the point group, whilst the products x^2, xy, . . . belong to g species. Thus the active *infrared* fundamentals are those whose normal coordinates belong to a u species, whereas the allowed *Raman* fundamentals have normal coordinates that belong to one of the g species. Consequently, if a molecule has a centre of symmetry, no fundamental frequency can occur in both the infrared and Raman spectrum. Conversely, when we can discover no coincidences among the fundamental frequencies in the two spectra, we have strong evidence that the molecule concerned is centrosymmetrical.

7.4 Depolarisation of Raman Lines

An important property observable in the vibrational Raman effect is the *polarisation* of the scattered light. As with the Rayleigh scattering Raman lines are in general found to be partially polarised, irrespective of whether the incident light is polarised or not.

Let us review briefly some of the conclusions reached in Section 7.1 in regard to the polarisation of the Rayleigh scattering. For spherically symmetrical molecules (the adjective 'spherical' refers to the symmetry of the polarisability ellipsoid) the scattering is quantitatively polarised in the plane perpendicular to the incident light. To understand this one recalls that the incident quantum can only produce a dipole moment perpendicular to its direction of motion; then the dipole is necessarily in a plane perpendicular to the incident beam and any light scattered at 90° to this beam will be completely polarised. For less symmetrical molecules, however, it is not expected that the induced dipole moment and the electric field will be oriented parallel. One anticipates some degree of depolarisation in these circumstances.

As to the Raman scattering, we consider the symmetry of the polarisability change, $(\partial\alpha/\partial Q_k)_0$, instead of the symmetry of the total polarisability. For this purpose the vibrations can be divided into the two classes of *totally symmetrical* and *non-totally symmetrical* vibrations. Totally symmetrical vibrations maintain the full symmetry of the molecule. Therefore the principal axes of the polarisability ellipsoid remain fixed in direction at every stage of the vibration. The lengths of the axes of the ellipsoid change during the vibration, because the polarisability of the molecule in the compressed phase is different from that in the extended phase. Thus the diagonal elements of the polarisability, α_{xx}, α_{yy}, and α_{zz} referred to a set of translational axes x, y, and z behave in the manner represented in curve (a), Fig. 7.4.1.

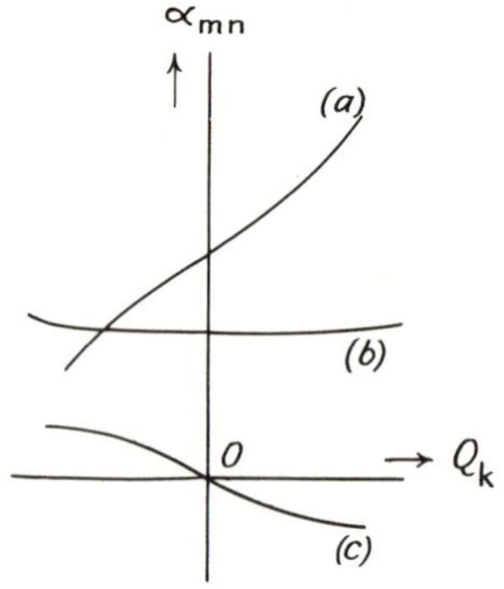

Fig. 7.4.1. Possible behavour of the components of the polarisability during vibration.

The derivatives $(\partial\alpha_{xx}/\partial Q_k)_0$, $(\partial\alpha_{yy}/\partial Q_k)_0$, $(\partial\alpha_{zz}/\partial Q_k)_0$ can all have values different from zero for totally symmetrical vibrations. If symmetry requires these three quantities to be equal (as in CCl_4), the Raman lines are completely polarised. Otherwise the derivatives have different values leading to an arbitrary, but often small, degree of depolarisation.

In a non-totally symmetrical vibration the axial lengths of the polarisability ellipsoid are constant. Therefore the derivatives of the diagonal elements of the polarisability are zero, as shown in curve (*b*), and are of no interest. When a non-totally symmetrical vibration occurs in the Raman effect it does so by virtue of the derivative of an off-diagonal element of the polarisability. Curve (*c*), Fig. 7.4.1, represents the possible behaviour of an off-diagonal element of α in the case that the axes xyz coincide with the principal axes of the ellipsoid when the nuclei are in their mean positions. Now, it can be shown that the Raman scattering due only to an off-diagonal component of the polarisability must be *completely depolarised*. The precise intensity ratio that corresponds to complete depolarisation depends upon the type of incident radiation and the geometry of the system used to record the scattering. With the arrangement illustrated in Fig. 7.1.1 the maximum value of the depolarisation factor, $\rho = I_\perp/I_\parallel$, is $\frac{3}{4}$ for incident polarised light and $\frac{6}{7}$ for incident unpolarised (natural) light. These values are attained by the Raman scattering associated with any non-totally symmetrical vibration whatsoever. Totally symmetrical vibrations may have any value of ρ less than 3/4 (or 6/7), except that the totally symmetrical vibrations of a tetrahedral or octahedral molecule must have $\rho = 0$.

Measurement of the depolarisation factors can thus give information about the symmetry of the normal vibrations involved in the scattering. Unfortunately a depolarisation factor is difficult to measure accurately; therefore it may not always be easy to distinguish a totally symmetrical vibration which happens to possess a high value of ρ from a non-totally symmetrical vibration for which ρ has necessarily the maximum value.

7.5 The Rotational Raman Effect

When a molecule rotates, the polarisability ellipsoid can be thought of as turning with the molecule. Consequently for non-isotropic molecules the polarisability, relative to a set of space-fixed axes, varies during the period of a rotation. In this sense the rotation of molecules is associated with a polarisability change, and hence the pure rotation frequency can be expected to occur in the Raman effect.

To observe a rotational Raman effect it is necessary to analyse the light scattered by gases, for the liquid state is not suitable for studying rotational transitions. This is a serious practical limitation because the Raman scattering from a gas or vapour is extremely weak, owing to the low concentration of molecules; and its observation meets with experimental difficulties that have only recently been overcome. The vapours of isotropic molecules, for which the polarisability ellipsoid is a sphere, give no rotational Raman spectrum because the polarisability along a fixed direction remains constant irrespective of the orientation.

Consider the case of a linear molecule. The polarisability ellipsoid is an ellipsoid of revolution about the internuclear axis (z). Rotation about an axis perpendicular to (z) brings the polarisability ellipsoid into a position indistinguishable from the original one twice in a single revolution. We may contrast this behaviour with that of a dipole moment in the molecule which returns to an indistinguishable position only once in a revolution. From a purely classical point of view this means we can expect the two-fold rotational frequency to appear in the Raman effect; quantum mechanically this corresponds to the selection rule $\Delta J = \pm 2$. An alternative approach is to consider Raman scattering to be a two step, or *two photon*, process as might be inferred from Fig. 7.3.1. For a linear molecule the rotational selection rule for each step is $\Delta J = \pm 1$, so that the possible end results for two such consecutive steps are

$$\Delta J = 0, \pm 2. \tag{1}$$

This is, indeed, the pure rotational selection rule for a linear molecule as a more rigorous quantum mechanical treatment confirms.

The selection rules for a symmetric top molecule may be similarly deduced. Again the polarisability ellipsoid is an ellipsoid of revolution, this time about the symmetry axis of the molecule. Rotation about this axis causes no change in polarisability and thus

$$\Delta K = 0. \tag{2}$$

The rotational selection rule appropriate to each step of the scattering process is in this case $\Delta J = 0, \pm 1$ so that the possible overall changes in J are

$$\Delta J = 0, \pm 1, \pm 2. \tag{3}$$

The benzene molecule provides a particular example. Because benzene lacks a permanent electric dipole moment its pure rotation spectrum cannot be studied in absorption; it can however be studied in the Raman effect. Since the molecule is an oblate symmetric top the appropriate selection rules are those of eqns. (2) and (3). $\Delta J = 0$ merely represents the Rayleigh scattering, and is of no further interest here.

The energy levels, assuming that benzene is a rigid symmetric top, are given by eqn. (5.2.15). The rotational Raman lines appear displaced from the exciting lines by a frequency $\Delta\nu$ given by,

for $\Delta J = \pm 1$,

$$|h\Delta\nu| = E(J+1, K) - E(J,K) = h(2BJ + 2B),\ J = 0, 1, \tag{4}$$

and for $\Delta J = \pm 2$,

$$|h\Delta\nu| = E(J+2, K) - E(J, K) = h(4BJ + 6B), J = 1, 2. \tag{5}$$

(4) is the equation of an evenly-spaced series of lines whose separation is $2B$, whilst (5) is an expression for a similar series wherein the line separation is $4B$. The lines of the second series therefore coincide with alternate members of the first series. As a large number of rotational levels is populated at ordinary temperature the series run to either side from the exciting line.

For benzene it happens that the $\Delta J = \pm 1$ transitions are exceedingly weak and that all the important lines belong to the series with $\Delta J = \pm 2$.* The line separation, $4B$, can be used to calculate the moment of inertia, I_b, perpendicular to the top axis. After applying a small correction for centrifugal distortion the results are:† for C_6H_6, $I_b^0 = 88{\cdot}91 \pm 0{\cdot}02$ uÅ^2 and for C_6D_6, $I_b^0 = 107{\cdot}50 \pm 0{\cdot}05$ uÅ^2.

On the assumption that the hexagonal structure of benzene is perfectly regular, the principal moment in the plane of the molecule is given by the expression

$$I_b = 3\{r(\text{C—C})^2 m_c + [r(\text{C—C}) + r(\text{C—H})]^2 m_H\}. \tag{6}$$

Taking the C—C and C—H bond lengths as identical in C_6H_6 and C_6D_6 one can introduce the observed moments into (4) and then solve the resulting pair of simultaneous equations for r(C—C) and r(C—H). The result is r_0(C—C) $= 1{\cdot}397_3 \pm 0{\cdot}001$, and r_0(C—H) $= 1{\cdot}084 \pm 0{\cdot}006$ Å. Some uncertainty should also be attributed to the residual, zero-point vibrational amplitude of the nuclei, but his has not been evaluated. The results are compared later (section 12.5) with the dimensions of the benzene molecule in the crystalline state.

Fig. 7.5.1 shows the S-branch of the pure rotational Raman spectrum of air. This spectrum was excited with the 5145 Å line of an argon ion laser and the assignment of the peaks to O_2 and N_2 is indicated in the figure. The presence of equivalent nuclei in these molecules produces striking effects in this spectrum. For O_2 (ground state $^3\Sigma_g^-$), in which the nuclei have zero spins, the even numbered rotational levels are entirely missing so the basic branch spacing is $8B$ and the first transition is displaced $10B$ from the exciting line (ν_0). In N_2 (ground state $^1\Sigma_g^+$) the nuclear spin quantum numbers are unity and there is a 2:1 statistical factor in the intensities in favour of the even numbered levels. This intensity alternation can be quite easily followed in Fig. 7.5.1. The branch spacing in N_2 is the normal $4B$ and the first transition has a separation $6B$ from ν_0.

The Raman scattering from gases contains, in addition to the pure rotation spectrum, a vibration-rotation spectrum which is due to the rotation of the polarisability change caused by the vibrations. Such spectra are analogous to vibration-rotation bands in the infrared region, but their structure has been studied in only a few examples.

* The lines for $\Delta J = +2$ constitute the S-branch; those for $\Delta J = -2$ make up the O-branch.
† Stoicheff, *Can. J. Phys.*, 1954, **32**, 339; Langseth and Stoicheff, *ibid.*, 1956, **34**, 350.

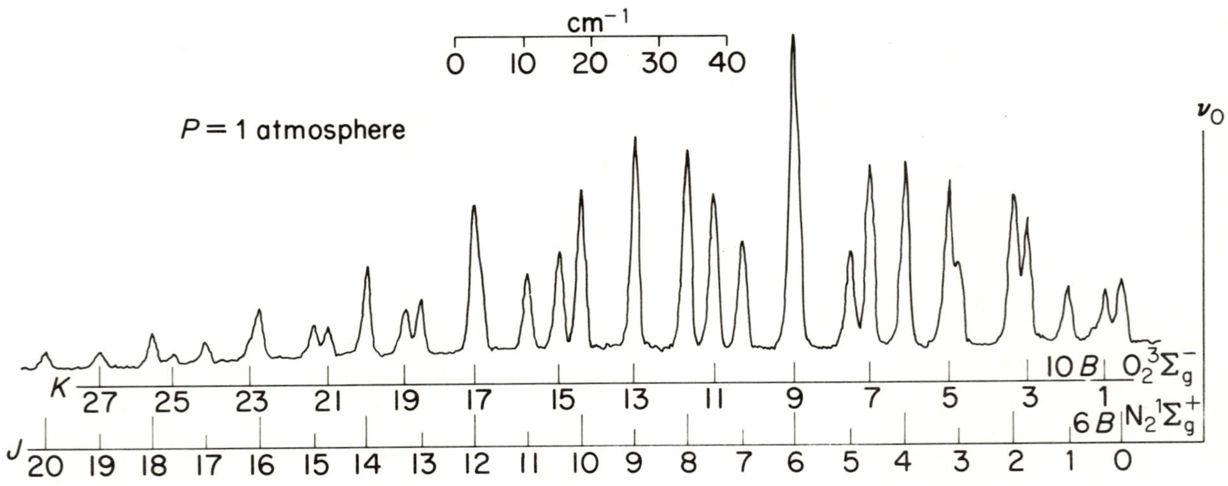

Fig. 7.5.1. The pure rotational Raman spectrum of air (A.P. Lane). *J* and *K* are the rotational quantum numbers for N_2 and O_2 respectively.

7.6 A Vibrational Analysis: Acetylene

The structure of acetylene is so well known that we shall use it directly in the interpretation of the spectrum. It should be emphasised, however, that the structure need not be taken for granted: that is to say, beginning with the observed spectrum, it would be possible swiftly to arrive at the equilibrium configuration of the molecule.

Let us enumerate the symmetry elements possessed by acetylene. They are illustrated in Fig. 7.6.1. The principal axis is, of course, the C_∞ axis, coincident with the line joining the nuclei. Perpendicular to C_∞ there are an infinite number of C_2 axes contained in a 'horizontal' symmetry plane, σ_h; while an infinite number of 'vertical' planes, σ_v, contain the axis C_∞. An improper axis S_∞ coincides with C_∞. Finally a

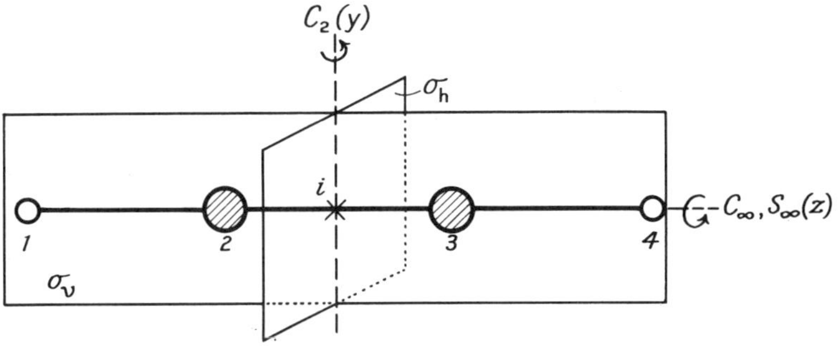

Fig. 7.6.1. Symmetry elements of acetylene.

cartesian set of axes can be fixed in the molecules so that the z axis is along C_∞, when the origin of the cartesian system is a centre of symmetry i. The operations that can be applied are, therefore: I, $2C_\phi$ (rotation about C_∞ by $\pm\,\phi$), $2C_\phi^2, \ldots,$ $\infty\,\sigma_v$, i, $2S_\phi = 2\sigma_h C_\phi, \ldots,$ and $\infty\, C_2$. These operations comprise the group $D_{\infty h}$.

We shall use internal coordinates to describe the distortion of the molecule by vibration. Seven $(3N-5)$ such coordinates are needed. The C—H bond extensions, r, provide two coordinates, while the C—C bond extension, r', gives a third. The $\angle$HCC bond angle deformations, θ, give the remaining four coordinates, since there are two such coordinates in each of two perpendicular planes. Table 7.6.1 contains, besides the character table of the group, the characters of the representation of the internal coordinates. The method by which the representations are obtained has been described in section 6.3. Take as an example the characters, $\chi(\theta)$, of the angular deformations, which are the least readily obtained. As there are four coordinates in the set, we have $\chi(I) = 4$. The operations i, S_ϕ, $S_{-\phi}$, and C_2 permute the coordinates, consequently $\chi(i)$, $\chi(S_\phi)$, and $\chi(C_2)$ are all zero. C_ϕ and $C_{-\phi}$ mix the coordinates, so that each degenerate pair gives the character $2\cos\phi$ (see section 6.3), whence $\chi(C_\phi) = 4\cos\phi$; whilst, for similar reasons, $\chi(\sigma_v) = 0$. The characters so obtained are collected in the final row of Table 7.6.1. The determination of the characters,

Table 7.6.1. Characters of the Internal Coordinates of C_2H_2

$D_{\infty h}$	I	$2C_\phi$	$\ldots$	$\infty\sigma_v$	i	$2S_\phi$	$\ldots$	∞C_2		
Σ_g^+	1	1	$\ldots$	1	1	1	$\ldots$	1		(x^2+y^2, z^2)
Σ_u^+	1	1	$\ldots$	1	-1	-1	$\ldots$	-1	z	
Σ_g^-	1	1	$\ldots$	-1	1	1	$\ldots$	-1		
Σ_u^-	1	1	$\ldots$	-1	-1	-1	$\ldots$	1		
Π_g	2	$2\cos\phi$	$\ldots$	0	2	$-2\cos\phi$	$\ldots$	0		(R_x, R_y) (xz, yz)
Π_u	2	$2\cos\phi$	$\ldots$	0	-2	$2\cos\phi$	$\ldots$	0	(x, y)	
	$\ldots$	$\ldots$	$\ldots$	$\ldots$	$\ldots$	$\ldots$	$\ldots$			
$\chi(r)$	2	2	$\ldots$	2	0	0	$\ldots$	0		
$\chi(r')$	1	1	$\ldots$	1	1	1	$\ldots$	1		
$\chi(\theta)$	4	$4\cos\phi$	$\ldots$	0	0	0	$\ldots$	0		

$\chi(r)$ and $\chi(r')$, of the C—H and C—C bond extension, respectively, will offer no serious difficulty to the reader who has studied section 6.3. The characters $\chi(r')$ are evidently those of an irreducible representation, namely, the totally symmetrical representation Σ_g^+. $\chi(r)$ and $\chi(\theta)$ form reducible representations but are easily broken down into their components, $\Sigma_g^+ + \Sigma_u^+$ and $\Pi_g + \Pi_u$, respectively. Thus the representation of the seven internal coordinates has the structure

$$\Gamma = 2\Sigma_g^+ + \Sigma_u^+ + \Pi_g + \Pi_u \tag{1}$$

As the normal coordinates form a representation having the same structure as the internal coordinates, Γ is also the representation which has the vibrational motions as its basis.

As an exercise the reader is recommended to write down the structure of the representation formed by the cartesian displacement coordinates of acetylene. After subtracting the structure of the representation of translation and rotation, which can be read directly from Table 7.6.1, he will obtain independently the structure, Γ, of the representation formed by the normal coordinates.

The qualitative form of the normal modes now becomes clear. There are two totally symmetrical modes, species Σ_g^+, involving the bond extensions r and r'; the angular deformation θ does not enter, so that these vibrations are pure stretching modes. Since the natural frequencies for C—H and C—C bond stretching are expected to be rather different, one of the totally symmetrical modes will be essentially a C—H stretching mode (with some C—C bond stretching character), while the second will be largely a C—C stretching mode (with some C—H character). The third stretching mode, Σ_u^+, the only vibration of this species, has the coordinate r as its basis and thus is a pure C—H stretching mode. The remaining four frequencies occur as two doubly degenerate pairs, one in each of the symmetry species Π_g

and Π_u: the coordinate involved is θ, hence they are pure angular deformation, or bending, modes. A schematic diagram of the normal modes, based on the foregoing analysis, is given in Fig. 7.6.2. Note that the vibrations ν_1 and ν_3 differ in that they are essentially in- and out-of-phase combinations, respectively, of the C—H bond extensions: likewise ν_4 and ν_5 are oppositely phased combinations of the ∠HCC deformation, θ.

The selection rules for the fundamental vibrations follow immediately from Table 7.6.1. The infrared active fundamentals have the same symmetry species as the cartesian axes $x, y\,(\Pi_u)$ and $z\,(\Sigma_u^+)$; therefore only the fundamental frequencies $\nu_5(\Pi_u)$ and $\nu_3(\Sigma_u^+)$ are expected to appear in absorption. Further, the vibration ν_5 involves a dipole moment change perpendicular to the internuclear axis z, so that the infrared vibration-rotation band will show P, Q, and R-branches; whereas in the ν_3 band,

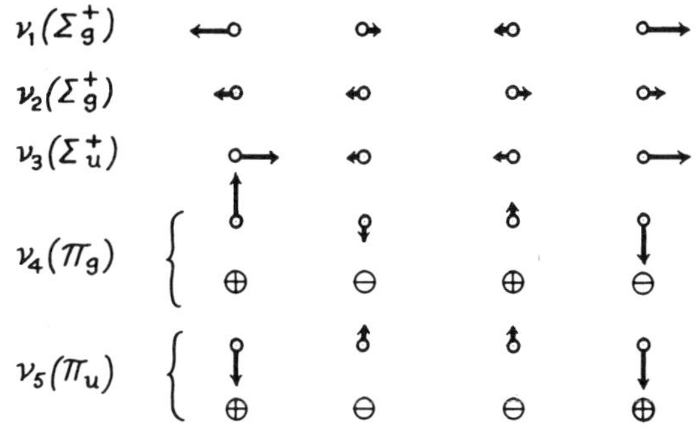

Fig. 7.6.2. Normal modes of acetylene.

where the moment change is parallel to z, the Q-branch will be missing. The Raman active fundamentals correspond to the species of products of coordinates, $x^2, xy, \ldots$ Thus the totally symmetrical vibrations ν_1 and ν_2 and the non-totally symmetrical vibration $\nu_4(\Pi_g)$ can be expected as fundamentals in the vibrational Raman effect. We see that all five fundamentals should occur, either in the infrared spectrum or in the Raman effect, but that no fundamental can appear in both spectra. All this information is summarised in Table 7.6.2. The non-occurrence of fundamentals in both infrared and Raman spectra is characteristic of a molecule that possesses a centre of symmetry (section 7.3).

The spectrum of gaseous acetylene below 3500 cm^{-1} is presented in Table 7.6.3. The Raman effect shows two strong bands,* centred at 1974 and 3374 cm^{-1}, which do not occur in the infrared spectrum. The displacements are expected to correspond to the totally symmetrical fundamentals ν_2 and ν_1, respectively; that is, to the essentially C—C and C—H stretching modes. It would be possible to check these assignments by measuring the degree of depolarisation of the Raman bands—the bands must be partially polarised if they represent totally symmetrical fundamentals—but this has not been done. The third Raman active fundamental $\nu_4(\Pi_g)$ has not been definitely observed; presumably it is intrinsically very weak, for the reason that it is

* Glocker and Morrell, *J. Chem. Phys.*, 1936, **4**, 15.

Table 7.6.2. Fundamental Vibrations of C_2H_2

Fundamental vibration		Infrared spectrum		Raman spectrum	
	Type	*Activity*	*Fine structure*	*Activity*	*Depolarisation factor (for natural light)*
$\nu_1(\Sigma_g^+)$	C—H stretching	No	—	Yes	< 6/7
$\nu_2(\Sigma_g^+)$	C≡C stretching	No	—	Yes	< 6/7
$\nu_3(\Sigma_u^+)$	C—H stretching	Yes	*P, R*	No	
$\nu_4(\Pi_g)$	∠HCC bending	No	—	Yes	6/7
$\nu_5(\Pi_u)$	∠HCC bending	Yes	*P, Q, R*	No	

allowed only by off-diagonal components, α_{yz} and α_{xz}, of the polarisability.

The infrared spectrum of acetylene* shows two very strong bands, at 731 and 3283 cm^{-1}, which are naturally attributed to the active fundamentals $\nu_5(\Pi_u)$ and $\nu_3(\Sigma_u^+)$. This is consistent with the rotational structure, for a *Q*-branch is present in the first band but absent from the second. The absorption spectrum also contains overtone and combination bands, two of which are noted in the table while many more occur in the frequency range above 3500 cm^{-1}. The combination bands enable one to evaluate the Raman active fundamental frequencies (including the 'missing' fundamental ν_4) rather more accurately than existing observations of the Raman effect, and the frequencies listed in Table 7.6.3 are values actually obtained from infrared measurements only. Observed frequencies of C_2D_2 are tabulated alongside those of C_2H_2 in the table: their interpretation follows precisely the same lines.

Table 7.6.3. Spectrum of Gaseous C_2H_2 and C_2D_2

	C_2H_2		C_2D_2	
Assignment	*σ, cm^{-1}*	*Intensity* and fine structure*	*σ, cm^{-1}*	*Intensity* and fine structure*
ν_4	613·3	—	511·4	—
ν_5	730·7	*I, vs: PQR*	538·7	*I, vs: PQR*
$\nu_4 + \nu_5$	1328·1	*I, s: PR*	1044	*I, m: PR*
ν_2	1974·0	*R, vs.*	1764·7	*R, s.*
$\nu_2 + \nu_5$	2701	*I, m: PQR*	—	—
ν_3	3282·5	*I, vs: PR*	2439·1	*I, s: PR*
ν_1	3373·2	*R, s.*	2703·6	*R, s.*

**vs* = very strong, *s* = strong, *m* = medium.

* Bell and Nielsen, *J. Chem. Phys.*, 1950, **18**, 1382 (C_2H_2); Saksena, *ibid.*, 1952, **20**, 95; Talley and Nielsen, *ibid.*, 1954, **22**, 2030 (C_2D_2).

Analysis of the rotational structure of the infrared bands gives

$$B_0(C_2H_2) = 1{\cdot}1769, \quad \text{and} \quad B_0(C_2D_2) = 0{\cdot}8479\ \text{cm}^{-1}$$

for the rotational constant in the lowest, zero-point vibrational state. It is interesting that the pure rotation spectrum of C_2H_2, observed in the Raman effect,† also yields $B_0 = 1{\cdot}1769\ \text{cm}^{-1}$, in perfect agreement with the infrared datum. The correction for the zero-point amplitude of the nuclei amends these results considerably, the figures adopted by Saksena for the hypothetical state in which the nuclei are at rest being

$$B_e(C_2H_2) = 1{\cdot}1838, \quad \text{and} \quad B_e(C_2D_2) = 0{\cdot}8507_5\ \text{cm}^{-1}. \tag{2}$$

Using the corrected moments of inertia, however, one is entitled to assume that the bond lengths are equal in the isotopic molecules. The expression for the moment of inertia,

$$I_b = 2\{[r(\text{C—H}) + \tfrac{1}{2}r(\text{C—C})]^2 m_H + [\tfrac{1}{2}r(\text{C—C})]^2 m_C\} \tag{3}$$

can then be solved for the two unknowns r(C—H) and r(C—C), yielding

$$r_e(\text{C—H}) = 1{\cdot}063_7, \quad \text{and} \quad r_e(\text{C—C}) = 1{\cdot}2010\ \text{Å}. \tag{4}$$

Force-Constants

The simplest assumption in regard to the potential energy change when vibration distorts the molecule is that the contribution to the energy from the C—H and C—C bond extensions, r and r', and from the ∠HCC deformations θ is additive. The potential can then be written

$$2V = 2Fr^2 + F'r'^2 + 2f\theta^2, \tag{5}$$

wherein F is the force-constant of a C—H bond, F' that of a C—C bond, and f that corresponding to the change of angle between H—C and C—C. The expressions for the normal frequencies (derived in section 6.12) are given below.

$$\Sigma_g^+: \lambda_1 + \lambda_2 = (\mu_H + \mu_C)F + 2\mu_C F' \tag{6}$$

$$\lambda_1\lambda_2 = 2\mu_H\mu_C FF' \tag{7}$$

$$\Sigma_u^+: \quad \lambda_3 = (\mu_H + \mu_C)F \tag{8}$$

$$\Pi_g: \quad \lambda_4 = \{\mu_H\rho_{CH}{}^2 + \mu_C(\rho_{CH} + 2\rho_{CC})^2\}f \tag{9}$$

$$\Pi_u: \quad \lambda_5 = (\mu_H + \mu_C)\rho_{CH}{}^2 f. \tag{10}$$

† Callomon and Stoicheff, *Can. J. Phys.*, 1957, **35**, 373

Here $\lambda_k = 4\pi^2 c^2 \sigma_k^2$, the μ's are reciprocal masses of the atoms and the ρ's are the reciprocal equilibrium bond distances. Before we attempt to find the numerical solution to these equations, let us note that they give rise to a number of relations between the normal frequencies of C_2H_2 and C_2D_2 which are *independent of the force-constants*. Take as an example the expression (7). Simple division gives

$$\frac{\lambda_1^{(i)}\lambda_2^{(i)}}{\lambda_1\lambda_2} = \frac{m_H}{m_D} \tag{11}$$

and hence

$$\frac{\sigma_1^{(i)}\sigma_2^{(i)}}{\sigma_1\sigma_2} = \left(\frac{m_H}{m_D}\right)^{\frac{1}{2}}, \tag{12}$$

in which the superscript (i) indicates the normal frequency of the isotopic molecule. Likewise, from eqns. (8) and (10) we have

$$\frac{\sigma_3^{(i)}}{\sigma_3} = \frac{\sigma_5^{(i)}}{\sigma_5} = \left(\frac{m_H M^{(i)}}{m_D M}\right)^{\frac{1}{2}}, \tag{13}$$

where $M = 2(m_H + m_C)$. To deal with eqn. (9) some preliminary reorganisation is needed. The expression (3) for the moment of inertia can be rearranged to give

$$2I_b = \{\mu_H\rho_{CH}^2 + \mu_C(\rho_{CH} + 2\rho_{CC})^2\}/\mu_H\mu_C\rho_{CH}^2\rho_{CC}^2. \tag{14}$$

Thus

$$\lambda_4 = 2\mu_H\mu_C\rho_{CH}^2\rho_{CC}^2 I_b f, \tag{15}$$

and therefore

$$\frac{\sigma_4^{(i)}}{\sigma_4} = \left(\frac{m_H I_b^{(i)}}{m_D I_b}\right)^{\frac{1}{2}}. \tag{16}$$

Eqns. (12), (13) and (16) may be derived in a more elegant way directly from the Teller–Redlich product rule (eqn.(6.3.19)) *without any prior calculation of the normal frequencies*. The indices appropriate to the product rule for acetylene are set out in Table 7.6.4 and the reader should have no difficulty in verifying these.

Table 7.6.4

Symmetry Species	n_j (H)	n_j (C)	t	$r(x,y)$
Σ_g^+	1	1	0	0
Σ_u^+	1	1	1	0
π_g	1	1	0	1
π_u	1	1	1	0

The observed and calculated product rule ratios for C_2H_2 and C_2D_2 are compared in Table 7.6.5. The small discrepancies apparent in this table are doubtless due to the vibrations not being strictly harmonic.

Table 7.6.5. Product Rule Ratios for C_2H_2 and C_2D_2

	Σ_g^+	Σ_u^+	Π_g	Π_u
	$\dfrac{\sigma_1^{(i)}\sigma_2^{(i)}}{\sigma_1\sigma_2}$	$\dfrac{\sigma_3^{(i)}}{\sigma_3}$	$\dfrac{\sigma_4^{(i)}}{\sigma_4}$	$\dfrac{\sigma_5^{(i)}}{\sigma_5}$
Observed	0·717	0·743	0·843	0·737
Calculated	0·706	0·732	0·832	0·732

As to the calculation of force-constants, there are five equations for the normal frequencies but only three independent constants. Thus F can be obtained from the Σ_g^+ frequencies as well as from $\nu_3(\Sigma_u^+)$, whilst f can be calculated from ν_4 and from ν_5. The quality of the agreement then serves as an indication of the quality of the assumed form (5) of the potential energy. Values calculated from the normal frequencies (Table 7.6.3) and the equilibrium bond lengths (eqn. (4)) are given below.

Force-constant	F dyne cm^{-1} × 10^{-5}		F' dyne cm^{-1} × 10^{-5}	f dyne cm rad^{-1} × 10^{12}	
Normal frequency	σ_1,σ_2	σ_3	σ_1,σ_2	σ_4	σ_5
Isotope:					
C_2H_2	5·9	5·9	15·5	1·53	3·28
C_2D_2	6·0	6·0	15·7	1·53	3·31

Looking forst at the stretching force-constants, it is evident that the eqns. (6)–(8) are fulfilled quite well; hence the longitudinal part of the expression (5) gives a good representation of the potential energy of the extended molecule. As to the bending force-constant f, the consistency of the results for C_2H_2 and C_2D_2 is of no significance, for it follows from the product rule and thus is independent of the value of the force-constant. The poor agreement between the constants calculated from ν_4 and ν_5 then indicates the assumption that deformation of the first $\angle$HCC does not affect the force necessary to produce a deformation at the second angle is a mistaken one. What the results tell us is that twice as much energy is necessary to encompass the in-phase bending ν_5, which distorts the molecule into a *cis*-configuration, as to develop the equivalent out-of-phase deformation, ν_4, into a *trans*-structure. This is undoubtedly a manifestation of the generally greater stability of a *trans*-configuration in comparison with the corresponding *cis*.

BIBLIOGRAPHY

Herzberg, *Infrared and Raman Spectra of Polyatomic Molecules*, Van Nostrand, New York, 1945.
Stoicheff, *Advances in Spectroscopy, Vol. 1*, Interscience, New York and London, 1959.

Chapter 8

Nuclear resonance spectra

8.1 Nuclear Magnetic Moments

Nuclei with spin greater than zero invariably possess a magnetic moment μ; that is to say, the nucleus behaves like a small bar magnet whose axis coincides with the axis of spin. According to quantum electrodynamics the absolute magnetic moment μ_P of the *proton* is proportional to the product of the angular momentum $I = (\sqrt{3}/2)h/2\pi$ times $e/2m_Pc$, the factor of proportionality being denoted by g_P. Thus

$$\mu_P = g_P \frac{\sqrt{3}}{2} \frac{e}{2m_Pc} \frac{h}{2\pi}. \tag{1}$$

Here m_P is the mass and e the charge of the proton; and g_P is an irrational number which from experiment is known to have the value 5·585. The group of universal constants $\frac{e}{2m_Pc}\frac{h}{2\pi}$ is known as the *nuclear magneton* μ_n, its magnitude being $5{\cdot}050 \times 10^{-24}$ erg gauss^{-1}. For nuclei other than the proton the absolute magnetic moment is given by the general expression

$$\mu = g\sqrt{I(I+1)}\frac{e}{2m_Pc}\frac{h}{2\pi} \tag{2}$$

or in terms of the nuclear magneton by

$$\mu = g\mu_n\sqrt{I(I+1)}. \tag{3}$$

Nuclei with the spin $I = 0$ therefore have zero magnetic moment. The dimensionless quantity $g \times I$ is often referred to as the 'magnetic moment' of the nucleus and is listed as such in Table 5.5.1. Clearly $g \times I$ is not equal to the absolute nuclear magnetic moment μ, though μ can be calculated from it by use of eqn. (3). We shall show later that $g \times I$ is actually the maximum component of the absolute nuclear moment in the direction of the lines of magnetic force of an external magnetic field. At the present time nuclear g factors can be found by experiment only; they are not calculable by any theory. Some nuclei have negative g (and hence negative moments) which signifies that the magnetic moment and angular momentum vectors, while they both coincide with the spin axis, actually point in opposite directions.

8.2 Spatial Quantisation of the Nuclear Angular Momentum

The orientation of the angular momentum vector $\boldsymbol{I} = \sqrt{I(I+1)}h/2\pi$ is quantised with respect to an axis fixed in space. The general principles of spatial quantisation of this kind were described in section 3.7. Theory allows I_z, the component of $\boldsymbol{I}$ along a space-fixed axis z, to assume any one of the values given by the expression

$$I_z = M_I h/2\pi. \qquad (1)$$

Here M_I is a quantum number that may adopt certain positive or negative integral or half-integral values, zero included. Numerically, M_I equals the component of angular momentum I_z in units of $h/2\pi$. Since I_z obviously cannot be greater than $\boldsymbol{I}$ itself, the limits of M_I are $\pm I$ and the spectrum of allowed values of M_I is

$$M_I = I,\ I-1, \ldots -(I-1), -I. \qquad (2)$$

(Note that the values of M_I are *either* all integral *or* all half-integral, depending on whether I is integral or half-integral). The total number of possible values of M_I is $2I+1$ and hence the vector $\boldsymbol{I}$ is allowed $2I+1$ discrete orientations with respect to a space-fixed axis. Let the angle between $\boldsymbol{I}$ and the fixed axis be θ_i, then

$$\cos\theta_i = \boldsymbol{I}_z/\boldsymbol{I} = M_I/\sqrt{I(I+1)}. \qquad (3)$$

The situation is analogous to the spatial quantisation of the total angular momentum $\boldsymbol{J}$ of a molecule due to its overall rotation (section 5.8).

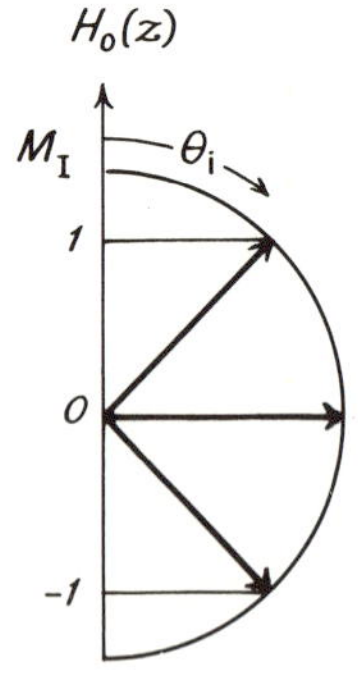

Fig. 8.2.1. Spatial quantisation of $\boldsymbol{I}$ for a nucleus with $I = 1$. The length of the vector $\boldsymbol{I}$ is proportional to $[I(I+1)]^{\frac{1}{2}} = 2^{\frac{1}{2}}$.

Fig. 8.2.1 shows the allowed orientations in space of the angular momentum of a nucleus with a spin $I = 1$. The components I_z are the projections of $\boldsymbol{I}$ on the fixed axis, set vertically in the figure. As the nuclear moments coincide with the axis of spin, their axes are also inclined at θ_i to the fixed axis z. Using eqns. (3) and (8.1.3), the possible values of the z-component μ_z of the absolute magnetic moment are given by

$$\mu_z = \mu \cos \theta_i = gM_I\mu_n. \tag{3a}$$

The limits of M_I are $\pm I$: hence the *maximum value* of μ_z is $gI\mu_n$, *i.e.*, the product of the 'magnetic moment' $g \times I$ times the nuclear magneton μ_n. Put another way, the 'magnetic moment' gI is the maximum value, in units of the nuclear magneton, of the component of the absolute moment parallel to a space-fixed direction.

The $2I + 1$ orientations of a nucleus may differ in energy for either of two quite different reasons. First, if $I \geqslant \frac{1}{2}$, a nucleus possesses a magnetic moment μ whose axis coincides with the vector $\boldsymbol{I}$ (though it may be opposite in sign). Hence if the nucleus is placed in a uniform magnetic field there is a contribution E_H to the potential energy of the nucleus that is given by

$$E_H = -\mu H_0 \cos \theta_i. \tag{4}$$

Here E_H is the potential energy of a magnetic dipole μ set at an angle θ_i to a field of strength H_0. The lines of magnetic force define a space-fixed direction and consequently we may substitute for $\cos \theta_i$ from (3); introducing also the expression (8.1.3) for μ, (4) becomes

$$E_H = -g\mu_n M_I H_0. \tag{5}$$

Eqn. (5) tells us that each allowed orientation of a nucleus in a magnetic field corresponds to a different value of E_H; or, in other words, that the $(2I + 1)$-fold degeneracy of the nuclear energy level is lifted in favour of a set of $2I + 1$ sub-levels of different energy. The process is represented for a nucleus with $I = \frac{1}{2}$ in Fig. 8.2.2.

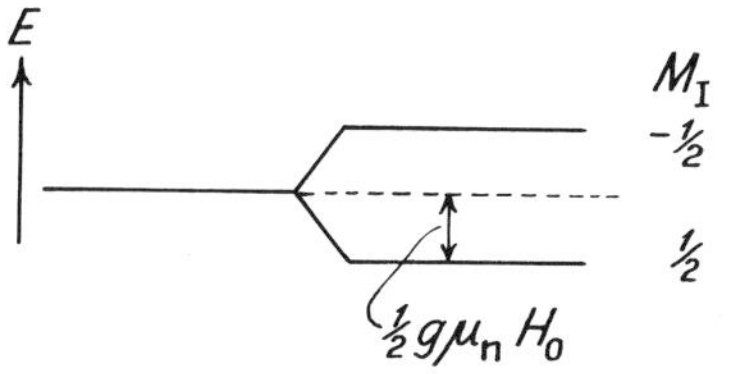

Fig. 8.2.2. Energy sub-levels of a nucleus with $I = \frac{1}{2}$ in a pure magnetic field. For nuclei with μ negative the sub-levels occur in the reverse order to that shown.

Transitions between these sub-levels involve a reorientation of the nucleus with respect to the lines of magnetic force of the applied field. It turns out that such transitions can be observed: they constitute the *nuclear magnetic resonance* spectra of atoms and molecules.

Secondly, the nuclear energy may depend on orientation if the nucleus possesses a quadrupole moment. The potential energy of an electric quadrupole in a non-uniform electric field varies according to the angle between the axis of the quadrupole and the line of electric force that passes through it. (In a *uniform* electric field, however, all orientations are of equal energy (see section 5.7)). Now, at every nucleus there is an electric field due to all other charged particles in the molecule: hence, if this field happens to be non-uniform, and the nucleus is quadrupolar, we have a situation in which the degeneracy of the nuclear energy level is partly lifted without the application of any external field whatever. Fig. 8.6.1 illustrates the splitting that occurs for a nucleus with spin $I = 3/2$. Transitions between the sub-levels involve the reorientation of the nucleus with respect to the internal electric field of the molecule. Such transitions can be observed in the crystalline state; they comprise the *pure nuclear quadrupole resonance* spectra of solids.

8.3 Nuclear Magnetic Resonance

As a nuclear quadrupole interacts with the internal electric field of a molecule (unless, exceptionally, the field happens to be uniform), a nucleus of this kind placed in a magnetic field has a two-fold contribution to its potential energy: from the orientation of the electric quadrupole in the electric field, and from the orientation of the magnetic dipole in the magnetic field. It then becomes necessary to take account of the quadrupole moment in the interpretation of the magnetic resonance spectrum. In the present section we shall consider only the magnetic resonance of nuclei of spin $I = \frac{1}{2}$ which have no quadrupole moment. The restriction is a practical one, for a majority of observations relate to nuclei of this type, precisely in order to avoid the complication of a quadrupole moment.

Common nuclei of spin $\frac{1}{2}$ are ^{1}H, ^{19}F, and ^{31}P, though the natural isotopes ^{13}C, ^{15}N, and ^{29}Si are also in this class. Unfortunately ^{12}C and ^{16}O, together with other nuclei of even atomic number and even mass (comprising about 60% of the total of stable nuclei) have $I = 0$ and so exhibit no nuclear resonance of any kind.

Eqn. (8.2.5) gives the relative energy E_H of the nuclear sub-levels in a pure magnetic field. From this equation the energy jump and transition frequency are readily calculated. Where $I = \frac{1}{2}$, M_I can be $+\frac{1}{2}$, or $-\frac{1}{2}$ and so we have

$$h\nu = E_H' - E_H'' = g\mu_n H_0. \tag{1}$$

Thus the resonant frequency ν is proportional to the field strength H_0. The proton has $g = 5{\cdot}585$ ($= 2 \times 2{\cdot}793$): therefore in a field of 10 000 gauss (using the numerical value, $\mu_n = 5{\cdot}050 \times 10^{-24}$ erg gauss^{-1}, for the nuclear magneton) we find

$$\nu = 42{\cdot}57 \text{ MHz.} \tag{2}$$

The frequency is that of a short radio wave. For fluorine in the same field ν is about 40·05, and for phosphorus it is 17·24 MHz. The inclination θ_i of the magnetic dipole with respect to the field is the same for all nuclei of spin $\frac{1}{2}$ and is given by (eqn. 8.2.3)

$$\cos \theta_i = \pm 1/\sqrt{3} \tag{3}$$

whence

$$\theta_i = 54°44' \quad \text{or} \quad 125°16'. \tag{4}$$

The orientations are shown in Fig. 8.3.1. Owing to the spin angular momentum of the nucleus, the nuclear axis, and hence the nuclear magnetic dipole, precesses about the direction of the lines of magnetic force, in the same manner as a top precesses about the direction of the earth's gravitational field.

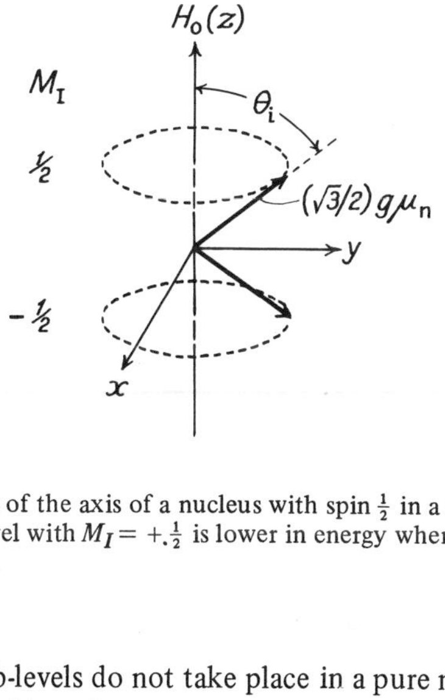

Fig. 8.3.1. Orientations of the axis of a nucleus with spin $\frac{1}{2}$ in a magnetic field. In this diagram the sub-level with $M_I = +\frac{1}{2}$ is lower in energy when μ is positive (see footnote to Fig. 8.2.2).

Transitions between sub-levels do not take place in a pure magnetic field. If, however, an electromagnetic field oscillating at the resonant frequency ν is superimposed, we may expect the oscillating field to induce the transitions to occur. Actually the probabilities of absorption and induced emission are equal, and consequently the observation of a net absorption of energy from the oscillating field depends on there being an excess population of nuclei in the lower energy sub-level. According to the Boltzmann law, the respective populations of the lower and upper sub-levels are proportional to $e^{-E_H''/kT}$ and $e^{-E_H'/kT}$. As the exponentials differ only insignificantly from unity the partition function, f_N, equals 2·0 for a nucleus with spin $\frac{1}{2}$. The excess relative population of the lower state is then

$$\tfrac{1}{2}(e^{-E_H''/kT} - e^{-E_H'/kT}) \simeq (E_H' - E_H'')/2kT$$
$$= g\mu_n H_0/2kT. \qquad (5)$$

In a field of 10 000 gauss, the expression (5) amounts to $3{\cdot}4 \times 10^{-6}$ for protons at 300 K: that is, the net absorption is due to between 3 and 4 nuclei out of every million. As the energy absorbed depends upon this small excess a large number of nuclei are necessary before the energy loss by absorption becomes detectable. 0·1—1 cm^3 samples of a liquid or solid are frequently used.

Experimental Method

In the arrangement shown in Fig. 8.3.2 the *source* is a split coil $T_{1,2}$, energised by a radio-frequency oscillator. The *detector* is a second coil R with its axis perpendicular to the source coil: and the magnetic field is applied at right angles to the axes of both coils. The windings of the electromagnet provide that the magnetic field strength can be varied at will. To describe the working of the spectrometer let us imagine a simple experiment. Let the sample be water, so that the observable transitions are those of a proton. The oscillator is adjusted to 42·57 MHz and the field strength H_0 to a little under 10 000 gauss. H_0 is then increased slowly, thus increasing the separation of the nuclear sub-levels, until at 10 000 gauss exactly, the energy separation coincides with the frequency of the source and the system comes into resonance. The transitions of the nuclear magnets from the lower to the higher energy level induce an e.m.f. in the detector coil R. The resultant voltage is minute

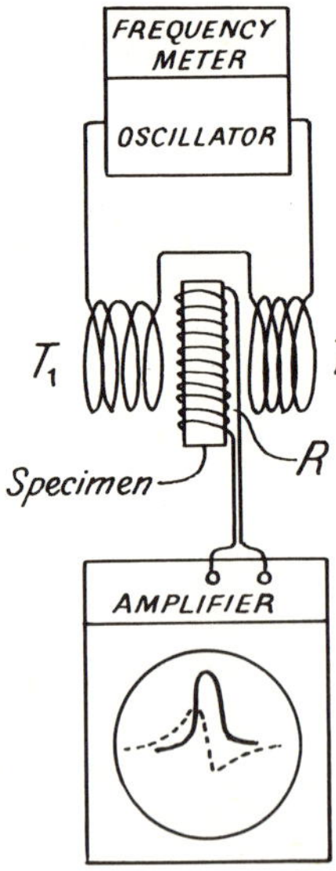

Fig. 8.3.2. Nuclear resonance spectrometer (induction type). The applied magnetic field is perpendicular to the plane of the paper.

but can be observed, after considerable amplification, by connection to the Y-plates of an oscilloscope. It is usual to sweep the magnetic field repeatedly through the resonance condition, with the same frequency applied to the X-plates of the oscilloscope. Then if H_0 is increased linearly with time the horizontal axis on the screen is proportional to the change of H_0, while the vertical axis, if certain precautions are observed, records the intensity of absorption from the radio-frequency field of the coils T_1 and T_2. Under these conditions the oscilloscope screen displays the nuclear magnetic resonance spectrum of the protons in the sample. For weak signals, H_0 is sometimes raised gradually and the output from the detector amplified in such a manner that the first derivative of the absorption curve is drawn by a pen-recorder. A derivative curve is shown dotted in Fig. 8.3.2: it can be transformed into the normal absorption curve by numerical or graphical integration.

The practice of recording spectra by gradually raising the field, keeping the radiofrequency constant, is adopted for purely experimental reasons: the same results would be obtained by keeping H_0 constant and changing the radiofrequency. The second method involves some practical difficulties and so is seldom used, but it corresponds more closely with the procedure in optical regions of the spectrum, and nuclear spectra are therefore often described as if they had been obtained by varying the frequency at constant field. The conversion is straightforward since field and frequency are proportional at the resonance condition. Suppose, for instance, we wish to discuss the width of a proton signal at one-half its maximum height. Let the corresponding field sweep be dH gauss: in frequency units the signal width is $d\nu = (g\mu_n/h)dH = 4257\,dH$ Hz, since for protons $g\mu_n/h = 4257$ Hz gauss^{-1}. For each field increment of 1 milligauss (10^{-3} gauss) the equivalent frequency interval is 4·257 Hz.

Proton resonance signals from molecules in the liquid state or in solution are as a rule extremely sharp; so sharp, in fact, that their width is usually determined by the extent to which the laboratory field varies across the sample. Unless all nuclei are in the same field they cannot be in resonance simultaneously, so that a non-homogeneous field gives rise to broad signals. Proton spectroscopy demands a very homogeneous field indeed since inhomogeneities greater than about one hundred-millionth part of the applied field lead to unacceptably broad signals. Very large pole faces are used and the sample kept as small as possible to minimise field changes over its area.

8.4 Proton Resonance Spectra of Liquids and Solutions

The behaviour of a given nucleus, such as ^1H or ^{19}F, is slightly different in different molecules; that is to say, the point of resonance (energy absorption) of protons in, for example, water and benzene samples does not occur at precisely the same field. The effect is in no way due to the protons themselves, for all protons have identical values of g and μ_n. It arises because the protons are in different environments whose interaction with the external field causes the field *at the proton* to vary very slightly from one case to another. Protons in different molecules, and even at different positions in the same molecule, therefore experience slightly different fields although the field applied externally is the same for all. The effect can be broken down into a part due to the surrounding electrons and another part due to neighbouring magnetic nuclei, and these two factors will now be considered in turn. Similar considerations apply to all magnetic nuclei, but to avoid unnecessary duplication the following discussion refers to protons only.

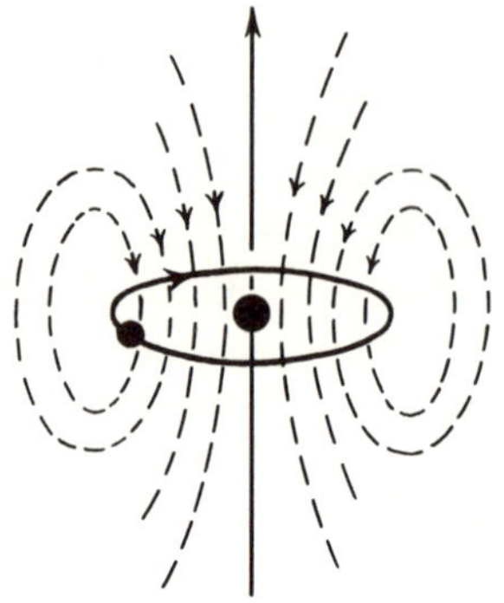

Fig. 8.4.1. The induced electron current sets up a local field (dotted lines) which opposes H_0 at the position of the nucleus.

Consider first a small symmetrical molecule such as methane, CH_4. The magnetic field exerts a force on the electrons so as to induce rotation about the axis of the field, as illustrated for a representative electron in Fig. 8.4.1. The circulating charges constitute a solenoid whose field is cast in opposition to H_0 so that the net field at the proton is slightly less than the field applied externally. In this sense the electrons partly *shield*, or *screen*, the protons from the laboratory field. In methane the four protons are symmetrically equivalent and so shielded equally, thus the resonance consists of a single line. The resonance of other simple molecules containing equivalent hydrogen atoms only is shown in Fig. 8.4.2. Each molecule gives rise to a single peak whose position varies from one molecule to another. The greater the shielding the more the external field must be raised in order to bring the field at the proton to its critical value. In the figure, H_0 increases from left to right; therefore, of the examples shown, shielding is greatest for the tetramethylsilane protons and least for the proton of trifluoracetic acid.

The resonance equation (8.3.1) needs to be amended slightly to take account of electronic shielding. Since the field, H', associated with the induced electron circulation is much smaller than H_0 we can assume H' is proportional to H_0 and so write,

$$H' = \sigma H_0 \qquad (1)$$

in which σ is a dimensionless constant, known as the *shielding constant*. (For atoms and some very simple molecules σ can be calculated approximately from the electronic wavefunction. Its value for the hydrogen atom works out to be $\approx 2 \times 10^{-5}$, representing the shielding of the proton by a single $1s$ electron). The resonance condition for a chemically-bound proton then becomes,

$$h\nu = g\mu_n(1 - \sigma)H_0 \tag{2}$$

Differences in resonance frequency for protons in different molecules, or in different positions in the same molecule, are known as *chemical shifts* and result from changes in the shielding constant. In practice, the magnitude of a chemical shift is always

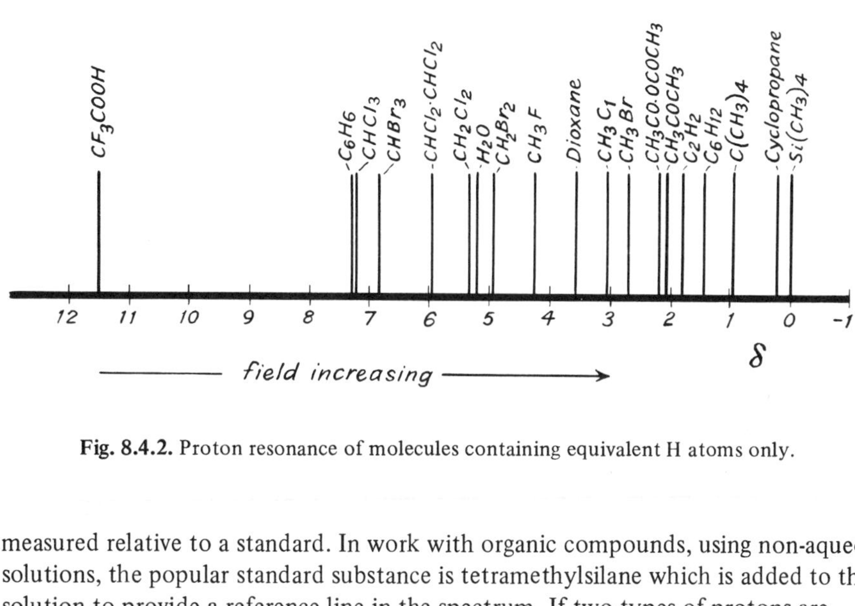

Fig. 8.4.2. Proton resonance of molecules containing equivalent H atoms only.

measured relative to a standard. In work with organic compounds, using non-aqueous solutions, the popular standard substance is tetramethylsilane which is added to the solution to provide a reference line in the spectrum. If two types of protons are present in the same solution, say a mixture of acetone (proton i) and tetramethylsilane (proton r = reference), resonance will occur at the frequencies ν_i and ν_r given by,

$$h(\nu_i - \nu_r) = g\mu_n(\sigma_r - \sigma_i)H_0\,. \tag{3}$$

Since shielding constants are very small numbers we can take $g\mu_n H_0/h$ equal to the radiofrequency ν_0 and so write,

$$(\nu_i - \nu_r)/\nu_0 = \sigma_r - \sigma_i\,. \tag{4}$$

$(\nu_i - \nu_r)/\nu_0$ is a measure of the shift of the acetone protons i relative to the standard r, but its numerical magnitude is extremely small, of the order 10^{-6}. The *chemical shift*, δ_i, is therefore defined to be,

$$\delta_i = 10^6 (\nu_i - \nu_r)/\nu_0 = 10^6 (\sigma_r - \sigma_i) \cdot \tag{5}$$

In proton spectroscopy, frequency is measured by superimposing on the radio-frequency a much lower frequency, ν_a, of roughly the same magnitude as $(\nu_i - \nu_r)$. ν_a is then in the audiofrequency range. The source now consists essentially of three frequencies, ν_0 and $\nu_0 \pm \nu_a$, so that when the spectrum is recorded as described in section 8.3 each line appears three times, as shown in Fig. 8.4.3. In effect, the 'side-bands' at $\pm \nu_a$ provide a frequency calibration from which the separation of peaks is found by simple proportion. Since the measured separation of acetone and tetramethylsilane signals is 125 Hz when $\nu_0 = 60$ MHz, the chemical shift of the acetone protons is $10^6 (125/60 \times 10^6) = 2{\cdot}08$. Eqn. (5) shows that the chemical shift is independent of the external field. This is especially useful in that not all spectrometers operate at the same value of H_0.

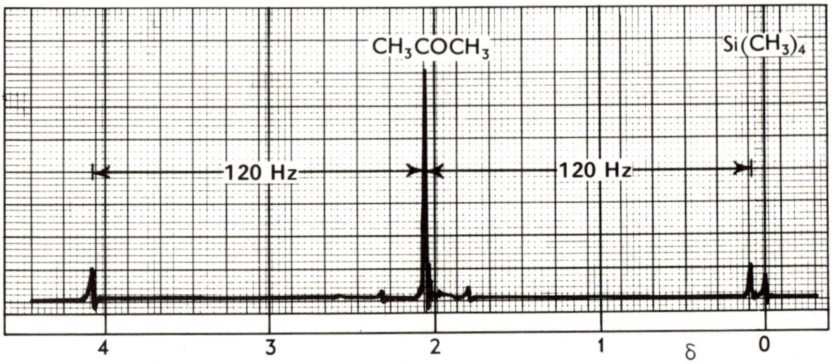

Fig. 8.4.3. Proton spectrum, with side bands, of acetone in presence of tetramethylsilane ($\nu_0 = 60$ MHz, $\nu_a = 120$ Hz).

In molecules which contain non-equivalent hydrogen atoms it should be possible to observe a separate resonance for each chemically distinct hydrogen. In fact, it is often practicable to do so even for quite large molecules. The spectrum of diacetone alcohol (Fig. 8.4.8(*a*)) provides a simple example. Evidently the method can be used to get information about an unknown structure from the chemical shifts observed for its various hydrogen atoms. Comprehensive tables of δ are available for protons and other magnetic nuclei in typical molecules.

In any proton spectrum the transitions are essentially similar in that they excite nuclei from the favoured to the unfavoured energy level. In consequence, all transitions are equally probable and their intensities are proportional simply to the numbers of protons involved. The relative intensity of the —OH, $—CH_2—$, $—CH_3$, and $(—CH_3)_2$ resonances of diacetone alcohol is therefore 1 : 2 : 3 : 6, which helps considerably with the assignment of peaks. Intensity should be understood as the total area enclosed by the resonance, a quantity less easily judged by eye than the peak height. Unfortunately peak heights are themselves not a reliable guide to area because the breadth of the resonance is determined by parameters which may vary even for different hydrogens in the same molecule. Quantitative intensities are obtained by integration of the signal. The integration may encompass more than one peak if the spectrum contains fine structure arising from spin coupling (see later).

To this point it might be thought that proton shielding constants are determined by the density of the valency electrons that bind the proton to the rest of the molecule. If this were correct there would be a general correlation between the acidity of a given proton and its chemical shift, since an acidic hydrogen is deprived of electrons and so should be relatively unshielded. One would expect acidic protons to occur at low field, and this is to some extent borne out by the position of the trifluoracetic acid proton in Fig. 8.4.2. But in other respects the compounds shown in the figure do not conform to this simple picture. For example, acetylene is certainly more acidic than chloroform yet its resonance occurs at higher field, implying greater shielding; and aromatic protons are not noticeably acidic but are comparatively unshielded. Evidently there is some other factor at least as important as the electron density immediately surrounding the proton. This is not surprising when one recalls the charge density around a proton is much smaller than that surrounding the carbon, nitrogen or oxygen atom to which hydrogen is chemically bound. The electron currents at these neighbouring atoms or groups, especially unsaturated groups and atoms carrying unshared electron pairs, are sufficiently large that their field at the proton often exceeds any effect resulting from the valency electrons of the proton itself.

Consider, for example, the effect of the aromatic ring on the chemical shift of the protons in benzene. The molecular π-orbitals of benzene can be thought of as a conductor around which the mobile electrons circulate under the influence of the applied field. The lines of magnetic force associated with this ring current are shown in Fig. 8.4.4 for the case when the ring plane is perpendicular to the external field.

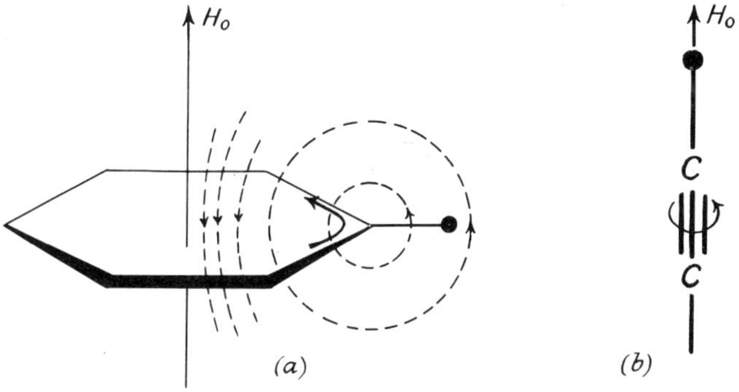

Fig. 8.4.4. π-Electron circulation in (*a*) benzene and (*b*) acetylene.

The field cast by the ring current supplements H_0 at the proton positions so that its effect is to deshield the protons, that is, to cause a shift to low field. Qualitatively, this accords with the large shift shown for benzene in Fig. 8.4.2. In a quantitative argument one must also take account of the fact that, owing to rotation in the liquid state, the particular orientation of molecule and field shown in Fig. 8.4.4 is held only momentarily. In general, it must be considered that the ring current is induced by the component of H_0 perpendicular to the ring plane. While the local field then varies in intensity according to the orientation of the molecule relative to H_0 its sense remains unaltered; that is to say, deshielding is greatest when the ring plane is perpendicular to H_0, least when the ring plane contains H_0, and varies in between according to a $\cos 2\theta$ function. The average contribution to deshielding is that calculated for the perpendicular configuration (shown in Fig. 8.4.4) multiplied by the average value ($\frac{1}{3}$)* of $\cos 2\theta$, corresponding to a random orientation of molecules in the sample as a whole. Calculations along these lines show that the ring current contributes about 2 units of chemical shift towards deshielding the benzene protons, that is, their resonance would occur near $\delta = 5$ if this current were somehow arrested.

A similar argument explains why the resonance of acetylene occurs at relatively high field. The four acetylenic π-electrons form a cylindrically-symmetrical shell so that these electrons circulate freely around the axis of the triple bond. The field set up by the circulation opposes the external field at the position of the proton, whose signal therefore appears at relatively high field in the spectrum since the external field must be raised in order to bring the field at the proton to its critical value. Other unsaturated groups and atoms carrying unshared electron pairs behave analogously, though the orientation of the electron currents is usually less obvious than in the two preceding examples.

It should be stressed that shielding or deshielding by a neighbour group only occurs when the electron circulation has a favoured orientation in that group. Take, for example, some unspecified group G whose electrons are equally free to circulate about any axis: the electrons then always circulate in the plane perpendicular to H_0 and the shielding experienced by a proton changes sign in the orientations (*a*) and (*b*), Fig. 8.4.5. Over a complete rotation the contribution of the group G towards shielding the proton averages to zero. No net shielding or deshielding is felt unless the neighbouring group is *anisotropic*, that is, unless its charge circulation has a preferred orientation in the group.

Spin Coupling (qualitative treatment)

Nearly all proton spectra contain more peaks than are accounted for by the number of non-equivalent hydrogen atoms in a molecule. The reason is that the protons are themselves magnetic, so that a given proton is exposed to small magnetic fields cast

* θ is the angle between the ring plane and the lines of magnetic force of the laboratory field. The average value of $\cos 2\theta$ is $\int_0^\pi \cos 2\theta \, d\Omega \, / \int_0^\pi d\Omega$ where $d\Omega = 2 \sin\theta \, d\theta$ is an element of solid angle (see Fig. 11.1.1). The integrals can be obtained from tables and give $(\cos 2\theta)_{av} = \frac{1}{3}$.

by neighbouring protons. The effect is reciprocal in the sense that a given proton in a set of non-equivalent protons both influences and is influenced by other protons in the set. The interaction is referred to as the *spin–spin coupling* of magnetic nuclei, and is recognised by the division of a signal into a sub-structure comprising two or more peaks. For reasons explained in section 8.5, spin coupling between *equivalent* protons does not as a rule lead to sub-division into separate peaks.

Spin coupling assumes its simplest form in molecules which contain just two magnetic nuclei. A molecule which meets this specification is H—F, where both nuclei have $I = \frac{1}{2}$, though the example is artificial in that the effects to be described would be difficult to observe owing to the intense chemical reactivity. For the sake of argument, however, consider a single molecule of HF and take the F nucleus to be at the origin of a set of cartesian coordinates so oriented that the z axis coincides with H_0. The fluorine nuclear dipole then precesses about z as shown in Fig. 8.3.1.

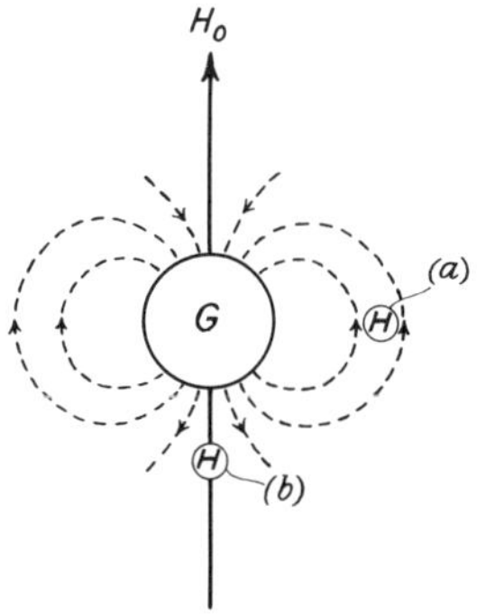

Fig. 8.4.5

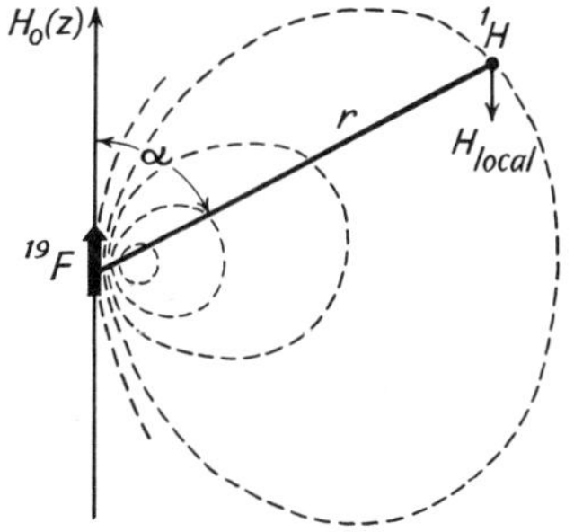

Fig. 8.4.6. Local field of a magnetic dipole.

We resolve its magnetic moment into components μ_z and μ_{xy} parallel and perpendicular to z. Owing to the precession, μ_{xy} rotates in the xy plane and so its magnitude along any fixed direction in that plane fluctuates sinusoidally: therefore its time average, and hence the time average of its local field, is zero. But μ_z gives rise to an axially-symmetrical local field which does not fluctuate with time: the component of its local field parallel to H_0 at the position of the proton (see Fig. 8.4.6) is calculable from classical electrostatic theory and is given by,

$$H_{\text{local}} = (\mu_z/r^3)(3\cos^2\alpha - 1). \tag{6}$$

A nucleus of spin $\frac{1}{2}$ has $\mu_z = \mu\cos\theta_i = \pm\frac{1}{2}g\mu_n$, the sign being determined by whether the nucleus occupies the lower (+) or higher (—) energy level. The field sensed by the proton is therefore

$$H_0(1-\sigma) \pm (g\mu_n/2r^3)(3\cos^2\alpha - 1). \tag{7}$$

This means the net field at the proton is either greater or less than $H_0(1-\sigma)$, depending on the orientation of the fluorine nucleus. In the sample as a whole the fluorine nuclei will be almost equally divided between the two orientations. If we consider a hypothetical single crystal in which all molecules are parallel, aligned at the same angle α to the field, the proton resonance will consist of two peaks of equal intensity whose field separation $\Delta H = 2H_{\text{local}}$ will be given by,

$$\Delta H = (g\mu_n/r^3)(3\cos^2\alpha - 1). \tag{8}$$

To get rough magnitudes, suppose the molecules are all oriented so that the H—F bonds are parallel to z ($\alpha = 0°$). ^{19}F nuclei have $g_F = 2 \times 2{\cdot}627$ (Table 5.5.1) and the bond distance r is approximately 1 Å; thus the components of the proton resonance are separated by,

$$\Delta H = 2g_F\mu_n/r^3 = 2(2 \times 2{\cdot}627) \times 5{\cdot}05 = 53 \text{ gauss}. \tag{9}$$

Likewise, the ^{19}F resonance divides into two components whose separation can be calculated from eqn. (8) using the value of g appropriate to protons. Since the splittings result from intramolecular fields their magnitude is independent of H_0.

In spite of its artificiality this analysis does indicate broadly the effects encountered with crystalline samples. Direct magnetic interaction of nuclei then dominates the spectrum, since its magnitude is immensely greater than the chemical shift. Solid state spectra form an important field of study though we shall not discuss them further. Proton spectra are ordinarily observed in the liquid state where the molecules are freely rotating. To estimate H_{local} under these conditions we substitute in eqn. (6) a mean value of $\cos^2\alpha$ corresponding to random tumbling of the H—F axis. The mean value is obtained in the manner described in the footnote on p. 230. It turns out $(\cos^2\alpha)_{\text{av}} = \frac{1}{3}$, hence $3(\cos^2\alpha)_{\text{av}} - 1 = 0$. In other words, the direct interaction of one magnetic nucleus with another vanishes in the liquid state. The same conclusion follows from consideration of the Uncertainty Principle (see section 3.8). The field at the HF proton has either of two values whose magnitude depends on the inclination α of the molecular axis relative to H_0. To this field difference ΔH there corresponds an energy difference $\Delta E = g\mu_n\Delta H$. We know the proton occupies one or other of these two states but we do not know which, thus the uncertainty in energy is ΔE. ΔE is associated with an uncertainty in time, τ, by the equation

$$\tau . \Delta E \approx h/2\pi. \tag{10}$$

For the orientation $\alpha = 0°$, $\Delta E/h = (g\mu_n/h)\Delta H = 4257 \times 53 = 225 \times 10^3$ Hz, hence $\tau = 0{\cdot}706 \times 10^{-6}$ sec. The orientation must be held at least as long as this in order to distinguish the two states. But collisions occur about every 10^{-10} sec in the liquid state and each collision must lead to reorientation (change of α); therefore the lifetime of a given orientation is several orders of magnitude less than the critical life time that allows the two states to be observed separately.

The proton spectrum of liquid HF nevertheless still shows two peaks though their separation is several hundred times smaller than calculated for the solid state. Nor is the residual splitting the result of incomplete averaging by rotation, for it is found to be independent of temperature. It results from a second order effect transmitted through the bonding electrons. These electrons are magnetic by virtue of their spin, but are compelled to adopt opposite spin orientations since they occupy the same orbital. Suppose the F nucleus has $M_I = \frac{1}{2}$; the electron most frequently in its neighbourhood then has $M_S = -\frac{1}{2}$, since the pairing of nuclear and electron magnets corresponds to a reduction of potential energy. In consequence, the proton sees a net field due to an electron of spin $+\frac{1}{2}$ (see Fig. 8.4.7). Likewise, when the F nucleus has $M_I = -\frac{1}{2}$, the proton senses the field of a slight excess of electron of spin $-\frac{1}{2}$. The two possibilities are not equivalent and so the proton spectrum contains two lines of equal intensity, corresponding to the equal probability that the F nucleus has $M_I = \frac{1}{2}$ or $-\frac{1}{2}$; and the effect is small because the bias of an electron towards a nucleus of opposite spin represents only a very slight unbalance in the otherwise

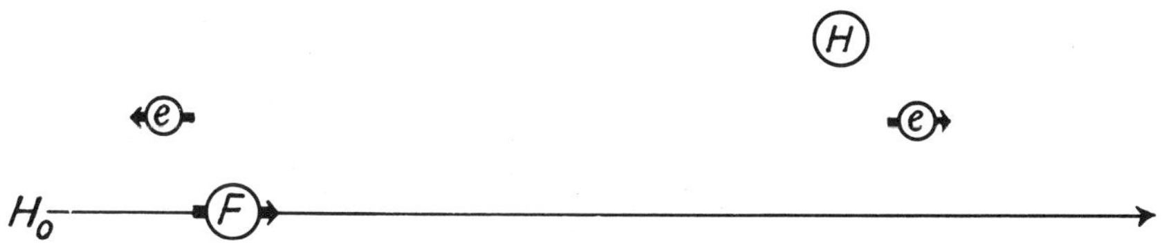

Fig. 8.4.7. Electron-coupled interaction of magnetic nuclei.

even flow of electrons around the nuclei. The phenomenon is known as the *electron-coupled spin–spin interaction* of magnetic nuclei. Such coupling is observable across more than one bond but dies away rather rapidly as the number of bonds increases.

A straightforward example of spin coupling occurs in the spectrum of ethyl acetate, $CH_3.CO.O.CH_2.CH_3$ [Fig. 8.4.8(*b*)]. As the carbon and oxygen nuclei are non-magnetic only proton–proton interactions need be considered. There are three groups of equivalent hydrogens, thus we expect three chemically shifted signals. The acyl-CH_3 protons are too distant to affect, or be affected by, the C_2H_5 protons so that their resonance consists of a single peak. What coupling there is occurs within the —$CH_2.CH_3$ group. Here, the CH_3 signal divides in consequence of interaction with the CH_2 protons (labelled 1 and 2), which go both into one or one into each energy level. Four arrangements are possible, namely,

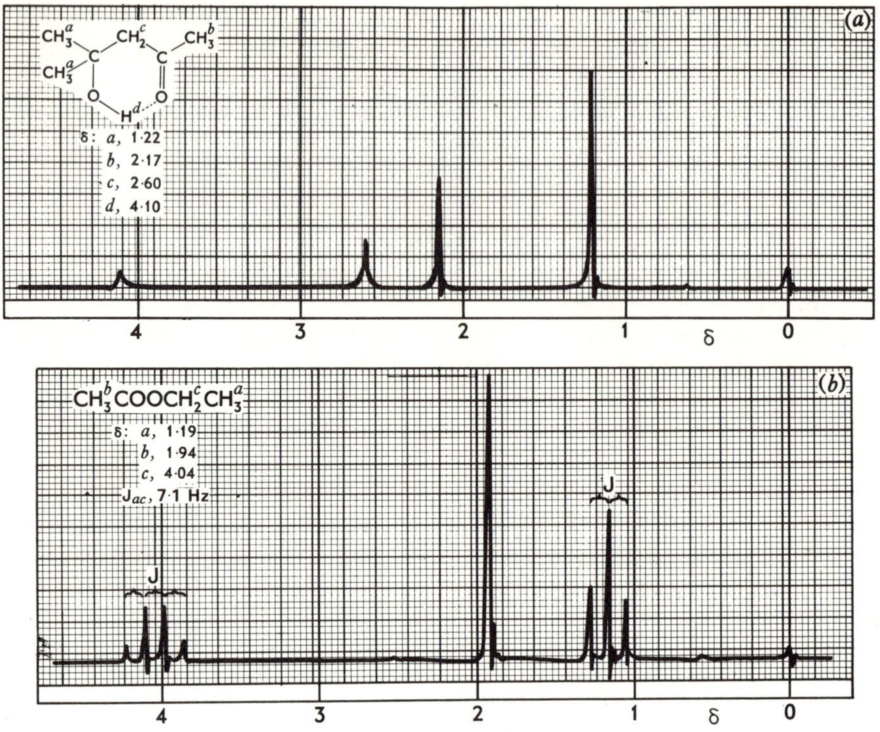

Fig. 8.4.8. Proton spectra (at 60 MHz) of (*a*) diacetone alcohol and (*b*) ethyl acetate.

(1)	(2)		(1)	(2)		(1)	(2)
$\frac{1}{2}$	$\frac{1}{2}$		$\frac{1}{2}$	$-\frac{1}{2}$		$-\frac{1}{2}$	$-\frac{1}{2}$
			$-\frac{1}{2}$	$\frac{1}{2}$			

In the sample as a whole, very nearly equal numbers of molecules belong to each arrangement. But as the CH_2 protons are equivalent the two arrangements $\frac{1}{2}, -\frac{1}{2}$ and $-\frac{1}{2}, \frac{1}{2}$ are indistinguishable at the position of the methyl protons; therefore two components of the split CH_3 resonance coincide and the signal consists of a group of three lines (that is, a triplet) with relative intensities 1 : 2 : 1.

The CH_2 resonance records the number of distinguishable spin orientations within the CH_3 group, namely,

$\frac{1}{2}$	$\frac{1}{2}$	$\frac{1}{2}$		$\frac{1}{2}$	$\frac{1}{2}$	$-\frac{1}{2}$		$\frac{1}{2}$	$-\frac{1}{2}$	$-\frac{1}{2}$		$-\frac{1}{2}$	$-\frac{1}{2}$	$-\frac{1}{2}$
				$\frac{1}{2}$	$-\frac{1}{2}$	$\frac{1}{2}$		$-\frac{1}{2}$	$-\frac{1}{2}$	$\frac{1}{2}$				
				$-\frac{1}{2}$	$\frac{1}{2}$	$\frac{1}{2}$		$-\frac{1}{2}$	$\frac{1}{2}$	$-\frac{1}{2}$				

Those arrangements which merely permute the spins among equivalent nuclei (*e.g.*,

$\frac{1}{2}\ \frac{1}{2} -\frac{1}{2}$, $\frac{1}{2} -\frac{1}{2}\ \frac{1}{2}$, *etc.*) give rise to the same field at the CH_2 group which therefore appears as a 1 : 3 : 3 : 1 quartet. Each interval in either group of lines records the effect on one set of protons of the reorientation of one proton in the other set. These intervals are therefore all equal, and are characteristic of the number and type of intermediate bonds (in this example, three single bonds in an acyclic structure). Measured in Hz, the splitting is known as the *coupling constant* and is denoted by the symbol J. The chemically-shifted position of the CH_2 and CH_3 signals corresponds to the mid-point of each spin multiplet. The reader is recommended to check for himself the values of J and δ recorded for ethyl acetate in Fig. 8.4.8. In the spectrum of diacetone alcohol non-equivalent protons are separated by a minimum of four σ-bonds: coupling is negligibly small under these conditions and only chemically-shifted signals are observed.

There is a fundamental difference between spin multiplets and chemically-shifted signals which can be used to distinguish them. Spin multiplets result from intramolecular fields and thus the coupling constant J is unaffected by changes in H_0. Shielding, on the other hand, is field-dependent [see eqn. (1)] and hence the separation in Hz of chemically-shifted signals is directly proportional to H_0. A measurement at two different fields will therefore discriminate immediately between them. For similar reasons J is quoted in frequency units, whereas the dimensionless δ is preferred for the chemical shift.

8.5 Quantitative Treatment of Spin–Spin Coupling (Two Nuclei)

Here we consider the quantitative treatment of a simple system comprising two protons separated by two or three saturated bonds so that their interaction is appreciable. Such interaction occurs in, for instance, CH_2Cl_2 (equivalent protons separated by two single bonds) $Cl_2CH.CHO$ (non-equivalent protons separated by three single bonds), as well as in larger structures like $C_6H_5.CH{=}CH.COOH$ where the two ethylenic protons couple with one another but are isolated (by four intermediate bonds) from interaction with other protons in the molecule. Svstems of three or more coupled protons are outside the scope of this book and the interested reader should consult the monograph by Pople, Schneider, and Bernstein.* The treatment, though more complex, does not demand new principles so that the extension is comparatively straightforward.

It is useful first to summarise the types of spectrum encountered. When two coupled protons are distinguished by a large chemical shift their spectrum contains two groups of two lines—that is, two doublets—of equal or nearly equal intensity. The situation corresponds to the top of Fig. 8.5.1. Within each group the line separation (denoted by J) is equal. J is known as the *coupling constant* and has the units of Hz. When the chemical shift is small the spectrum still contains four lines but the inner lines are more intense than the outer lines, corresponding to the centre part of Fig. 8.5.1. The pattern is symmetrical about its mid-point, the separation of the outer pairs of lines being equal (J). Finally, two equivalent protons give rise to a spectrum of just one line, as shown at the foot of the figure. These two-proton systems are referred to as AX (large shift), AB (small shift), and A_2 (equivalent), respectively.

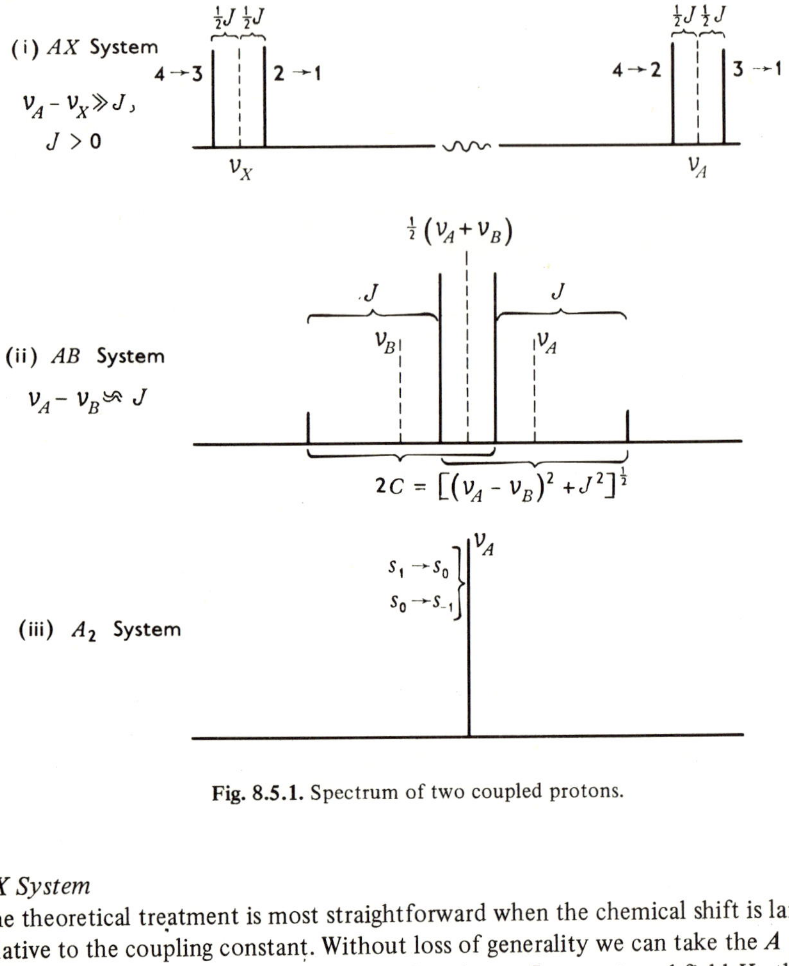

Fig. 8.5.1. Spectrum of two coupled protons.

AX System

The theoretical treatment is most straightforward when the chemical shift is large relative to the coupling constant. Without loss of generality we can take the A proton to be at higher field, thus $\nu_A > \nu_X$ and $\sigma_A > \sigma_X$. In an external field H_0 the field at the two protons is then $H_0(1 - \sigma_A)$ and $H_0(1 - \sigma_X)$. There are four possible spin orientations of the composite system, since both nuclei may go into the favoured energy level ($M_I = \frac{1}{2}$), or both into the unfavoured level ($M_I = -\frac{1}{2}$), or one into each. Assuming for the moment there is no interaction between the nuclei, their total energy when both occupy the unfavoured energy level is simply the sum of the energies of two isolated nuclei with $M_I = -\frac{1}{2}$, namely,

*Pople, Schneider, and Bernstein, *High-Resolution Nuclear Magnetic Resonance*, McGraw-Hill, New York, 1959.

$$E_1 = E_A + E_X = \tfrac{1}{2} g\mu_n [(1-\sigma_A) + (1-\sigma_X)] H_0 \tag{1}$$

But $g\mu_n(1-\sigma_A)H_0 = h\nu_A$ and $g\mu_n(1-\sigma_X)H_0 = h\nu_X$, so eqn. (1) can be written,

$$E_1 = \tfrac{1}{2} h(\nu_A + \nu_X) \tag{2}$$

Likewise, when A is in the unfavoured and X in the favoured level the energy is,

$$E_2 = E_A + E_X = \tfrac{1}{2} g\mu_n [(1-\sigma_A) - (1-\sigma_X)] H_0 = \tfrac{1}{2} h(\nu_A - \nu_X). \tag{3}$$

The student should check for himself that the two remaining states have the energy E_3 and E_4 shown in the sixth column of Table 8.5.1, and that the states 1 to 4 are arranged in order of decreasing energy.

Table 8.5.1. Energy levels and Wavefunctions of the AX System

State	$M_I(A)$	$M_I(X)$	$F_z = M_I(A) + M_I(X)$	Wavefunction	Energy $J = 0$	Energy $J > 0$
1	$-\frac{1}{2}$	$-\frac{1}{2}$	-1	$\psi_1 = \alpha_A\alpha_X$	$E_1 = \frac{1}{2}h(\nu_A + \nu_X)$	$E_1 = \frac{1}{2}h(\nu_A + \nu_X) + \frac{1}{4}Jh$
2	$-\frac{1}{2}$	$\frac{1}{2}$	0	$\psi_2 = \alpha_A\beta_X$	$E_2 = \frac{1}{2}h(\nu_A - \nu_X)$	$E_2 = \frac{1}{2}h(\nu_A - \nu_X) - \frac{1}{4}Jh$
3	$\frac{1}{2}$	$-\frac{1}{2}$	0	$\psi_3 = \beta_A\alpha_X$	$E_3 = \frac{1}{2}h(-\nu_A + \nu_X)$	$E_3 = \frac{1}{2}h(-\nu_A + \nu_X) - \frac{1}{4}Jh$
4	$\frac{1}{2}$	$\frac{1}{2}$	1	$\psi_4 = \beta_A\beta_X$	$E_4 = \frac{1}{2}h(-\nu_A - \nu_X)$	$E_4 = \frac{1}{2}h(-\nu_A - \nu_X) + \frac{1}{4}Jh$

In any one of these states the total spin angular momentum along the z direction is $[M_I(A) + M_I(X)]\,h/2\pi$. Writing $M_I(A) + M_I(X) = F_z$, the value of F_z associated with states 1, 2, 3, and 4 is -1, 0, 0, and $+1$.* Transitions between the four energy levels are governed by a selection rule which permits only those transitions in which F_z changes by ± 1. Physically, this means an allowed transition reorients one or other of the nuclei, not both. The rule disallows the transitions $4 \rightarrow 1$ (corresponding to the excitation of both nuclei simultaneously) and $3 \rightarrow 2$ (a simultaneous upward transition of X and downward transition of A) and so simplifies the spectrum considerably. The allowed transitions and frequencies are,

$$\left.\begin{array}{ll} 4 \rightarrow 3 & (E_3 - E_4)/h = \tfrac{1}{2}(-\nu_A + \nu_X) - \tfrac{1}{2}(-\nu_A - \nu_X) = \nu_X \\ 4 \rightarrow 2 & (E_2 - E_4)/h = \nu_A \\ 3 \rightarrow 1 & (E_1 - E_3)/h = \nu_A \\ 2 \rightarrow 1 & (E_1 - E_2)/h = \nu_X \end{array}\right\}. \tag{4}$$

* In the monograph by Pople, Schneider, and Bernstein (see footnote on p. 236) H_0 is taken to coincide with the *negative* z direction rather than (as here) the positive z direction. Their convention reverses the signs of M_I and F_z.

Evidently the first and last are transitions of the X nucleus while the second and third belong to the A nucleus. The result, of course, is one that would have been expected intuitively from two non-interacting nuclei, that is, a spectrum containing two lines of frequency ν_A and ν_X.

To calculate the spectrum of two interacting protons we have to introduce wavefunctions to describe the magnetic states of the nuclei. Consider first a single, isolated proton. To it we assign a wavefunction α if $M_I = -\frac{1}{2}$ and a wavefunction β if $M_I = +\frac{1}{2}$. It is not necessary to know in detail the form of α and β beyond that, like the wavefunctions of the rotator discussed in section 3.7, they are functions of a variable ϕ which can be visualised as the angle of rotation of the spin about its own axis. α (and β) must satisfy the Schroedinger equation,

$$\mathcal{H}\alpha = E\alpha, \tag{5}$$

where $\mathcal{H}$ is the Hamiltonian operator and E is the energy of the proton in the magnetic field. (We use $\mathcal{H}$ for the Hamiltonian operator to avoid confusion with H meaning magnetic field intensity). α is taken to be properly normalised,

$$\int \alpha^2 \text{đ}\phi = 1 \tag{6}$$

where the bar in $\text{đ}\phi$ indicates that integration is over the full range of values of ϕ, that is, from $\phi = 0$ to 2π. Duplicate expressions (7) and (8) can be written for the wavefunction β,

$$\mathcal{H}\beta = E\beta, \tag{7}$$

$$\int \beta^2 \text{đ}\phi = 1. \tag{8}$$

Finally, α and β must be orthogonal,

$$\int \alpha\beta \text{đ}\phi = 0. \tag{9}$$

Eqn. (9) states mathematically that a given nucleus cannot have $M_I = +\frac{1}{2}$ and $-\frac{1}{2}$ at the same time.

As α and β are functions of ϕ it will be useful to recast eqns. (5) and (7) so as to separate the energy from the variable. To do this, multiply (5) from the left by α,

$$\alpha\mathcal{H}\alpha = E\alpha^2 \tag{10}$$

and integrate from 0 to 2π,

$$\int \alpha\mathcal{H}\alpha \text{đ}\phi = E\int \alpha^2 \text{đ}\phi \tag{11}$$

$$E = \int \alpha\mathcal{H}\alpha \text{đ}\phi \Big/ \int \alpha^2 \text{đ}\phi = \alpha\mathcal{H}\alpha \text{đ}\phi \tag{12}$$

since α is normalised. Turning to the Hamiltonian operator itself, we know the energy of an isolated nucleus, say A, is $E_A = -M_I h\nu_A$. Therefore eqn. (5) is satisfied if we take $\mathcal{H}_A = -M_I h\nu_A$; for if the nucleus in the higher energy state ($M_I = -\frac{1}{2}$) we have,

$$E_A = \int \alpha_A \mathcal{H}_A \alpha_A \, d\phi = \tfrac{1}{2}h\nu_A \int \alpha_A{}^2 d\phi = \tfrac{1}{2}h\nu_A \qquad (13)$$

since $\mathcal{H}_A$ is a constant and so commutes with α_A.

The next step is to choose wavefunctions to describe the system of two non-interacting protons. When $M_I(A) = M_I(X) = -\frac{1}{2}$ the Schroedinger eqns. for the isolated nuclei are,

$$\mathcal{H}_A \alpha_A = E_A \alpha_A; \qquad \mathcal{H}_X \alpha_X = E_X \alpha_X. \qquad (14)$$

Multiply the first eqn. from the left by α_X, the second by α_A, and add them together. Then we have,

$$\alpha_X \mathcal{H}_A \alpha_A + \alpha_A \mathcal{H}_X \alpha_X = \alpha_X E_A \alpha_A + \alpha_A E_X \alpha_X. \qquad (15)$$

But $\mathcal{H}_A$ does not operate on α_X, or $\mathcal{H}_X$ on α_A; hence the left-hand side of (15) equals $(\mathcal{H}_A + \mathcal{H}_X)\alpha_A \alpha_X$. E_A and E_X are constants and so commute with α_A and α_X, so that the right-hand side equals $(E_A + E_X)\alpha_A \alpha_X$. Eqn. (15) can therefore be written,

$$(\mathcal{H}_A + \mathcal{H}_X)\alpha_A \alpha_X = (E_A + E_X)\alpha_A \alpha_X. \qquad (16)$$

Thus $\psi_1 = \alpha_A \alpha_X$ is a solution of the composite equation. Similar considerations for the states 2, 3, and 4 lead to the wavefunctions shown in the fifth column of Table 8.5.1. The reasoning employed in eqns. (10) to (13) can then be used to obtain the energy of the system in each of the various states. When $M_I(A) = M_I(X) = -\frac{1}{2}$ (state 1), we have

$$E = \int \psi_1 \mathcal{H} \psi_1 \, d\phi / \int \psi_1^2 d\phi. \qquad (17)$$

The denominator equals unity since the spin wavefunctions are normalised. Therefore,

$$\begin{aligned} E &= \int \alpha_A \alpha_X (\mathcal{H}_A + \mathcal{H}_X) \alpha_A \alpha_X \, d\phi \\ &= \int \alpha_A \mathcal{H}_A \alpha_A \, d\phi \int \alpha_X{}^2 d\phi + \int \alpha_X \mathcal{H}_X \alpha_X \, d\phi \int \alpha_A{}^2 d\phi \\ &= \tfrac{1}{2}h(\nu_A + \nu_X) \end{aligned} \qquad (18)$$

which accords with the result obtained earlier [eqn. (4)].

When the AX nuclei interact a further term must be added to $\mathcal{H}$ to express their interaction. Instead of $\mathcal{H} = \mathcal{H}_A + \mathcal{H}_X$, we take

$$\mathcal{H} = \mathcal{H}_A + \mathcal{H}_X + \mathcal{H}_{AX}, \tag{19}$$

in which $\mathcal{H}_{AX}$ is the interaction operator. The selection of $\mathcal{H}_{AX}$ is semi-empirical in that the final criterion is its success in explaining the observed spectra. For two nuclei, $\mathcal{H}_{AX}$ is assumed to have the form,

$$\mathcal{H}_{AX} = \tfrac{1}{4}Jh(2P_{AX} - 1). \tag{20}$$

Here, J is a constant with the dimensions Hz, so that Jh has the dimensions of energy; and P_{AX} is an operator that interchanges the nuclei A and X. The energy of the highest state (state 1) in the manifold is then,

$$\begin{aligned} E &= \int \psi_1(\mathcal{H}_A + \mathcal{H}_X + \mathcal{H}_{AX})\psi_1 \, d\phi \\ &= \int [\psi_1(\mathcal{H}_A + \mathcal{H}_X)\psi_1 + \psi_1\mathcal{H}_{AX}\psi_1] \, d\phi. \end{aligned} \tag{21}$$

According to (18), the first term equals $\tfrac{1}{2}h(\nu_A + \nu_X)$. As to the second term,

$$P_{AX}\psi_1 = P_{AX}\alpha_A\alpha_X = \alpha_X\alpha_A = \psi_1. \tag{22}$$

$$\text{Therefore } \tfrac{1}{4}Jh\int \psi_1(2P_{AX} - 1)\psi_1 \, d\phi = \tfrac{1}{4}Jh\int [2\psi_1^2 - \psi_1^2] \, d\phi = \tfrac{1}{4}Jh \tag{23}$$

Hence the energy of state 1 is,

$$E_1 = \tfrac{1}{2}h(\nu_A + \nu_X) + \tfrac{1}{4}Jh. \tag{24}$$

For the state 2 we have $\psi_2 = \alpha_A\beta_X$. But $P_{AX}\psi_2 = P_{AX}\alpha_A\beta_X = \beta_A\alpha_X = \psi_3$, and so $\int \psi_2 P_{AX}\psi_2 \, d\phi = \int \psi_2\psi_3 \, d\phi = 0$ owing to the orthogonality of the wavefunctions. The energy of state 2 is therefore,

$$\begin{aligned} E_2 &= \int [\psi_2(\mathcal{H}_A + \mathcal{H}_X)\psi_2 + \psi_2\mathcal{H}_{AX}\psi_2] \, d\phi \\ &= \tfrac{1}{2}h(\nu_A - \nu_X) + \tfrac{1}{4}Jh(0 - 1) \\ &= \tfrac{1}{2}h(\nu_A - \nu_X) - \tfrac{1}{4}Jh. \end{aligned} \tag{25}$$

The reader should be able to obtain for himself the energy of states 3 and 4, given in the right-hand column of Table 8.5.1.

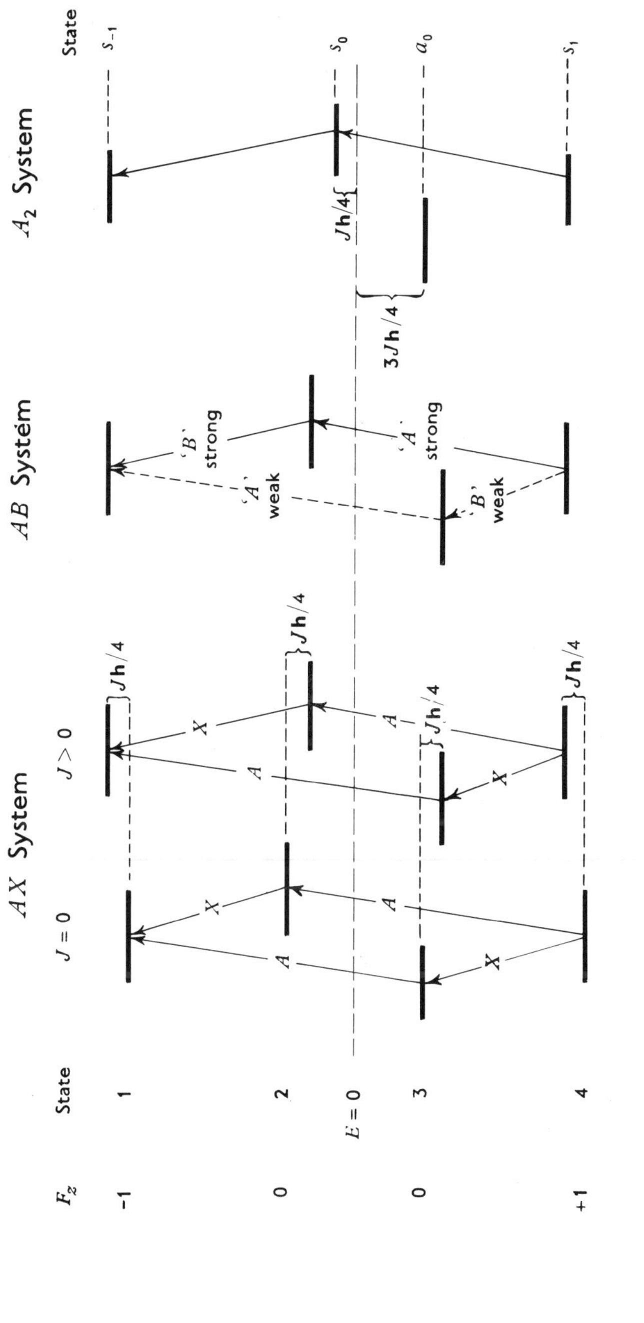

Fig. 8.5.2. Transitions of two coupled protons.

The energy levels of the coupled AX system are shown in Fig. 8.5.2. Four transitions are allowed by the F_z rule, namely,

$$\left.\begin{array}{ll} 4 \rightarrow 3 & (E_3 - E_4)/h = \nu_X - \frac{1}{2}J \\ 4 \rightarrow 2 & (E_2 - E_4)/h = \nu_A - \frac{1}{2}J \\ 3 \rightarrow 1 & (E_1 - E_3)/h = \nu_A + \frac{1}{2}J \\ 2 \rightarrow 1 & (E_1 - E_2)/h = \nu_X + \frac{1}{2}J \end{array}\right\} .$$

These transitions represent two doublets whose mid-points occur at the frequencies ν_A and ν_X. Each component of a doublet is displaced by $+\frac{1}{2}J$ from the mean position so that the doublet splitting (in Hz) is J. The spectrum corresponds to that illustrated at the top of Fig. 8.5.1.

It is useful to reconsider these results from a qualitative standpoint. When the A and X nuclei have the same spin orientation, that is, when both have $M_I = +\frac{1}{2}$ or both $-\frac{1}{2}$, one might expect that the field of one nucleus transmitted to the other will tend to raise its energy, since a parallel arrangement of dipoles normally corresponds to an increase of potential energy. In agreement with this, states 1 and 4 are found to have higher energy than in absence of interaction. Likewise, the interaction of opposite spins should tend to reduce the energy in the manner shown for states 2 and 3 in Fig. 8.5.2. In either case this intuitive argument explains the facts *provided the coupling constant J is positive*. But the interaction is transmitted through the bonding electrons, not directly through space, and therefore may not operate in the same sense as we have assumed. This consideration implies that J may be positive or negative in sign. Fig. 8.5.2 is drawn up on the assumption that J is positive. It is left to the reader to confirm that the appearance of the spectrum is unaffected by reversing the sign of J. This means the analysis of an AX spectrum can give no information as to whether J is positive or negative. A decision is sometimes possible in systems of more than two nuclei.

Two Equivalent Nuclei (A_2 System)

The system of two equivalent nuclei introduces a complication for those two states which have $F_z = 0$. Following the treatment of the AX problem, one might expect their wavefunctions to be $\psi_2 = \alpha_A \beta_A'$ and $\psi_3 = \beta_A \alpha_A'$ (primes are used merely to distinguish the nuclei, not to imply any non-equivalence) but this violates the principle that when two non-distinguishable nuclei have opposite spin it is impossible to tell which spin is attached to which nucleus. What we require are wavefunctions that are either left unchanged or change only in sign when the equivalent nuclei are switched about. The wavefunctions describing the highest and lowest states, $\psi_1 = \alpha_A \alpha_A'$ and $\psi_4 = \beta_A \beta_A'$, already meet this requirement, since permutation of the nuclei leaves both these wavefunctions unchanged.

The solution is to write new wavefunctions as the sum and difference of the old, namely,

$$\psi_s = 2^{-\frac{1}{2}}(\psi_2 + \psi_3) \quad (27)$$

and

$$\psi_a = 2^{-\frac{1}{2}}(\psi_2 - \psi_3). \quad (28)$$

The factor $2^{-\frac{1}{2}}$ enters in order that ψ_s and ψ_a shall be properly normalised. On permutation of the nuclei ψ_s is left unchanged, while ψ_a changes only in sign. This explains the subscripts s (= symmetric) and a (= anti-symmetric to exchange of the nuclei). The four states can be distinguished as s_{-1}, s_0, a_0 and s_1, according to whether the wavefunction is s or a and whether $F_z = -1, 0,$ or $+1$ (see Table 8.5.2).

Table 8.5.2. Energy Levels and Wavefunctions of A_2 and AB Systems

A_2 *System*

F_z	*State*	*Wavefunction*	*Energy* $J = 0$	*Energy* $J > 0$
-1	s_{-1}	$\alpha_A\alpha_{A'}$	$h\nu_A$	$h\nu_A + \frac{1}{4}Jh$
0	s_0	$2^{-\frac{1}{2}}(\alpha_A\beta_{A'} + \beta_A\beta_{A'})$	0	$+\frac{1}{4}Jh$
0	a_0	$2^{-\frac{1}{2}}(\alpha_A\beta_{A'} + \beta_A\alpha_{A'})$	0	$-\frac{3}{4}Jh$
1	s_1	$\beta_A\beta_{A'}$	$-h\nu_A$	$-h\nu_A + \frac{1}{4}Jh$

AB *System*

F_z	*Wavefunction*	*Energy* $J > 0$
-1	$\alpha_A\alpha_B$	$\frac{1}{2}h(\nu_A + \nu_B) + \frac{1}{4}Jh$
0	$a\alpha_A\beta_B + b\beta_A\alpha_B$	$\{[\frac{1}{2}h(\nu_A - \nu_B)]^2 + [\frac{1}{2}Jh)^2\}^{\frac{1}{2}} - \frac{1}{4}Jh$
0	$a\alpha_A\beta_B - b\beta_A\alpha_B$	$-\{[\frac{1}{2}h(\nu_A - \nu_B)]^2 + [\frac{1}{2}Jh]^2\}^{\frac{1}{2}} - \frac{1}{4}Jh$
1	$\beta_A\beta_B$	$-\frac{1}{2}h(\nu_A + \nu_B) + \frac{1}{4}Jh$

We next determine the energy of the four states. The highest and lowest, s_{-1} and s_1, present no problem since the wavefunctions are analogous to those for the AX system: in fact, all that is necessary is to substitute ν_A for ν_X in the formulae given on the extreme right in Table 8.5.1. As to the states s_0 and a_0, both correspond to a configuration with opposite spins so that unless there is interaction between the nuclei they are degenerate with $E(s_0) = E(a_0) = 0$. Put another way, only the interaction between nuclei makes any net contribution to the energy. Thus,

$$E(s_0) = \int \psi_s \mathcal{H}_{AA'} \psi_s \, d\phi, \quad (29)$$

and

$$E(a_0) = \int \psi_a \mathcal{H}_{AA'} \psi_a \, d\phi, \quad (30)$$

in which the interaction operator $\mathcal{H}_{AA'} = \frac{1}{4}Jh(2P_{AA'} - 1)$. Since $P_{AA'}\psi_s = P_{AA'}\psi_a = -\psi_a$, the expressions (29) and (30) give,

$$E(s_0) = \tfrac{1}{4}Jh \int [2\psi_s^2 - \psi_s^2]\, d\phi = \tfrac{1}{4}Jh \tag{31}$$

and

$$E(a_0) = \tfrac{1}{4}Jh \int [-2\psi_a^2 - \psi_a^2]\, d\phi = -\tfrac{3}{4}Jh. \tag{32}$$

The results to this point are summarised in Table 8.5.2 and Fig. 8.5.2.

Besides the selection rule for F_z, another rule operates which forbids transitions between states of different symmetry. The transitions $s_1 \to a_0$ and $a_0 \to s_{-1}$ are therefore disallowed and fail to appear in the spectrum. The allowed transitions $s_1 \to s_0$ and $s_0 \to s_1$ occur at the same frequency

$$[E(s_0) - E(s_1)]/h = [E(s_{-1}) - E(s_0)]/h = \nu_A\,, \tag{33}$$

so that the A_2 system gives rise to just one peak at the same frequency as an isolated A proton. It is impossible to tell anything from the spectrum about the value of J, even to decide if J is different from zero.

Nearly Equivalent Nuclei (AB *System*)

There is a gradual transition from the A_2 to the AX case as the nuclei become non-equivalent. When the shift is small, in place of the wavefunctions (27) and (28) for the states with $F_z = 0$, we write

$$\psi_+ = a\psi_2 + b\psi_3 \tag{34}$$

and

$$\psi_- = a\psi_2 - b\psi_3. \tag{35}$$

Here, a and b are constants which, however, must satisfy the condition $a^2 + b^2 = 1$ for ψ_+ and ψ_- to be normalised. As the AB system approaches A_2, ψ_+ correlates with the symmetric wavefunction ψ_s, and ψ_- with ψ_a; while at the same time a and b tend to become equal so that in the A_2 limit we have $a = b = 2^{-\frac{1}{2}}$. The states described by ψ_+ and ψ_- are therefore not strictly symmetric or antisymmetric with respect to exchange of nuclei; but ψ_+ contains more s than a character, ψ_- conversely.

It can be shown* that the energy of the states described by ψ_+ and ψ_- are given by the roots, E, of the determinantal equation,

$$\begin{vmatrix} E_2 - E & E_{23} \\ E_{23} & E_3 - E \end{vmatrix} = 0. \tag{36}$$

* Eyring, Walter, and Kimball, *Quantum Chemistry*, Wiley, New York, 1944, section 7*b*.

In this equation E_2 and E_3 are the eigenvalues of ψ_2 and ψ_3 and can be taken directly from Table 8.5.1; and,

$$E_{23} = \int \psi_2 \mathcal{H} \psi_3 d\phi \tag{37}$$

where $\mathcal{H} = \mathcal{H}_A + \mathcal{H}_B + \mathcal{H}_{AB}$. Therefore,

$$E_{23} = \int [\psi_2 \mathcal{H}_A \psi_3 + \psi_2 \mathcal{H}_B \psi_3 + \psi_2 \mathcal{H}_{AB} \psi_3] d\phi. \tag{38}$$

The first term in (38) equals

$$\int \alpha_A \beta_B \mathcal{H}_A \beta_A \alpha_B \, d\phi = \int \alpha_A \mathcal{H}_A \beta_A \, d\phi \int \beta_B \alpha_B \, d\phi = 0,$$

and the same result is obtained for the second term. The third term becomes,

$$\tfrac{1}{4}Jh \int [2\psi_2\psi_2 - \psi_2\psi_3] d\phi = \tfrac{1}{4}Jh[2-0] = \tfrac{1}{2}Jh. \tag{39}$$

Thus the energy of the two middle states of the AB system are the roots of (36) with $E_2 = \frac{1}{2}h(\nu_A - \nu_B) - \frac{1}{4}Jh$, $E_3 = \frac{1}{2}h(-\nu_A + \nu_B) - \frac{1}{4}Jh$, and $E_{23} = \frac{1}{2}Jh$.

Multiply out the determinant and we obtain,

$$\begin{aligned} E &= \tfrac{1}{2}(E_2 + E_3) \pm \{[\tfrac{1}{2}(E_2 - E_3)]^2 + E_{23}{}^2\}^{\frac{1}{2}} \\ &= -\tfrac{1}{4}Jh \pm \{[\tfrac{1}{2}h(\nu_A - \nu_B)]^2 + [\tfrac{1}{2}Jh]^2\}^{\frac{1}{2}}. \end{aligned} \tag{40}$$

We can test this equation in the A_2 limit by putting $\nu_A = \nu_B$, which gives $E = \frac{1}{4}Jh$ and $-\frac{3}{4}Jh$ [compare (31) and (32)]. Likewise, in the limit when $(\nu_A - \nu_B) \gg J$, $E = \frac{1}{2}h(\nu_A - \nu_B) - \frac{1}{4}Jh$ and $\frac{1}{2}h(-\nu_A + \nu_B) - \frac{1}{4}Jh$, corresponding to the energies E_2 and E_3 of the AX system. The energies of the four states of the AB system are collected together in the last column of Table 8.5.2.

The work of writing down the line positions in the AB spectrum is simplified if we make use of the fact that frequency measurements in nuclear resonance are not absolute; they are taken from a reference point in the spectrum. A convenient reference for the present purpose is halfway between A and B resonances. We can then write $\frac{1}{2}(\nu_A + \nu_B) = 0$ and relate the line positions to this origin. The four transitions allowed by the F_z rule occur at the relative frequencies given in the table below, where C is used as shorthand for $\{[\frac{1}{2}(\nu_A - \nu_B)]^2 + [\frac{1}{2}J]^2\}^{\frac{1}{2}}$. The transitions are arranged in order of increasing frequency, provided that J is positive in sign. A negative J alters the assignments but not the appearance of the spectrum.

Three aspects of the analysis deserve special attention. First, the separation of the first and second, and also the third and fourth, transitions in the table is equal to $|J|$, so that the magnitude of the coupling constant can be obtained directly from the spectrum. Secondly, the separation of the first and third transitions, as well as the second and fourth, is $2C = \{(\nu_A - \nu_B)^2 + J^2\}^{\frac{1}{2}}$; thus the interval between next neighbours, combined with $|J|$, gives the A–B chemical shift, $\nu_A - \nu_B$. Thirdly, the first and last transitions in the table correlate in the A_2 limit with transitions that are forbidden by the $s \nleftrightarrow a$ selection rule. This prohibition is not strictly enforced

Transition A_2 *limit*	AX *limit*	*Relative frequency*	*Intensity*
$s_1 \to a_0$	$4 \to 3$	$-C - \frac{1}{2}J$	weak
$s_0 \to s_{-1}$	$2 \to 1$	$-C + \frac{1}{2}J$	strong
$s_1 \to s_0$	$4 \to 2$	$C - \frac{1}{2}J$	strong
$a_0 \to s_{-1}$	$3 \to 1$	$C + \frac{1}{2}J$	weak

$$C = \tfrac{1}{2}\{(\nu_A - \nu_B)^2 + J^2\}^{\frac{1}{2}}$$

in an AB system, but transitions to and from the state described by ψ_- are allowed only in proportion to the s character in the wavefunction and so are weak relative to the transitions to and from the largely symmetric state (ψ_+). Accordingly, the AB spectrum consists of two pairs of lines of unequal intensity, the inner lines being more intense. As the A–B shift becomes progressively smaller the inner lines intensify at the expense of the outer lines. Eventually the inner lines coalesce and the outer lines vanish, corresponding to the single peak observed in the A_2 limit.

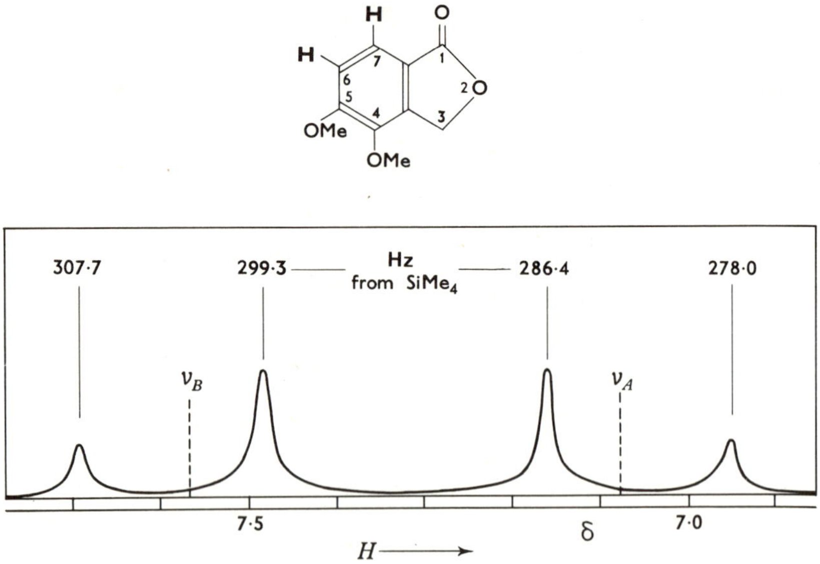

Fig. 8.5.3. Spectrum of the aromatic protons (H_6, H_7) of 4,5-dimethoxyphthalide.

In an AB spectrum, it can be shown that the ratio of intensities of inner and outer lines is given by the expression,

$$\frac{I(\text{inner})}{I(\text{outer})} = \frac{2C+J}{2C-J}. \tag{41}$$

Although intensity cannot be measured as accurately as frequency, eqn. (41) is useful as a check on the analysis.

The aromatic hydrogens of 4,5-dimethoxylphthalide form a typical AB spin system. A record of their spectrum measured on an instrument operating at 40 MHz is shown in Fig. 8.5.3 (other protons in the molecule give rise to signals lying outside the region contained in the figure). The displacement (in Hz) of each peak from the tetramethylsilane resonance is given in the body of the diagram. It is left to the student to show that:

$$\delta_A = 7{\cdot}07\,(A = \mathrm{H}_6) \qquad |J| = 8{\cdot}4\ \mathrm{Hz}$$

$$\delta_B = 7{\cdot}56\,(B = \mathrm{H}_7) \qquad I(\text{inner})/I(\text{outer}) = 2{\cdot}3\ (\text{calc.})$$

The analysis gives the chemical shift for the two protons without distinguishing between them. From the spectra of related compounds one may determine that H_7 (deshielded by the neighbouring carbonyl group) occurs at the lower field.

8.6 Pure Nuclear Quadrupole Resonance in Solids

The second type of nuclear resonance arises when a nucleus possesses an electric quadrupole moment. In section 5.7 the classical electrostatic orientation energy of a quadrupolar nucleus in a field of axial symmetry was seen to be

$$E_Q = \frac{1}{8}eQq\,(3\cos^2\theta - 1). \tag{5.7.7}$$

The complication of free molecular rotation with its associated angular momentum does not arise for molecules in the solid state as it does for gaseous molecules. We can then, in a quantum mechanical approach consider the nuclear spin I to be directly quantised with respect to the axis of the field. The allowed values for the component I_z of the spin angular momentum along the field axis (z) are then

$$I_z = M_I h/2\pi. \tag{8.2.1}$$

with M_I given by eqn. (8.2.2), and the quantum mechanical expression for the quadrupole interaction energy E_Q can be shown to be

$$E_Q = eQq\,\frac{3M_I^2 - I(I+1)}{4I(2I-1)}\,. \tag{1}$$

The scope of eqn. (1) needs to be clearly understood. It applies to a nucleus in an axially symmetric electric field; because in a field of lower symmetry the single parameter $q = \partial^2 V/\partial z^2$ is not sufficient to describe the field gradient. The case when the field symmetry is less than axial need not concern us, however. On the other hand, we shall on occasions meet a situation when the environment of the given nucleus is spherically symmetrical. In this event the three components of the field gradient must all be equal, that is,

$$\partial^2 V/\partial x^2 = \partial^2 V/\partial y^2 = \partial^2 V/\partial z^2,$$

and so we have, from the Laplace equation,

$$\partial^2 V/\partial x^2 = \partial^2 V/\partial y^2 = \partial^2 V/\partial z^2 = 0. \tag{2}$$

Hence the components of the field gradient are all zero, which means that the field at the nucleus is uniform. All orientations of the nucleus then have the same energy, the energy level being $(2I+1)$-fold degenerate.

The frequency required to induce the reorientation of a nucleus in an axial field is calculated from eqn. (1). Suppose that the nucleus has $I = \frac{3}{2}$; therefore $M_I = \pm\frac{1}{2}$, $\pm\frac{3}{2}$, and hence

$$E_Q = \pm\tfrac{1}{4}eQq. \tag{3}$$

The positive sign applies when $M_I = \pm\frac{3}{2}$, the negative sign when $M_I = \pm\frac{1}{2}$. The interaction of the nuclear quadrupole with the electric field splits the four-fold degenerate nuclear energy level into two sub-levels each doubly degenerate. Which sub-level is higher in energy depends upon the sign of eQq. In Fig. 8.6.1 we show the splitting that occurs when eQq is positive: if eQq is negative, the sub-levels become interchanged. In either case, the frequency of the transition is given by

$$h\nu = E_Q' - E_Q'' = \tfrac{1}{2}\,|eQq|. \tag{4}$$

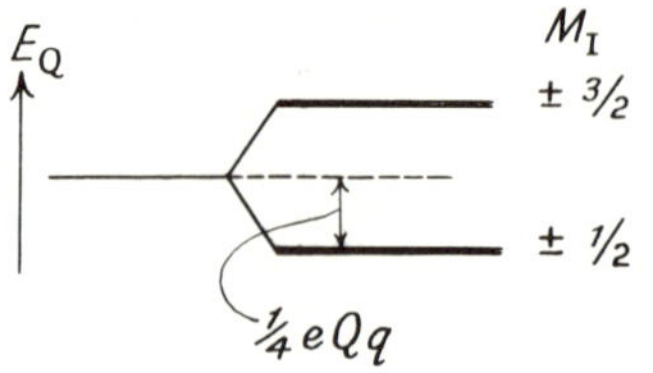

Fig. 8.6.1. Energy sub-levels of a nucleus with $I = \frac{3}{2}$ in an axially symmetric electric field.

Therefore an experimental measurement of ν yields $|eQq|$, *i.e.*, the magnitude of the product eQq but not its sign.

The absorption frequency of a nucleus in a given compound is observed by exposing a sample to a radio-frequency field. The sample is placed in the tuned coil of an oscillator whose output frequency can be varied over a small range. Absorption of energy from the field of the coil occurs when the oscillator sweeps through the resonant frequency ν: this absorption can be detected and displayed visually on an oscilloscope screen. The present theory does not apply to gases and so gas samples cannot be used. It turns out that crystalline solids are suitable (though with a reservation, mentioned later), but not liquids. This might have been foreseen, for in liquids the haphazard movement of a molecule with respect to its neighbours causes slight variations in the field gradient q which broaden the absorption line into a band too weak to be detected.

The coupling constants obtained from rotation spectra of gases (section 5.6) are often several per cent greater in magnitude than the values determined from pure nuclear quadrupole resonance in the solid state. Table 8.6.1 cites examples, all relating to nuclei in an axially symmetrical location within the molecule, where eQq has been determined by both methods. In part, the explanation of the difference is that the environment of a molecule in the crystal lattice has lower symmetry than the molecule itself. Neighbouring molecules therefore disturb the pure axial symmetry of the field within an isolated linear molecule or symmetric top, and a defect arises from the use of eqn. (1) in a situation where it is not strictly valid. Fortunately the effect is small enough not to be too troublesome. The magnitude of the defect relative to the crystal structure is an interesting field for exploration.

Table 8.6.1. Coupling Constants from Rotational Hyperfine Structure and from Pure Nuclear Quadrupole Resonance

Nucleus	*Substance*	eQq *vapour MHz*	$\|eQq\|$ *solid state MHz*
^{14}N	ICN	−3·80	3·39
	BrCN	−3·83	3·28
^{35}Cl	CH_3Cl	−74·8	68·4
	CF_3Cl	−78·0	77·6
^{79}Br	CH_3Br	577·3	528·9
^{81}Br	CH_3Br	482·4	441·8
^{127}I	ICl	−2930	3037
	ICN	−2420	2549
	CH_3I	−1934	1753

Neither pure quadrupole spectra nor the hyperfine structure of rotational transitions afford means for separating the quadrupole coupling constant eQq into its factors. For common nuclei the quadrupole moment eQ is known approximately from atomic beam experiments so that q (or $|q|$) may be estimated. For isotopic molecules, where the field gradient q is unchanged, the coupling constants eQq are in the same ratio as the nuclear quadrupole moments eQ (see Tables 8.6.1, 5.5.1 and 5.7.2).

8.7 Nuclear Quadrupole Coupling and the Chemical Bond

The electric field gradient q at a nucleus is an indication of the electron configuration in the vicinity of the nucleus. In the present section we shall examine this relationship more closely and consider some of the information that can be deduced from it. We shall not attempt to give the theory, but will simply describe the results.

Consider first a single atom or ion. If the electrons form a closed shell—as, for instance, in argon, or as in the chloride or potassium ions—the environment of the nucleus is spherically symmetrical and thus the electric field is uniform: in this case, then, $q = 0$. The same argument applies to each inner shell of electrons taken separately: if a shell is filled the charge distribution is spherical, and therefore the electrons within the shell make no contribution to q. Thus we can ignore all electrons except those in the outermost valency shell in assessing the contribution to q. Take as an example the chlorine atom where the electronic configuration is $(1s)^2(2s)^2(2p)^6(3s)^2(3p)^5$. The orbitals up to and including $3s$ are filled and the electrons in them give rise to a uniform field at the nucleus. However in the set $3p$ there is an unbalance of one electron and, moreover, the configuration $(3p)^5$ is one that must establish a field which is symmetrical about the axis of the singly-occupied $3p$ orbital. Hence we can expect the $(3p)^5$ configuration to give rise to a field gradient whose axial component at the nucleus will differ from zero. Likewise the valence shell configurations p^4, p^2, and p will be associated with q's that are different from zero, though it should be remembered that q is only observable for those isotopes which possess a quadrupole moment. The physical basis for the dependence of the energy of a nucleus upon its orientation in the field of a p-orbital is illustrated schematically in Fig. 8.7.1.

Two configurations which do not give rise to a field gradient at the nucleus may be mentioned here. s orbitals have spherical symmetry: therefore an atom that has a single s electron in its valence shell (*e.g.*, Na, K) must have a uniform electric field at the nucleus. Secondly, the configuration p^3 will give rise to a uniform field, if, but only if, one electron goes into each of three available p-orbitals; for then the unbalance is equal for each orbital, yielding a symmetrical structure. This situation obtains in the ^4S ground state of the nitrogen atom.

Turning to polyatomic systems, the gradient q at a given nucleus must be attributed in principle to all other charges in the molecule. Not all charges are equal in importance, however. Eqn. (5.7.3) gives the field gradient due to a charge e at a fixed distance z as

$$q = 2e/z^3. \qquad (5.7.3)$$

This equation requires modification before it will apply to the internal situation in a molecule, for the motion of the electrons in their orbits means that the distance from an electron to a nucleus is a fluctuating one. Theory shows that the contribution to q of any charge in the molecule is proportional to the *average value of the inverse cube of its distance from the nucleus in question*. In consequence, q depends in first approximation upon the electrons encircling the nucleus, the effect of the remaining nuclei and electrons being small in comparison.

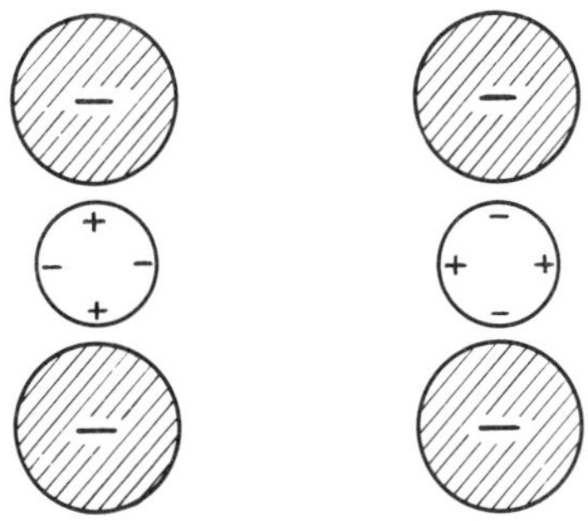

Fig. 8.7.1. Ground state (*a*) and excited (*b*) configuration of a nucleus in the field of a *p*-orbital electron (schematic).

The chlorine compounds in Table 8.7.1 provide an interesting series of results. The fourth column of the table gives the observed coupling constants for the ^{35}Cl nucleus. The chloride ion has a filled shell of electrons; therefore inorganic chlorides should have $q = 0$ at the chlorine nucleus, provided the interatomic 'bond' is a pure electrovalency. The ionic chlorides actually have very small or zero coupling constants, as expected. The second fixed point in the discussion is the chlorine atom whose coupling constant, – 109·7 MHz, can be taken as characteristic of the state in which there is an unbalance of one $3p$ electron. Two electrons are shared equally between the nuclei in the chlorine molecule, implying that on an average one electron is missing from each nucleus. The fact that the coupling constant for the molecule is nearly equal to that for the atom Cl thus indicates that the bond in Cl_2 is a pure p bond, so that the mean defect amounts to one p electron at each atom. The coupling constant for methyl chloride is considerably less than for the chlorine atom, as would be expected if the environment of the chlorine atom in methyl chloride had shifted towards that in the chloride ion. Put another way, the sense of the change is explained if the unbalance at the chlorine nucleus in methyl chloride amounts to less than one p electron. One may suppose, therefore, that the chlorine retains control over more than one-half of the electrons bonding it to the methyl

Table 8.7.1. Quadrupole Coupling Constants for ^{35}Cl

Substance	*State*	*Structures*	*eQq* MHz	*% Ionic Character* from (*eQq*) atomic	Pauling's scale
Cl (atom)	gas		−109·7		
KCl	gas	K^+Cl^-	<0·04	100	70
NaCl	gas	Na^+Cl^-	<1	>99	66
TlCl	gas	Tl^+Cl^-, Tl—Cl	−15·8	86	
SiH_3Cl	solid	$SiH_3^+Cl^-$, SiH_3—Cl	−40·0	64	30
HCl	gas	H^+Cl^-, H—Cl	−53·4	51	19
CH_3Cl	gas	CH_3—Cl, $CH_3^+Cl^-$	−75·3	31	5
ICl	gas	I—Cl, I^+Cl^-	−82·5	25	5
BrCl	solid	Br—Cl, Br^+Cl^-	−103·6	5	1
Cl_2	gas	Cl—Cl	−109·0	<1	0
FCl		F—Cl, F^-Cl^+	−146·0		

group. This view associates the low value of eQq in methyl chloride and related molecules with the *inductive effect* of the chlorine atom: thus it is in line with evidence from dipole moments (Section 11.3) and with a wealth of observations on the influence of a chlorine substitutent upon equilibria and chemical reactivity. In the alternative language of resonance theory, methyl chloride is a hybrid of the hypothetical pure covalent and ionic forms, CH_3—Cl and $CH_3^+Cl^-$. The remaining compounds in the table bear out this point of view. In FCl, where Cl tends to be positive (resonance between F—Cl and F^-Cl^+), eQq is numerically greater than for atomic chlorine, indicating an average unbalance of more than one p electron. In other substances the p orbital defect is less than unity, showing that the chlorine atom has some negative character.

It has been suggested* that the quantity $(1 - (eQq)/(eQq)_{\text{atomic}}$ is an index of the fractional *ionic character* of the bond to chlorine in compounds where the chlorine atom tends to be negative. Such values, expressed as percentages, are given in the fifth column of Table 8.7.1. But it should be remarked that this assumption leads merely to one of several semi-empirical scales of ionic character. Historically the first such scale was deduced by Pauling† from consideration of bond energies, and several of Pauling's values are given in the final column of the table. Although there is a measure of agreement as to the relative amounts of ionic character, the numerical estimates differ quite seriously. The scale based on $(eQq)_{\text{atomic}}$ has the advantage that it rests on quantities which are accurately measurable: but a final verdict on the quantitative aspects of ionic character is still awaited.

In other molecules the coupling constant can give information on the hybridisation of atomic orbitals used in bond formation. Ammonia is a case in point. It is known that the molecule is pyramidal with the interbond angle, $\angle$HNH,

* Gordy, *Discuss. Faraday Soc.*, 1955, **19**, 14: compare, however, Dailey, *ibid.*, p. 255.
† Pauling, *Nature of the Chemical Bond*, Cornell University Press, 1945, sec. 12.

approximately equal to 107°. As to the bond structure, there are two possibilities. *Either* the N—H bonding orbitals are formed by the overlap of a pure $2p$ nitrogen orbital with $1s$ atomic orbitals of hydrogen, the bonds being bent outwards by some repulsive force between the hydrogen atoms so that the ∠HNH is greater than the natural angle, 90°, subtended by p-orbitals: *or* hybridisation at the nitrogen atom gives orbitals of mixed s and p character, which then overlap with the hydrogen atom orbitals in bond formation. Whichever description applies, approximately one-half the electron probability in a bond orbital will be attached to nitrogen, and one-half to the hydrogen atom. If the first description is correct, each of the three $2p$ orbitals must be equally occupied leading to a symmetrical structure (the non-bonding electrons fill the $2s$ orbital), and hence $eQq = 0$. On the other hand, if the orbitals are hybridised, the equal occupation of the bonding orbitals does not establish spherical symmetry of charge because the orbital carrying the unshared pair is now a hybrid with p-character and so will unbalance the p-electron distribution. In fact, the quadrupole coupling constant for the nitrogen nucleus in ammonia is −4·10 MHz and thus there must be some p character in the orbital holding the unshared pair. Evidently sp^3 hybridisation (leading to ∠HNH ≃ 109·5°) would be in fair agreement with the known geometry; but unfortunately the coupling constant cannot give a quantitative ratio for the hybridisation, because the unbalance at the nitrogen nucleus associated with a single $2p$ electron is as yet undiscovered.

BIBLIOGRAPHY

Pople, Schneider, and Bernstein, *High-Resolution Nuclear Mangetic Resonance*, McGraw-Hill, New York, 1959.

Roberts, *Introduction to the Analysis of Spin–Spin Splitting in Nuclear Magnetic Resonance*, Benjamin, New York, 1961.

Chapter 9

Crystal structure analysis

9.1 Introduction

Diffraction is an essential property of any form of wave-motion. Under appropriate conditions two beams of waves can interact so as to reinforce, or interfere with, one another. Light *plus* light may produce darkness.

In particular when a beam of radiation interacts with matter, the rays scattered by different atoms will give rise to diffraction effects, which can be used to get information about the relative positions of the atoms. A number of techniques for the study of molecular structure have been based on this principle. The radiation may take different forms, provided only that its wavelength is of the same order of size as the distance between neighbouring atoms. The material may be studied in the gaseous, liquid, or amorphous forms; but the method is most powerful when crystalline solids are used. In this chapter the methods of crystal structure analysis will be described, with all the emphasis on the X-ray method. X-rays were used exclusively in the development of these methods and X-rays still dominate the field.

In a crystal some unit of structure is repeated by the symmetry elements embodied in the space-group (see section 2.13) so as to give a three-dimensional pattern. This asymmetric unit may be a single atom (as in metallic iron) or an ion-pair (as in the alkali halides), or a part of a molecule (as in naphthalene), or a whole molecule (as in a sugar), or some larger and more complex group (as in $p:p'$-dimethoxybenzophenone, in which two crystallographically distinct molecules are repeated). This regularity of array facilitates study, even without recourse to diffraction. For example, it is sometimes found that the refractive index of a crystal is relatively high for light polarised so that its electrical vector vibrates in a particular crystallographic direction; this implies that the electrons are more free to move in that direction than in another. With an aromatic compound this may indicate that the benzenoid rings are orientated with their planes in that direction. Anisotropy of magnetic properties can be similarly interpreted in molecular terms.

Fullest advantage of the ordered arrangement of units is taken when the crystal is used as a diffraction medium. In favourable cases it is possible to determine the positions of all the atoms in the unit cell. When this has been done for a molecular crystal, the molecular structure is known; and all the questions asked by classical stereochemistry are answered, as well as many others.

Necessarily the molecular structure found will be that obtaining in the crystal. It is possible that a different one may occur in the gaseous state or in solution. (An extreme example is phosphorous pentachloride: see section 12.5). With most covalent compounds, however, and especially with typical organic crystals, the valency forces holding the atoms together within the molecule are much stronger than the van der Waals forces between molecules. The effect of the crystalline environment on the molecular structure will then be slight. A small difference of conformation is sometimes detected: the benzenoid rings in diphenyl are probably coplanar in the solid state; in the vapour, one ring is rotated through some 40° about the C—C bond linking it to the other. Usually, when a comparison has been made, the structures appear to be almost identical.

Even when an X-ray analysis has been pushed to the highest attainable state of refinement, the accuracy with which molecular dimensions can be measured is considerably less than that attainable by spectroscopic methods. At the time of writing, few bond lengths have been determined by X-rays with an accuracy as good as ±0·002 A. But the X-ray method can be applied to a far wider range of materials, including those with molecules of great complexity.

9.2 Principles of X-ray Diffraction by Crystals

At the beginning of 1912 it was not known whether X-rays consisted of particles or waves, and nothing was known about the internal structures of crystals. A suggestion made by von Laue elucidated both these problems, and in doing so originated the whole subject of X-ray crystallography. He pointed out that the wavelength (~1 Å) then roughly estimated for X-rays, on the assumption that they were indeed a wave-motion akin to ordinary light, was approximately equal to the interatomic distances in crystals implied by the known value of the Avogadro number; and hence that a crystal might act as a three-dimensional diffraction grating for X-rays. After overcoming certain difficulties, Friedrich and Knipping demonstrated the occurrence of diffraction when a beam of X-rays was passed through a crystal of copper sulphate pentahydrate. They obtained what would now be recognised as the first *Laue photograph*. The detailed interpretation of such diffraction patterns was not straightforward, because the X-rays used were 'white'–*i.e.*, their wave-lengths covered a considerable range–and this was not fully realised at the time. von Laue did however state a fundamental equation of X-ray diffraction from a crystal. When rays of wavelength λ are scattered from a row of points regularly spaced a distance a apart, and when ϕ_0 is the angle the direction of the incident rays makes with the row and ϕ_1 that of the scattered ray, then reinforcement will occur provided that the difference between the paths traversed by rays scattered from successive points is equal to a whole number (n) of wavelengths; *i.e.*, when

$$a(\cos\phi_0 - \cos\phi_1) = n\lambda, \ldots \qquad (1)$$

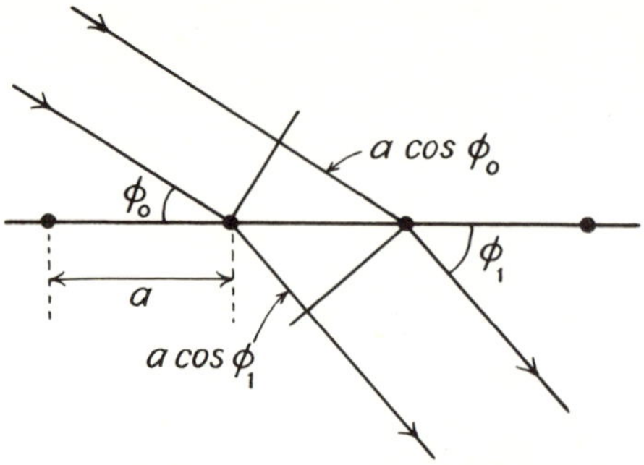

Fig. 9.2.1. Diagram to illustrate the Laue condition for reinforcement of X-rays scattered from a row of equidistant points.

as can easily be seen from Fig. 9.2.1.

A crystal can be regarded as composed of parallel rows of points along the directions of the three principal axes. The condition for a reflexion is that three equations of the form of (1) should be satisfied simultaneously. In the Laue photograph, ϕ_0 is held constant, but a continuous range of wavelengths is available, so that reflexions occur in certain directions when ϕ_1 and λ have appropriate values.

In the method initiated by W. L. Bragg, monochromatic X-rays are used, and a simpler treatment then becomes possible. With λ fixed as well as ϕ_0, the Laue conditions would not in general be fulfilled. When ϕ_0 is varied continuously by rotating the crystal, reflexions do occur at certain orientations. The conditions for such reflexions are defined by the Bragg equation (2), which is a special case of (1). The crystal is now regarded as built from parallel layers of scattering points, with a spacing d between the layers; reflexion from these layers, as from a mirror, takes place when the angles of incidence and reflexion, θ, are equal, and when the difference in path between rays scattered from successive layers equals a whole number (n) of wavelengths. From Fig. 9.2.2 it should be evident that

$$n\lambda = 2d \sin \theta. \qquad (2)$$

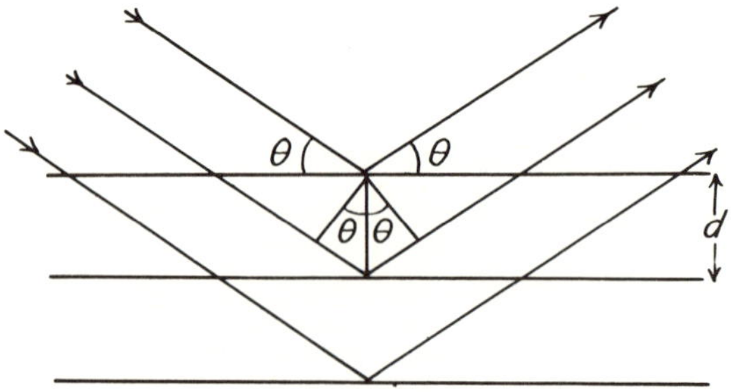

Fig. 9.2.2. Diagram to illustrate the Bragg condition for reinforcement of X-rays scattered from equidistant, parallel layers of scattering material.

As the crystal represented in Fig. 9.2.2 is rotated counter-clockwise, θ will increase, and reflexions will occur every time the Bragg equation is satisfied, with $n = 1, 2, 3$, etc. Since $\sin\theta$ cannot exceed unity—the value reached when the beam is reflected back perpendicularly—only a limited number of orders can be observed. When the spacing is 10 Å, for instance, and X-rays from a copper target are used ($\lambda = 1{\cdot}54$ Å),

$$\sin\theta = (1{\cdot}54n)/20 = 0{\cdot}077n,$$

and only twelve orders are accessible. Should it be desired to observe a wider range of reflexions, X-rays of shorter wavelength may be substituted—*e.g.*, those from molybdenum. But there is always a limit implicit in the Bragg equation, quite apart from another phenomenon which restricts the observation of high-order reflexions (see section 9.11). The importance of these limitations will be discussed later.

The layers of lattice points shown in Fig. 9.2.2 will be parallel to some (actual or potential) external face of the crystal. Suppose the face were (100). The successive reflexions from this face are usually denoted by substituting the Millerian indices for n; so that the orders would correspond to $h00 = 100, 200, 300$, etc. The second order ($n = 2$) can thus be legitimately considered as a first-order reflexion from planes (200) of spacing, $\frac{1}{2}d$. This is as if additional scattering-points were so placed as to interleave those lying on the layers explicitly shown in Fig. 9.2.2. In an actual crystal, whose unit cell comprises atoms in various positions, this situation often obtains, to a greater or lesser degree. A similar interpretation can be given to higher orders: h-th order from 100 is equivalent to 1st order from $h00$, with spacing $(d(100))/h$.

Besides the layers of points parallel to the obvious, external faces of the crystal, an infinite number of general (hkl) planes can be discerned. A few are noted for the two-dimensional net represented in Fig. 9.2.3. In principle, each such set of layers gives rise to a series of Bragg reflexions, subject always to the limitation mentioned

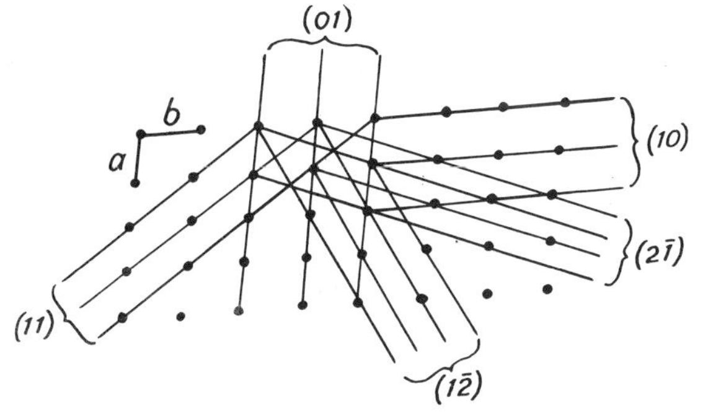

Fig. 9.2.3. Two-dimensional illustration of some of the (infinite number of) layers of points that exist in an infinite lattice.

earlier; when the layer-spacing is less than $\lambda/2$, the reflexion is no longer observable. The h-, k-, and l-indices of the particular set of layers turn out to be the value of n in the three Laue equations that must be satisfied.

9.3 The Reciprocal Lattice

When a point of light is viewed through a piece of woven fabric, a two-dimensional diffraction pattern can be observed, whose geometry depends upon that of the fabric. In an analogous way the totality of possible reflexions from a given crystal can be regarded as a three-dimensional X-ray *diffraction pattern*. The reflexions are usually, but not necessarily, recorded photographically; and various types of X-ray photographs can be taken. Their technical differences will not be described in any detail in this book. But they can all be considered as derived in some way from the diffraction pattern.

Crystallographers find it convenient to discuss the X-ray pattern in terms of the *reciprocal lattice*. The concept can be explained in terms of its two-dimensional analogue. Fig. 9.3.1(i) represents a net in real space. Two prominent rows of points are selected as principal axes, a and b, thus defining a unit cell, one of which is indicated. The horizontal rows correspond to the (10) planes, which would give rise to the 10 Bragg reflexion. This reflexion, in its first order, is represented in *reciprocal space* by taking an arbitrary origin and drawing from it a vector of direction perpendicular to the (10) planes and of length inversely proportional to $d(10)$, the 10-spacing. In Fig. 9.3.1(ii) this vector is represented by the point labelled 10, with respect to the origin 00. The second order from the same set of planes corresponds to half the spacing $d(10)$ and hence is represented by the point 20. Negative indices such as $\bar{1}0$ denote reflexions from the opposite sides of the (10) planes. Similarly

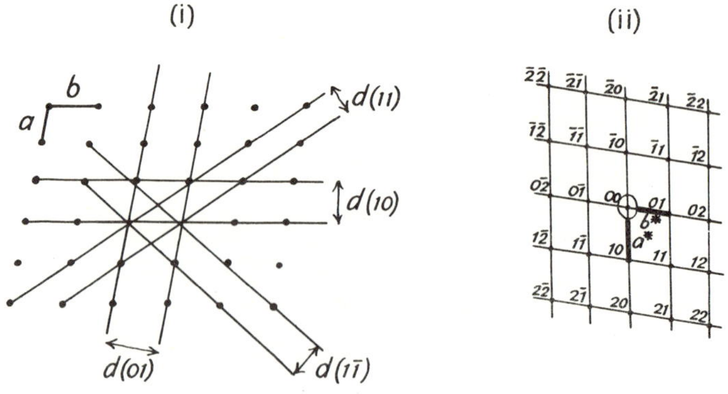

Fig. 9.3.1. Two-dimensional illustration of the relationship between lattices in (i) real, and (ii) reciprocal, space.

Bragg reflexions from the more widely spaced (01) planes are represented by the more closely spaced points in reciprocal space, $0\bar{1}$, 01, 02, etc.; and those from the diagonal (11) planes by $\bar{1}\bar{1}$, 11, 22, etc. In fact, Fig. 9.3.1(*ii*) is the reciprocal net derived from 1(*i*). The same construction can be applied in three dimensions to yield the reciprocal lattice appropriate to any particular crystal lattice.

Two matters require some emphasis. First, any single reciprocal lattice point, such as that indexed 10, corresponds to the *set* of planes; not merely to the two planes specifically marked *d*(10) in Fig. 9.3.1(*i*), but to the infinite set of such planes which give rise to the 10-reflexion. Secondly, the reciprocal of the reciprocal lattice is the original lattice in real space. Real (or direct) and reciprocal spaces are inversely related.

If each point in the reciprocal lattice is now weighted in proportion to the intensity of the corresponding X-ray reflexion, then the reciprocal lattice is a complete representation of the behaviour of the crystal in X-ray diffraction. Incidentally the weighted reciprocal lattice is an unique property of any crystalline material.

An X-ray diffraction photograph shows a pattern of 'spots'. This pattern is always derived from the weighted reciprocal lattice; usually only a section of the lattice is included in the photograph, and in most types of photograph there is more or less geometric distortion. Nevertheless, from a study of suitable X-ray photographs it is possible to reconstruct the reciprocal lattice; and to do this–at least in part– is always a preliminary to the analysis of a crystal structure.

Of the relation between photographic record and reciprocal lattice only one example will be given here. A crystal can be set up to rotate about an axis parallel to one of its principal crystallographic axes; let it be *a*. While the crystal is rotating, a fine beam of monochromatic X-rays is allowed to fall on the crystal in a direction normal to the axis; and the diffraction pattern is recorded on a photographic film bent into a cylindrical shape about the rotation-axis. On the developed film (*e.g.*, see Plate IA) spots appear along a number of parallel *layer lines*, which characterise the *single-crystal rotation photograph*.

The concept of reciprocal space may be developed by introducing the related concept of the *Ewald sphere*. The two-dimensional case is illustrated by Fig. 9.3.2, in which (*i*) merely repeats the essentials of Fig. 9.2.2 for the Bragg reflexion from a set of planes indexed as *hkl*. In (*ii*) we draw a circle – a section of the Ewald sphere; it has a radius $1/\lambda$ and is constrained so that one point on its circumference always passes through the origin, O, of reciprocal space. Following the rules above, we draw OB in a direction perpendicular to the *hkl* planes, and of length $1/d(hkl)$; B is therefore the reciprocal lattice point corresponding to the planes drawn in (*i*). We have orientated the circle so that it also passes through B. We now suppose the direction of the incident X-ray beam, E′C′ of (*i*), to lie along that diameter of the circle which passes through the origin, ECO of (*ii*), and the direction of the reflected beam, along C′B′ in (*i*), to be parallel to the line from the centre of the circle, C, towards the reciprocal lattice point, B. We now have to demonstrate that the construction we have described is equivalent to the Bragg condition explained in section 9.2.

In Fig. 9.3.2(*ii*) draw a perpendicular from C to bisect OB at A. It is easily seen that the angles OCA and ACB are each equal to θ, and therefore that

$$OA = AB = CO.\sin\theta; \quad \text{and hence}$$

$$1/d(hkl) = OB = 2.CO.\sin\theta = 2(1/\lambda)\sin\theta.$$

Rewritten as

$$\lambda = 2d\sin\theta,$$

this is seen to be the Bragg condition for the first-order reflexion from the *hkl* set of planes.

Though a formal proof in three dimensions is rather more difficult, the same law holds. Rotating a crystal in the X-ray beam is equivalent to rotating the Ewald sphere, of radius $1/\lambda$, with respect to the reciprocal lattice, whilst maintaining the constraint that a particular point of the surface of the sphere must be fixed at the reciprocal-lattice origin. The Bragg condition will be satisfied, and a reflexion may result, whenever any reciprocal lattice point comes to some other point on the surface of the sphere. And corresponding directional relationships hold.

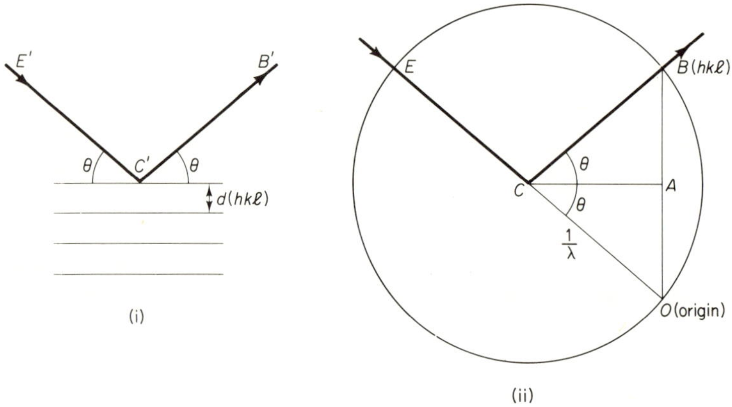

Fig. 9.3.2. Interpretation of the Bragg condition in terms of the Ewald sphere, which becomes a circle in this two-dimensional case. The reciprocal-lattice point, *B*, corresponds to the set of reflecting planes (*hkl*).

The disposition of spots on the single-crystal rotation photograph can now be explained. The Ewald sphere is set in reciprocal space as sketched in Fig. 9.3.3(*i*). This follows because the crystal has been set to rotate about its *a*-axis in direct space. From the account given above, as the crystal rotates points on the upper reciprocal-lattice net will cut the surface of the sphere along the small circle drawn in (*i*). Moving to (*ii*), we see that the directions of all consequent reflected rays must cut the

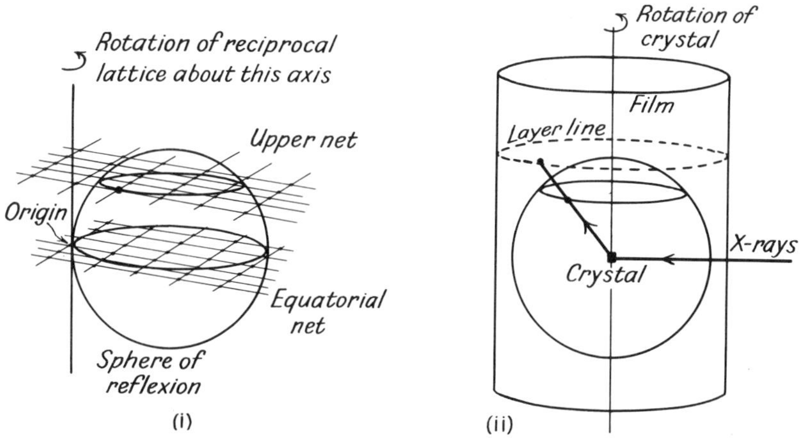

Fig. 9.3.3. Sphere of reflexion in reciprocal space, and formation of layer line pattern in a single-crystal rotation X-ray diagram.

enveloping cylindrical film at points on the (upper) layer-line, which is represented by the broken line. For each reciprocal lattice net that is within reach of the sphere of reflexion, there will be such a layer line. It should be evident that, when the crystal is rotated about its *a*-axis (as here), the equatorial layer line will comprise spots of the type $0kl$; the upper lines those of types $1kl$, $2kl$, etc.; and the lower those of types $\bar{1}kl$, $\bar{2}kl$, etc.

The spacing of the layer lines on the film readily leads to the primitive translation (*a* in the present instance) along the crystal axis parallel to the axis of rotation. One way of determining the dimensions of the unit cell (*a, b,* and *c*; and, where necessary, α, β, and γ) is to take a series of these photographs about a sufficient number of independent axes. A rather complicated type of X-ray camera developed by Buerger yields an undistorted picture of a selected reciprocal lattice net; this is the *precession photograph*, an example of which is shown in Plate IB.

9.4 Information to be derived from Unit-Cell Dimensions

Useful information can sometimes be derived from a knowledge of the unit-cell parameters. For example, they constitute a rather reliable means of identifying a crystalline substance. They also enable a good estimate to be made of its molecular weight (*M*). From the cell dimensions the unit-cell volume (U_0) can be calculated; an accuracy better than ± 1% is readily attainable. By flotation in a suitable medium, the density (*D*) of the crystal can be found with a similar accuracy. There must be a whole number (*Z*) of molecules in the cell: and for many substances this number is 1, 2, 4 or 8, with 2 or 4 commonest. We can therefore write

$$D = (ZM)/(N_A U_0),$$

N_A being Avogadro's number. If we know M even roughly, we can recognise the value of Z, and thence obtain an accurate value of M. For example, the acid potassium salt of *p*-hydroxybenzoic acid had been described by the formula $KH(C_7H_5O_3)_2$, which would require a molecular weight of 314·3. It was found later that the monoclinic unit cell has $a = 16{\cdot}50(1)$, $b = 3{\cdot}846(5)$, $c = 11{\cdot}20(1)$ Å, $\beta = 92{\cdot}8(2)°$, which leads to $U_0 = 709{\cdot}9$ Å^3, and that the density is 1·555 g cm^{-3}. The contents of the cell therefore correspond to $709{\cdot}9 \times 1{\cdot}555 \times 6{\cdot}023 \times 10^{-1} = 665$ (at. wt. units). The difference between this number and $2 \times 314{\cdot}3$ is significant. The explanation is that there are indeed two molecules in the cell, but that the molecule is hydrated; the compound should be given the formula $KH(C_7H_5O_3)_2.H_2O$ (= 332·3), as was borne out by the detailed structure analysis. (The difference between 665 and $2 \times 332{\cdot}3$ is not significant).

Occasionally the dimensions of the unit cell give a clue to the shape of the molecules and their arrangement. Corresponding polymorphic forms of the straight-chain fatty acids have unit cells two of whose translations remain virtually constant as the homologous series is ascended ($a \approx 9\frac{1}{2}$, $b = 4{\cdot}9_5$ Å, for the 'C' forms of the even members), but the third is much longer (*e.g.*, $c \approx 51$ Å for the C_{18}-acid) and increases by about $2{\cdot}5_4$ Å per additional CH_2—group. This is just the distance between alternate carbon atoms in the extended, planar zigzag adopted by the normal paraffin chain. The molecules of the acids therefore lie almost exactly along the *c*-direction, and there are two such molecules within the primitive translation; in fact, they occur as dimers, linked by hydrogen bonds between contiguous carboxyl groups. A celebrated structural deduction was made by Bernal in 1932 from a study of the unit cell data for a number of compounds in the steroid group. He found that their cells were all monoclinic, with $c \approx 7{\cdot}2$ Å, $a \approx 10$, 20 or 30 Å, and $b \approx 20$ or 40 Å. Taken with the respective numbers of molecules in the unit cells—and some other physical data—these translations suggested that the molecules had dimensions of about $7{\cdot}2 \times 5 \times 20$ Å^3. This was irreconcilable with the structure then accepted for the steroid skeleton, but was in accord with a formulation based on phenanthrene. Bernal's work had much to do with the general acceptance of the new formulation. However, it should be emphasised that attempts to deduce molecular structures from unit-cell data alone can lead to errors.

9.5 Determination of the Space Group and its Significance

By the methods of classical crystallography it was easy to decide to which of the seven systems a crystal belongs;* and, by a more detailed study, it was often possible to decide between the 32 point groups. But, though space-group theory had

* The crystal system is now almost always found directly by X-rays.

been developed at the end of the nineteenth century, there was no practical way of assigning any crystal to its space group until after the discovery of X-ray diffraction. The 230 space groups (see section 2.13) derive from the crystallographic point-groups because of the possibility of using translational symmetry operations in three-dimensional repeat patterns, and, in particular, screw-axes and glide-planes. These translational symmetry elements can be detected by X-rays. The detailed technicalities do not concern us here, but we may illustrate the principle by reference to a three-fold screw axis (3_1) in the c-direction. This symmetry element means that if any atom, or group of atoms, in the structure is rotated through $360°/3 = 120°$ about the screw-axis, and then translated by $c/3$, it will come into coincidence with an equivalent atom or group, as is suggested in Fig. 9.5.1(*i*). Now, were this group projected on to the c-axis, as shown in Fig. 9.5.1(*ii*), the effective primitive translation is reduced to $c/3$. Reflexions of the type $00l$ are from planes perpendicular to the c-axis; and, since layers of atoms which in fact differ, but which look identical when thus projected on to the axis, occur at intervals of $c/3$, $00l$ reflexions will be missing unless l is divisible by three; 001, 002, 004, 005, etc., are *systematically absent*. Putting the matter in an alternative way, the intensities of reflexions from the $00l$ set of planes depend only on the z-coordinates of the atoms (see section 9.6). Now the 3_1-axis generates from each atom at z two others at $z + 1/3$ and $z + 2/3$. Hence the effective length of the c-axis is reduced to one third of its true length, though *only* for the set of reflexions $00l$. For all other (hkl) reflexions the x- and y-coordinates are relevant.

A twofold screw-axis likewise causes 'halving' of the axial reflexions. (This effect is evident along the central, horizontal row of spots in Plate IB of the frontispiece). In a similar way, a glide of $a/2$, in a plane parallel to (010), results in a halving of the a-axis when—and only when—it is viewed in its projection on to the (010)-plane. This should be clear from a consideration of Fig. 2.13.1(*a*). Reflexions of the type $h0l$ then occur only when h is even.

There are 70 space groups that can be uniquely recognised from the systematic absences. (Amongst them is $P2_1/a$ ($\equiv P2_1/c$), which is very common amongst the simpler molecular crystals; see section 2.13). Unfortunately it is not generally possible to make an unambiguous diagnosis. This is because a centre of symmetry appears in the X-ray diffraction pattern whether one is present in the space group or not. The reason for this will be given later, along with an account of special conditions where the principle breaks down. For instance, the space group $P2_1$ lacks a centre; when one is added, the group becomes $P2_1/m$. These two groups give exactly the same set of absent reflexions, and hence they are indistinguishable on this basis.

The ambiguity can be overcome by other methods. For instance, if a sufficiently large number of external faces is developed by the crystal, then an application of classical goniometry may suffice to decide the point group, a definite knowledge of which, together with the absences, will indicate the space group. Again the detection of certain pyro- or piezo-electric effects proves the absence of a centre of symmetry. But the converse may not be true: failure to detect such effects may mean that they are merely very weak. Useful, though not infallible, are certain statistical tests, which

(i)   (ii)

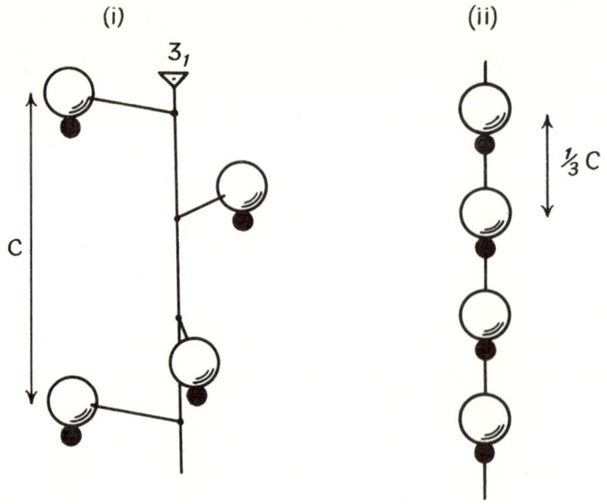

Fig. 9.5.1. Effect of screw-axis in causing certain orders of X-ray reflexions to be identically absent. (Here the 3_1-axis eliminates $h00$ reflexions, unless h is divisible by three).

depend on the distribution of intensities amongst the various types of reflexions.

A knowledge of the space-group sometimes gives information about molecular structure. According to an early investigation, diphenyl crystallises with two molecules in a cell belonging to the space group $P2_1/a$ (see section 2.13 and Fig. 2.13.1). The symmetry elements of this group are such that any atom placed in a general position, x, y, z in the cell must give rise to three others at $-x, -y, -z; \frac{1}{2} + x, \frac{1}{2} - y, z$; and $\frac{1}{2} - x, \frac{1}{2} + y, -z$. The general position is four-fold. Therefore, since there are only two diphenyl molecules in the cell, half a molecule must constitute the *asymmetric unit*—the minimum portion of matter from which the whole crystal would be developed by the symmetry operations. Two halves of the molecule must be in the relation of x, y, z to $-x, -y, -z$, so that the whole must possess a centre of symmetry. This is a valid deduction without any *a priori* chemical knowledge apart from the molecular formula. Given the further chemical information that the molecule consists of two benzenoid rings linked by a bond radial to each, we can deduce that the rings are co-planar. In the analysis of the vinylideneamine compound (I), there proved

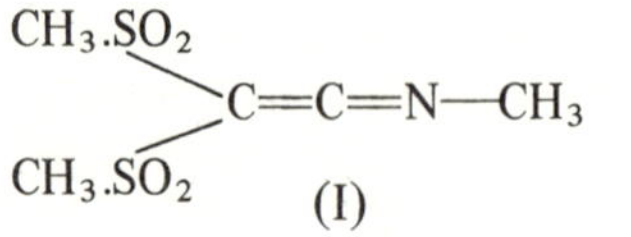

$$CH_3.SO_2$$
$$\diagdown$$
$$C{=}C{=}N{-}CH_3$$
$$\diagup$$
$$CH_3.SO_2 \qquad (I)$$

to be four molecules in a cell of orthorhombic space group *Pbcn*. This implies that the molecule possesses either a centre of symmetry or a two-fold axis. As the former can be rejected on chemical grounds, it must have the axis. Such an axis can run only along the C—C—N—C chain, which must therefore be linear. (The methyl group is presumably rotationally disordered with respect to its hydrogen atoms, as will be explained shortly). Space-group evidence enabled Carlisle and Crowfoot (1941) to allocate *meso*- and *racemic*-structures to certain pairs of dibenzyl derivatives of the type, C_6H_5.CHR.CHR.C_6H_5; those required to possess a molecular symmetry-centre must belong to the *meso*-series.

The full symmetry of the (idealised) gaseous molecule is rarely used in the crystal. The naphthalene molecule, for example, can be assigned to the point group mmm ($= D_{2h}$); it embodies a centre of symmetry and three two-fold axes as well as the three mirror-planes explicitly shown in the Hermann–Mauguin symbol. In the crystal the space-group is $P2_1/a$ and there are two molecules per cell, so that only the centre is used. The mean positions of the atoms are indeed very close to those needed to satisfy the requirements of the other symmetry elements; but they do not *necessarily* satisfy them. High refinement of the structure analysis (see section 9.15) in fact shows that the atoms do deviate slightly from the ideal positions, and that the deviations can be attributed to repulsions by atoms of neighbouring molecules. The symmetry of the molecular site is only $\bar{1}$; and this inevitably debases the mmm symmetry of the isolated molecule. In this sense the crystallographic symmetry of a molecule is always a minimum. Most aromatic hydrocarbons have planar molecules, for instance; but the molecular plane has rarely been found to lie in a crystallographic plane of symmetry.

On the other hand, crystals are known in which the molecule appears to have a symmetry higher than can be accepted on chemical grounds. *p*-Chlorobromobenzene crystallises with two molecules in a cell of space group $P2_1/c$, so that the molecules are formally centrosymmetric, like those in the isomorphous crystals of *p*-dichlorobenzene. But the molecule Cl.C_6H_4.Br cannot have a true centre of symmetry. The explanation is that these molecules are randomly orientated in the crystal with respect to the alternative aspects, Cl.C_6H_4.Br and Br.C_6H_4.Cl. In any macroscopic region of the crystal, the halogen sites are occupied by 'mean atoms', $\frac{1}{2}$(Cl + Br), and the intensities of the reflexions are best accounted for on this hypothesis. Enhanced symmetry of this kind, due to disorder, is fairly common; and the possibility must always be kept in mind when deductions about molecular symmetry are drawn from space group evidence.

A rather different kind of disorder is illustrated by the crystal symmetry of the ammonium ion. Its inherent symmetry is tetrahedral ($\bar{4}3m$), but it is rarely found in a tetrahedral crystalline environment. The low-temperature form of ammonium chloride is one of the few examples: each nitrogen atom has eight chloride ions surrounding it at the corners of a cube; the four hydrogen atoms lie on alternate N . . . Cl diagonals, so as to give an ordered tetrahedron of N—H . . . Cl hydrogen bonds. At −30°C there is a transition accompanied by an entropy change of 0·82 calorie degree^{-1} mole^{-1}, to give the ordinary form, in which the same arrangement obtains for the nitrogen and chlorine atoms, but the hydrogen positions are randomised amongst the eight N . . . Cl diagonals so as to yield an ammonium ion with effectively full cubic symmetry ($m3m$). Any individual ion must of course remain tetrahedral, but it can adopt alternative orientations with respect to the eight surrounding chloride ions. At 184° there is a second transition from this caesium chloride type of structure to one of rock salt type. Now the environment of the ammonium ion is six-fold octahedral, and there is a notable increase in disorder between a large number of possible orientations–a situation often, though not accurately, described as 'free rotation' of the ion. Rotational randomisation of the hydrogen atoms in a methyl group was mentioned above in relation to the molecule (I).

9.6 Structure Analysis and the Determination of Atomic Positions

In some very simple structures all the atoms are in explicit special positions, so that a determination of unit-cell dimensions and space group suffices to locate them definitively. Such are diamond and rock-salt. What are ordinarily regarded as molecular substances rarely crystallise in such simple patterns. Most of, if not all, their aroms lie in general positions, for each of which x-, y- and z-coordinates need to be found. To do this requires a full *structure analysis*. An analysis is by no means always practicable; it may require more effort than the importance of the problem justifies.

If, after considering the unit-cell dimensions and space group, the crystallographer decides to undertake the analysis, his first step is to measure the intensities of a large number of reflexions, preferably of all those that are accessible. This may be done simply by a photographic method; if greater accuracy is desired, the reflexions may be measured by some form of Geiger counter.

Simple considerations will show why intensity data are needed to find the atomic positions. Were the atoms which scatter the X-rays exactly located on the lattice planes (see Fig. 9.2.2), a very intense reflexion would result when the Bragg condition was fulfilled. Now suppose these atoms to be progressively 'smeared out', until finally the scattering matter is evenly distributed with no particular concentrations of it on the planes. The material has then ceased to be crystalline, and no Bragg reflexion would occur at all. The more the matter is concentrated on the planes, the stronger the reflexion. Applying this principle to a large number of different planes may make it feasible to deduce the atomic positions.

We must now consider the scattering of X-rays by a crystal in more detail. The kind of scattering that concerns us here is wholly due to the electrons. Each electron is caused to vibrate by the incident X-ray beam; and its vibration, at a frequency equal to that incident, causes it to emit scattered X-rays of the same wavelength. The proper measure of the scattering power per unit cell is a quantity known as the *structure factor*, represented by F, or by $F(hkl)$ when we wish to emphasise that F depends on the particular set of planes, hkl, involved. The *intensity*, I, of a given reflexion is related to F^2, and it depends, amongst other things, on the size of the crystal. It is a simple technical problem to convert experimental measurements of I

into F^2.

We can explore the nature of F most simply by reference to a one-dimensional crystal whose structure repeats itself indefinitely at intervals of c along the direction z. (We choose z here because we shall have other uses for x and y in the immediate context). First, suppose our crystal consists of only one stationary* electron in each cell. Our choice of an origin is free; so we simplify matters by taking the origin to be at the electron, whose fractional coordinate is therefore $z = 0$, with other electrons at $z = 1, 2 \ldots$, as we suggest in Fig. 9.6.1(*i*). The horizontal lines represent the infinite set of 'planes', indexed as $l = 1$, which is the one-dimensional analogue of the planes drawn in Fig. 9.2.2. X-rays will be reflected when the Bragg equation, (9.2.2), is satisfied, which implies that the waves scattered from each successive electron are exactly in phase. The structure factor for this reflexion, $F(1)$, is then taken to be $+1$. It is thus expressed in terms of a unit: one-electron-per-cell, stationary at the origin. (According to an alternative approach, the structure factor is taken to be a pure number. We are then taking it as a ratio of the scattering power of the one-electron case.)

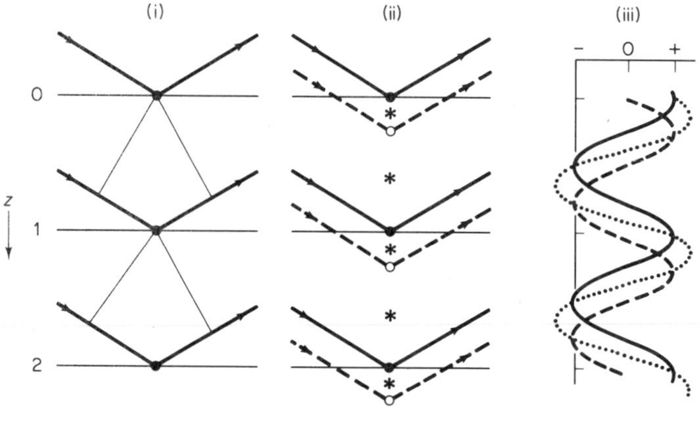

Fig. 9.6.1. A one-dimensional structure: (*i*) single electron in each cell at $z = 0, 1, 2 \ldots$; (*ii*) two electrons per cell, the second being at $z = 1/4, 5/4, 9/4 \ldots$; (*iii*) summation of waves, combining the partly out-of-phase scatterings from the two electrons, to yield the total structure factor represented by the dotted line. (This summation is alternatively represented in Fig. 9.6.2(*i*)).

Now let us add a second stationary electron to each cell, with a coordinate $z = \frac{1}{4}$ (Fig. 9.6.1(*ii*)). Whenever the Bragg equation is satisfied for the first set of electrons it will also be satisfied for the second set, because their spacing (c, in this case) is

* We can evade the objection that our electron, if it remained stationary, would not scatter the X-rays by supposing the amplitude of forced vibration to be small compared with the wavelength.

the same. But the respective, reinforced rays from the two sets are not in phase with one another. As suggested in (*iii*), the ray from the second set lags behind by a phase-angle of $\pi/2$. By inspection, the sum of the two waves (continuous and broken lines) gives the composite wave (dotted line) which has an amplitude of $\sqrt{2}$ and a phase-angle of $\pi/4$ with respect to the origin. Thus to define a structure factor in general, both an *amplitude* and a *phase-angle* are needed.

A situation of this sort is best handled by the methods of complex numbers. It is represented, for the case of Fig. 9.6.1(*iii*), by the Argand diagram of Fig. 9.6.2(*i*). The contribution of each electron is represented by a unit-vector measured anti-clockwise from the x-axis: that of the first at $z = 0$ is directed along the x-axis (of reals); that of the second at $z = \frac{1}{4}$ is directed $2\pi/4$ away, along the y-axis (of imaginaries). Vector addition yields F, with the amplitude and phase stated above, represented by the broken line.

Any number of electrons can be treated in the same way. For example, suppose we have single electrons at $z = 0{\cdot}1$ and $0{\cdot}2$ and a pair at $0{\cdot}4$. They correspond to

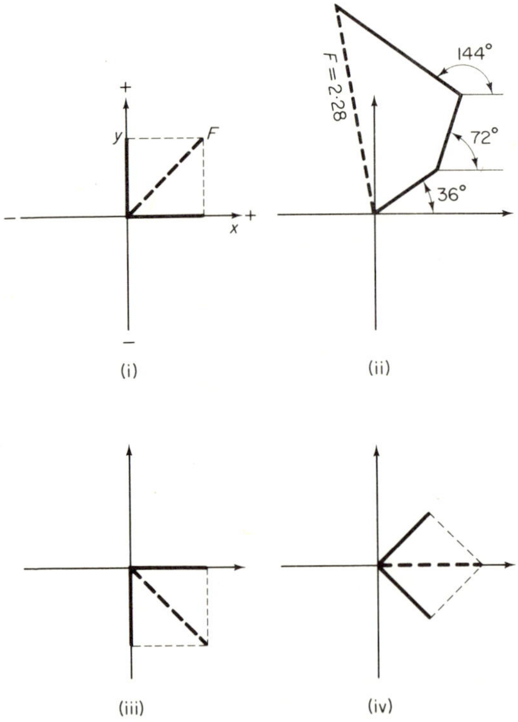

Fig. 9.6.2. Use of the Argand diagram to represent vector addition of X-rays scattered from electrons at different positions in the cell: (*i*) covers the case represented in Fig. 9.6.1(*iii*); (*ii*) combined effects of several electrons in a non-centrosymmetric structure; (*iii*) and (*iv*) alternative choices of origin in the situation of Fig. 9.6.1(*iii*).

vectors of amplitudes 1, 1 and 2, directed at phase-angles, $\alpha = 36°$, $72°$ and $144°$ respectively, as sketched in Fig. 9.6.2(*ii*). Summing these vectors, in any order, we arrive at the structure factor, shown by the broken line, with $|F| = 2{\cdot}3$ (electrons) and $\alpha = 110°$. Details of the trigonometric calculation are set out below; they should explain themselves by reference to (*ii*).

No. of electrons	z	α	$\cos\alpha$	$\sin\alpha$
1	0·1	36°	0·809	0·588
1	0·2	72°	0·309	0·951
2	0·4	144°	2(–0·951)	2(0·309)

$$|F| = \sqrt{\{(2{\cdot}157)^2 + (-0{\cdot}784)^2\}} = 2{\cdot}28;$$
$$\alpha = \tan^{-1}(-0{\cdot}784/2{\cdot}157) = 110°.$$

The choice of origin is arbitrary, and immaterial. In the two-electron structure (Figs. 9.6.1(*ii*) and 9.6.2(*i*)) we might have put it at the second electron. The co-ordinate of the first would then have been $z = -\frac{1}{4}$. From Fig. 9.6.2(*iii*) we see that the structure amplitude would still be $\sqrt{2}$. The phase-angle would be $-\pi/4$, but, as we shall emphasize later, only $|F|$ is directly observable, and α is not. Actually in this case it would be more convenient to choose one of the centres of symmetry as origin. If we choose the top one marked with an asterisk in Fig. 9.6.1(*ii*), the Argand diagram takes the form shown in Fig. 9.6.2(*iv*): $|F| = \sqrt{2}$; $\alpha = 0$. Or more simply, having got rid of the imaginary part, we can write $F = +\sqrt{2}$. For this reason we always take our origin at a centre of symmetry when there is one; for then the structure factors become real, their phase-angles being reduced to either 0 or π. F is either positive or negative.

We now move on to a situation which, though still one-dimensional, is one stage more realistic – the structure sketched in Fig. 9.6.3(*i*). An imaginary molecule, MN_2, with atomic numbers $M = 15$ and $N = 8$, is strung out along the x-direction, at intervals of a. There are two types of centres of symmetry in each unit cell. It would be normal to choose the centre at which atom M is situated as origin, but in fact we have chosen the other. The structure then consists of atom M at the *special position*, $x = \frac{1}{2}$, whilst N occupies the set of *general positions* x and $1 - x$. Thus one *parameter* suffices to define this structure: $x(N) = 0{\cdot}31$, which implies that the second N is at 0·69.

The representation of the molecule in Fig. 9.6.3(*i*) is conventional. Since the X-rays are scattered by the electrons, the relevant feature of the cell is its *electron-density distribution*, which is better represented by (*ii*). This is drawn on a larger scale, and it covers only the unique half-cell (or rather more) needed to describe the whole. The atoms are shown as symmetrical peaks, with the density falling to zero between them. This would not be strictly valid in an actual molecule whose atoms were covalently bonded together. Any such curve of electron density against x can be represented generally as $\rho(x)$; it is a repetitive function which assumes identical

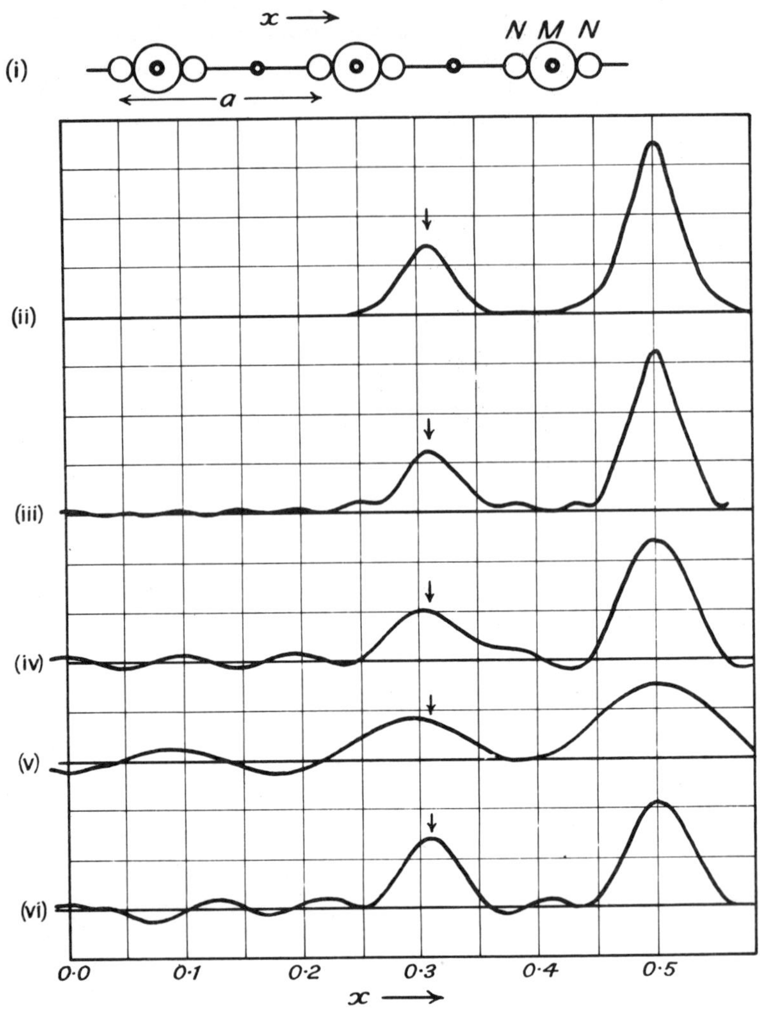

Fig. 9.6.3. An imaginary one-dimensional crystal (*i*), and the representation of its electron-density distribution (*ii*) by single Fourier series of various degrees of incompleteness (*iii-vi*).

values at $\ldots x-1, x+1, x+2 \ldots$. Our task is to determine $\rho(x)$ from the experimental diffraction-intensity measurements.

We start with the inverse problem, which has the advantage of being generally soluble: to calculate the intensities–or, more conveniently, the structure factors–of the various reflexions, $\rho(x)$ being known. Consider an element of the unit cell lying between x and $x + dx$. Its contribution to the X-ray structure factor will be proportional to the number of electrons it contains: *i.e.*, to $\rho(x)dx$. When the Bragg condition for a first-order reflexion is satisfied, this element scatters a wavelet of amplitude $\rho(x)\mathrm{d}x$; and similar wavelets from every other cell in the crystal will be in phase, and so reinforce one another. Its contribution to the structure factor will be $\rho(x)\mathrm{d}x.\cos(2\pi x)$. This follows from the treatment earlier of single electrons. The wavelets from all elements of the cell have to be integrated to yield the total structure factor.

There is a centre of symmetry at the origin, so that for every element at x there is an identical one at $-x$. Following the argument based on Fig. 9.6.2(*iv*), we see that these elements will give amplitudes that are equal but directed at opposite phase-angles, $+2\pi x$ and $-2\pi x$. Their sine components will cancel, whilst their cosine components have to be summed. The integration must cover a whole cell, (say) from $-\frac{1}{2}$ to $+\frac{1}{2}$; but because of the symmetry, it suffices to double the integral from 0 to $+\frac{1}{2}$: and thus the structure factor for the first-order reflexion is given by

$$F(1) = 2\int_0^{\frac{1}{2}} \rho(x).\cos(2\pi x)\,\mathrm{d}x;$$

and for any order (h)

$$F(h) = 2\int_0^{\frac{1}{2}} \rho(x).\cos(2\pi hx)\,\mathrm{d}x. \qquad (1)$$

This equation also gives us the meaning of the zero-order term, $F(0)$, which we cannot observe as it necessarily lies in the direction of the powerful, incident X-ray beam, but which we shall need later. Since cos(0) is 1,

$$F(0) = 2\int_0^{\frac{1}{2}} \rho(x).\,\mathrm{d}x,$$

which is simply the total number of electrons in the cell, or ΣZ, where Z is the atomic number of each atom. For our imaginary one-dimensional crystal, $F(0) = 15 + 2 \times 8 = 31$ (electron units).

Though the integration of an expression involving $\rho(x)$ is the most fundamental way of developing the structure factor, it is not the most convenient in practice. We are accustomed to regard matter as composed of discrete atoms with more-or-less spherical shapes, and this turns out to be a very good approximation in chemical crystallography. The scattering potency of an atom depends naturally on its atomic number, Z. For a non-vibrating point-atom, the *scattering factor, f,* would be equal to Z in all orders of reflexion.

In fact the electron density of an atom occupies a significant region of space compared with the wavelength of the X-rays. The scattered rays coming from different parts of the same atom may therefore be partly out-of-phase, and the scattering power of the atom will then be diminished. To allow for this an *atomic-scattering function* must be used. Represented by f, it starts from the atomic number, Z, when $\sin\theta$ = zero, and decreases to half of Z or less as $\sin\theta$ rises towards its upper limit of unity for the high-order reflexions. The function f can be calculated from our knowledge of the atomic wavefunction.

Atoms in crystals are vibrating. This effectively spreads the electron density still further, and hence causes an additional fall-off of scattering power with sin θ. We shall return to the importance of vibrations later.

We are now able to replace the integral (1) by an analogous summation:

$$F(h) = \sum_i f_i . \cos(2\pi h x_i), \tag{2}$$

in which x_i is the coordinate and f_i the scattering function of the ith atom, and the summation extends over all the atoms in the cell. With scattering functions appropriate to atoms of the shapes shown in Fig. 9.6.3(*ii*), we have used this equation to calculate the structure factors given in Table 9.6.1.

Table 9.6.1. Structure Factors for the One-dimensional Crystal in Fig. 9.6.3(*i*)

h	0	1	2	3	4	5	6	7	8	9	10
$F(h)$	31	−21	+4	−2	+14	−21	+16	−7	+4	−7	+12
h (*cont.*)	11	12	13	14	15	16	17	18	19	20	
$F(h)$ "	−10	+6	−3	+4	−6	+6	−4	+2	−2	+3	

A real crystal is periodic in three dimensions. If it is centrosymmetric, and provided our origin is taken at a centre, structure factors can be calculated by (3) which is wholly analogous to (2).

$$F(hkl) = \sum_i f_i \cos 2\pi(hx_i + ky_i + lz_i). \tag{3}$$

In the most general case of a non-centrosymmetric crystal the structure factor is best treated as a complex quantity (p. 268). It is then written as $F = A + iB$, where $i = \sqrt{(-1)}$. A is the real part of F, and it now corresponds to the right-hand side of equation (3); B is the imaginary part, and it may be calculated from a similar expression with sines instead of cosines. The *structure amplitude*, which should now be written $|F|$, is equal to $\sqrt{(A^2 + B^2)}$, whilst the phase-angle, α, is $\tan^{-1}(B/A)$. Since

$$e^{iw} = \cos w + i \sin w,$$

where w is any variable, the most concise way of writing the equation for the structure factor is

$$F(hkl) = \Sigma f_i \exp\{2\pi i(hx_i + ky_i + lz_i)\}, \tag{4}$$

in which we follow the convenient convention of representing e^w by $\exp(w)$.

9.7 The Fourier Series Method

We will now suppose that the structure factors of Table 9.6.1 are available, and that we wish to use them to find the structure of the crystal shown in Fig. 9.6.3(*i*). Though it is not the only approach, a Fourier series is the most important and is generally used. In any case this method serves well to illustrate the principles involved and the difficulties to be overcome.

Any function periodic in one dimension can be reproduced by combining a series of harmonic cosinusoidal waves. So the electron density, $\rho(x)$, in our crystal can be represented by the *Fourier series*:

$$\rho(x) = \sum_{n=-\infty}^{n=+\infty} C(n) \cos(2\pi n x + \alpha(n)), \tag{1}$$

where $C(n)$ is the amplitude of the n-th harmonic and $\alpha(n)$ its phase angle. When the structure is centrosymmetric, the series can be simplified to

$$\rho(x) = 2 \sum_{n=0}^{n=+\infty} (\pm)C(n) \cos(2\pi n x). \tag{2}$$

The factor 2 appears because the terms in n and $-n$ are identical and have now been combined together. (The zero-order term occurs only once however, and should be considered separately). If we substitute this equation for the electron density into (9.6.1) for the structure factor, we get

$$F(h) = 2\int_0^{\frac{1}{2}} \left[2 \sum_{n=0}^{n=\infty} C(n) \cos(2\pi n x)\right] \cos(2\pi h x)\,\mathrm{d}x.$$

For a reason which is given in Appendix III, the integral is zero unless n and h are numerically equal, when it assumes the value $aC(h)/2$. Therefore $C(h) = F(h)/a$; the coefficients needed in the Fourier series are proportional to the corresponding structure amplitudes, so that (2) becomes,

$$\rho(x) = 2(1/a) \sum_{h=0}^{h=\infty} F(h) \cos(2\pi h x). \tag{3}$$

This beautiful inverse relationship between eqns. (3) and (9.6.1) is of profound significance. Technically the expressions for structure factor and electron density are each the *Fourier transform* of the other; they are interrelated as are the pair of equations:

$$\psi(v) = \int \phi(w) \cos(vw)\,\mathrm{d}w$$

and

$$\phi(w) = \int \psi(v) \cos(vw)\,\mathrm{d}v,$$

in which ψ and ϕ are functions of the variables v and w, which exist in mutually reciprocal spaces. To be more precise, the transform of the electron density, $\rho(x)$, in a single unit cell would be a continuous function in h, corresponding to the fact that the diffraction pattern from a single cell–could it be observed–would consist of a continuum of gradually varying intensity. (A situation closely resembling this occurs in electron diffraction with gases, as is explained in Chapter 10). However, a crystal is composed of a vast number of unit cells arranged in a lattice. The transform of the lattice is the reciprocal lattice. The observed diffraction pattern of sharp spots is obtained by combining this transform of the lattice with the transform of the unit cell; in other words, the cell transform (F with h continuously variable) can be sampled only at the points of the reciprocal lattice (where h has integral values). This is equivalent to a statement of the Laue condition given in eqn. (9.2.1). Thus we arrive at the concept of the weighted reciprocal lattice, which we have already asserted to embody a complete account of the diffraction properties of the crystal. As Sayre has expressed it, 'it is not the function, $\rho(x)$, . . . but another function, derived from it and called its Fourier transform, which most directly presents the scattering properties of a structure. Hence experiments on scattering essentially measure the transform of the structure in question, and from the transform the structure is deduced.'*

Incidentally the atomic scattering function, $f(\sin\theta)$, and the expression for the electron-density distribution in the atom are Fourier transforms of one another. The latter exists in real space, the former in reciprocal space.

We are now in a position to make use of the structure factors in Table 9.6.1. For the moment we must accept the signs there given as correct, and then we can make the summation according to eqn. (3), remembering that $F(0)$ is to be taken once, but all the other terms twice, *i.e.*, for $F(h)$ and $F(\bar{h})$. We need compute the electron density only between $x = 0$ and $\frac{1}{2}$, since the symmetry of the cell then suffices to complete the pattern.

If an infinite number of terms could be summed in (3), we should exactly reproduce the ideal electron density shown in Fig. 9.6.1(*ii*). Since only a limited number of orders of reflexion is ever accessible, we must do with an incomplete series. When all 21 terms up to $h = 20$ are used, we get the result shown in (*iii*), where, despite a few 'ripples', the atoms are well reproduced. There is a progressive deterioration as more of the higher terms are omitted, as can be seen in (iv) and (v) where the series has been terminated as $h = 10$ and 5 respectively. The atomic peaks are less sharp, the ripples and negative regions are more pronounced, and errors are consequently more likely to affect the indicated atomic positions: atom M, in a special position, necessarily comes out at $x = \frac{1}{2}$, but N is displaced. *Termination-of-series errors* affect all such electron-density maps, and they become more serious as the number of orders used in the summation diminishes.

For a centrosymmetric, three-dimensional crystal the electron density can be

* Amongst other practical uses, the Fourier transform has recently become the basis of a method for the more rapid and more precise experimental recording of infrared and NMR spectra.

computed by a corresponding triple Fourier series:

$$\rho(x, y, z) = (1/U_0)\sum_h \sum_k \sum_l (\pm)F(hkl) \cos 2\pi (hx + ky + lz), \tag{4}$$

where U_0 is the volume of the unit cell. When this function has been successfully evaluated, atomic positions will be indicated by more-or-less spherical regions of high density. A double Fourier series enables us to derive the electron density projected down an axis of the cell. For instance, we may use a more limited set of structure factors, $F(hk0)$, to give us the structure projected down the c-axis:

$$\rho(x,y) = (1/A_0)\sum_h \sum_k F(hk0) \cos 2\pi (hx + ky), \tag{5}$$

where A_0 is the area of the projected cell. Similar equations hold for projections down other axes.

In dealing with a non-centrosymmetric structure we must introduce the phase-angles $\alpha(hkl)$; this can be done by using the Fourier series in the form:

$$\rho(x, y, z) = (1/U_0)\sum_h \sum_k \sum_l |F(hkl)| \cos \{2\pi (hx + ky + lz) - \alpha(hkl)\}; \tag{6}$$

or, in the most general form, when the electron density is written as the Fourier transform of (9.6.4),

$$\rho(x,y,z) = (1/U_0)\sum_h \sum_k \sum_l F(hkl) \exp \{-2\pi i(hx + kz + lz)\}, \tag{7}$$

which is independent of any particular choice of origin.

9.8 A Note on Experimental Methods

Though we are not concerned in this book with experimental methods, technical advances have recently had a profound effect on the practicabilities of crystal structure analysis, and hence on its strategy. Two aspects are especially important.

The first X-ray diffraction patterns were recorded on photographic plates, or films. At an early date W. H. Bragg used the ionisation spectrometer to measure the intensities of X-ray reflexions; but this method, though accurate, was slow and troublesome. Some form of visual estimation of the spots recorded on photographic films was more convenient, and this method was used in nearly all structural work until the middle 1960's, and is still used. However it is of limited accuracy, and the visual estimation of (say) two thousand spots is a laborious task. Most analyses now—and all with any pretensions to high accuracy—are based on intensities measured by an automatic diffractometer. The machine can be programmed to orientate the crystal to the proper position for each reflexion in turn, and the spot is scanned by some form of counter to yield the integrated intensity. Much higher speed and much greater accuracy are attained.

The advent of the digital electronic computer, from around 1960, has had an even more revolutionary effect. The scale of the problem can be illustrated by considering Fourier analysis. In the one-dimensional exercise featured in Fig. 9.6.3, we need to sample the electron density at, perhaps, 21 points at intervals of 0·025 from $x = 0{\cdot}00$ to 0·500, and we have at our disposal 20 structure factors. The amount of computation may then be measured by $20 \times 21 \approx 4 \times 10^2$. The double Fourier series, for a projected structure (equation 9.7.5), might be based on 200 terms, and the density might have to be calculated at a grid of points, 21 along the x-axis and 21 along y. The computational labour may then be measured by $21 \times 21 \times 200 \approx 8 \times 10^4$. In a three-dimensional synthesis we might have at least 2000 terms, and the need to calculate the density at a lattice of points along x, y and z; and the labour works out at $21 \times 21 \times 21 \times 2000 \approx 2 \times 10^7$. The relative efforts in 1, 2 and 3 dimensions are roughly in the proportion of 1 : 200 : 40 000; by hand calculation, half-an-hour: a fortnight: three years. Only with the general availability of computers have three-dimensional structure analyses become really practicable, though a few heroic studies by hand calculation are on record. Previously structure analysis had to depend on projections. This is effective only when the unit cell happens to have one axis short (say, <6 Å); otherwise the atoms of the molecule are not well enough resolved. This is one reason why most of the classical structure analyses achieved by J. M. Robertson between 1933 and 1950 were of simple, planar molecules, such as naphthalene, phthalocyanine and oxalic acid in its dihydrate.

9.9 The Phase Problem

The general strategy of an X-ray analysis can now be appreciated. After the preliminary routines of determining the unit cell and space group and of collecting adequate intensity data, the crucial difficulty is to solve the *Phase Problem*. Since we can record the values of $|F|$ only, and not those of α, an infinite number of solutions of (9.7.6) is mathematically possible. Any conceivable combination of phase-angles substituted in this expression would yield an electron-density distribution which would account for the observed intensities. Even with a centrosymmetric crystal, each structure factor could have alternative values of opposite sign, so that the number of solutions of (9.7.4) is still very large (2^N for N reflexions), though it is not infinite. The physical situation is much more restrictive however: there can be no regions of negative density—apart from such trivial ones as may be ascribed to termination-of-series errors—and the electron density must be so distributed as to comprise atoms, of the correct kinds and in the correct numbers, arranged in a chemically acceptable way. These requirements drastically curtail the number of possible solutions, and it is believed that a general solution of the Phase Problem ought, in many cases, to be accessible; that there is, implicit in the intensity measurements, information which might be used to determine the phases of the stronger reflexions. A method for doing this was developed during the period 1950–1960 by J. Karle and H. Hauptman, and it has since been extended and exploited by J.

and I. L. Karle in particular, as well as by Hauptman, Woolfson and others. We shall revert to these 'direct methods' later in this section.

Lacking any direct way of solving the Phase Problem, crystallographers had meantime to circumvent it by methods of limited applicability, which may, in favourable circumstances, indicate the signs or phase-angles of an adequate number of the larger structure amplitudes. When this has been achieved, the Fourier series method will give approximate positions for some of the atoms, if not all. For example, the six strongest terms ($|F| > 10$) of Table 9.6.1, taken with correct signs, lead to the electron-density trace shown in Fig. 9.6.1(*vi*), which reveals the structure rather well. Once a correct notion of the structure has been roughed out, refinement proceeds by a straightforward routine: from the approximate structure, approximate structure factors are calculated—amplitudes and phases; these *calculated* phases are then used with the *observed* amplitudes in a second synthesis, which is expected to yield a more exact structure. At any stage, only those terms can be included for which the agreement between observed and calculated amplitude is good enough to warrant adoption of the calculated phase (or sign). The cycle of operations (symbolised in Fig. 9.9.1) is repeated until all observed reflexions can be included in the series. Occasionally a pseudo-structure may be found; it gives some measure of agreement in the earlier stages, but refinement fails to proceed to a really satisfactory agreement; and suspicion is often aroused by the obtrusion of improbable bond lengths or intermolecular contacts. A fresh start has then to be made.

In the necessary initial attack on the Phase Problem a number of methods have been used. Most of the structures that have been successfully solved hitherto have been based on one of the following four methods—or sometimes on a combination of several of them.

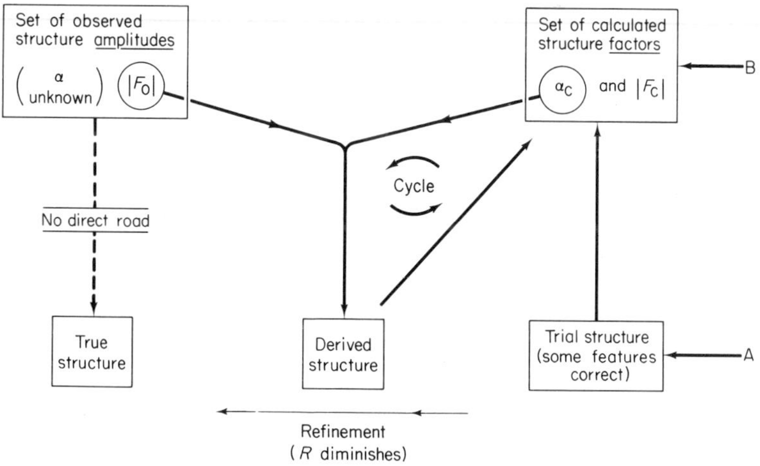

Fig. 9.9.1. The Phase Problem: possible strategies for its solution, and the tactics of structure refinement.

(1) *Trial and Error*
When the structure is a very simple one, and particularly, if the chemical structure is not in doubt, it may be practicable to try various sets of atomic positions in the unit cell until one is found giving an encouraging measure of agreement between observed and calculated structure amplitudes. If this trial structure is near enough, the iterative process described above can be trusted to lead to the correct one. In the absence of specially favourable circumstances, this method is not likely to succeed when more than six or eight atoms have to be located.

(2) *The Heavy-Atom Method*
When the molecule contains an atom of relatively high atomic number–*e.g.*, bromine or iodine in an organic compound–the X-rays scattered from this 'heavy' atom dominate the situation; and the sign of the structure factor is likely to be that which would obtain if only this atom were present. The position of the heavy atom can usually be found by method (4), and the signs required by this position can then be calculated. They are coupled with the observed amplitudes and used in a Fourier synthesis. In the resulting map it may be possible to recognise other atoms, which can be included in a second calculation of structure factors. The usual iteration follows.

This method becomes particularly elegant when the heavy atom is at a centre of symmetry, which is taken as origin. The signs imposed by this atom will then be all positive, and most of the structure factors will be positive also. Where the atom is sufficiently heavy, all structure factors will be positive. (In the simple one-dimensional case shown in Fig. 9.6.1, all signs are positive when the atom M is taken as origin). The Fourier series can then be used directly, with positive signs, and the structure appears in the resulting electron-density map, without any chemical assumptions. Such an absolute structure determination was first achieved by Robertson and Woodward (1939) for the platinum derivative of phthalocyanine.

When the crystal is not centrosymmetric this method is much more difficult. A simple choice between positive and negative signs can generally be made unambiguously; but the general phase angle can have any value from 0 to 360°, and the angle calculated for the heavy atom will at best, only be approximately correct for the whole structure. Nevertheless, the method was used successfully by Carlisle and Crowfoot (1946) in their analysis of cholesteryl iodide; and it played a major part in the elucidation of the much more complex structure of vitamin B_{12} (Hodgkin, White, *et al.*, 1957*).

When other methods of solving some difficult organic structure have failed, some effort may be profitably spent in searching for a heavy-atom derivative that is suitably crystalline. As a rough index, the prospects of success depend on the ratio of the square of the atomic number of the heavy atom to the sum of the squares of those of all the lighter atoms: for a fair chance of success, this ratio should exceed unity, though–after the expenditure of a correspondingly greater amount of labour –success is now often achieved when the ratio is less than unity.

* *Proc. Roy. Soc.*, 1957, **A.242**, 228.

(3) *Isomorphous Replacement*
Corresponding atoms in two isomorphous crystals occupy very nearly the same positions in the unit cells, but the interchangeable pair differ in atomic number–*i.e.*, in scattering power. To a good approximation for a centrosymmetric crystal, therefore, structure factors will differ simply by the difference between the contributions made to each particular reflexion by the replaceable atoms. The position of these atoms can usually be found, by some form of method (4); and then it becomes possible to calculate this *difference* between the F-values, both magnitude and sign. Provided that the observed amplitudes are both on an absolute scale, their signs can be decided by inspection. For instance, suppose the *amplitudes*, $|F|$, for a corresponding reflexion from isomorphous chloro- and bromo-compounds are observed to be 35 and 16; and that, from a knowledge of the halogen-atom position, the difference between the *structure factors* is calculated to be, $F_{Br} - F_{Cl} = -21$. These values can best be accounted for by attributing negative signs to both structure factors, since $-35-(-16) = -19$, which does not differ significantly from -21.

This method was developed by Robertson (1936) in his classical work on the phthalocyanines. It was found that phthalocyanine itself ($PhthH_2$) and its nickel derivative (PhthNi) crystallise isomorphously and with centrosymmetric molecules. Since the two hydrogens were negligible so far as X-rays are concerned, it followed that the 'hole' at the centre of symmetry of the molecule could be isomorphously replaced by a nickel atom. Advantage was taken of this situation to achieve most elegant determinations of both crystal structures.

Like the heavy-atom method, isomorphous replacement is much more difficult to apply to a non-centric crystal. The application was however made to the sulphate and selenate of strychnine (Bohkoven, Schoone, and Bijvoet, 1951) and, with spectacular success since 1960, to myoglobin and about a score (1971) of other crystalline proteins.

(4) *Patterson Methods*
A Fourier synthesis may be performed using F^2, instead of F, for the amplitude terms, and disregarding the problem of signs or phase angles. As was shown by Patterson in 1934, the resulting function, when plotted out as a map, has a maximum (a peak) for every pair of atoms in the structure; each maximum is situated at a distance, and in a direction, from the origin, corresponding to the vector between the corresponding pair of atoms in the real crystal. The *Patterson synthesis* thus leads to a *vector map*. The pattern shown in Fig. 9.9.2(*i*)–which some readers may recognise as derived from a natural object–gives rise to the vector map shown in (*ii*). When, as here, there are 5 points in the object, there are 5^2 vector points, 5 of which –between each point and itself–lie superposed to give the large peak at the origin, leaving $\frac{1}{2}(5 \times 4)$ pairs related by the centre of symmetry at the origin. In a crystal the unit-cell pattern is repetitive, so that the vector map is also repetitive. But it is still true that N atoms in the real cell give rise to N^2 peaks in the cell of the vector map–N of them superposed at the origin, and the remainder distributed over the vector cell.

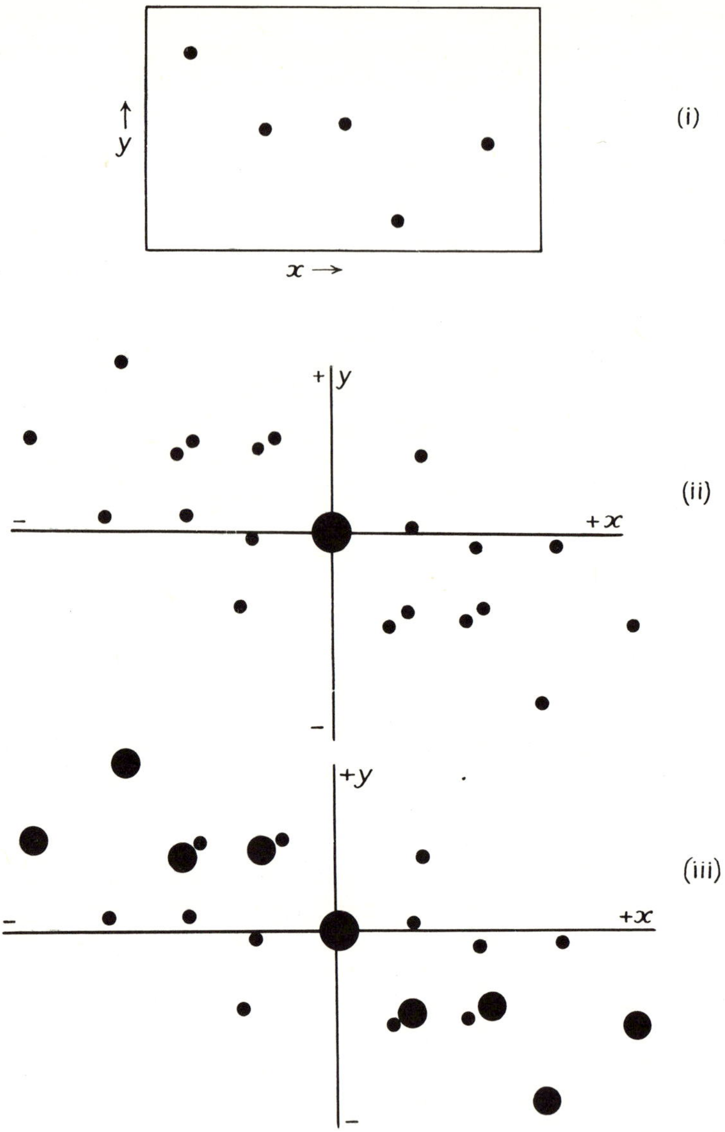

Fig. 9.9.2. Diagram to illustrate the Patterson, or vector, function: (*i*) a two-dimensional pattern in real space, (*ii*) the corresponding vector diagram, (*iii*) the vector diagram when the extreme left-hand point in (*i*) is given extra weight.

The problem is to deduce the unknown, real structure, such as (*i*), from the calculated Patterson function, such as (*ii*). In practice the Patterson function consists not of points, but of peaks with finite areas, which are more and more likely to overlap as their number increases. With a very simple pattern it may be possible to 'unscramble the egg', as it may also be when certain special conditions obtain. Generally it becomes difficult, and soon impossible, as the structure becomes more complex. Nevertheless, some form of Patterson function, and particularly when it can be calculated three-dimensionally, is a method for solving a structure that does not include heavy atoms.

When the molecule contains one heavy atom, or a small number of them, amongst a larger number of lighter atoms, the special conditions mentioned above obtain, and solution of the Patterson function may be greatly facilitated. If the cell contains just one heavy atom, the vectors between it and the other, lighter atoms will give rise to higher peaks than those between pairs of light atoms. The former class of vectors will then be enhanced. For example, if the 'atom' on the extreme left of the pattern in Fig. 9.9.2(*i*) were heavier than the others, it would give rise to eight non-origin peaks of greater height than the others, with the suggestive result shown in (*iii*). Again, if the cell contains two, or four, heavy atoms, the vector peaks between pairs of these will be outstanding, and it is usually possible to recognise these peaks and so to determine coordinates for the heavy atoms in the real cell. This can be a necessary preliminary to the application of methods (2) and (3).

(5) *'Direct Methods'*

We now revert to the possibility of a more general solution of the Phase Problem. We start with the relatively easier task of the centrosymmetric crystal, where we have to find *signs* for a reasonable number of the stronger observed structure amplitudes, $|F_o|$. Encouragement was drawn from the consideration that a crystal structure is potentially over-determined: with 20 atoms to locate, we need to know 60 coordinates; and from such a crystal we ought to have no difficulty in measuring 1000 or more $|F_o|$. There is a large excess of observations over parameters, and this mass of information may carry implications about signs.

It has been recognised since about 1950 that, when we have three strong reflexions whose Millerian indices are in the relationship, $hkl : h'k'l' : h+h'\ k+k'\ l+l'$, then their signs ($s$) are usually connected by an equation which is variously associated with the names of Sayre, Cochran and Zachariasen:

$$s(hkl).s(h'k'l').s(h+h'\ k+k'\ l+l') = +1;$$

or, written in shorter form:

$$s(h).s(h') = s(h+h'). \tag{1}$$

(An explicit example would be the triplet of reflexions, 324, $21\bar{2}$, and 532).

Some idea of the physical reasons behind this equation in particular, and behind direct methods in general, can be gleaned from Fig. 9.9.3. In (*i*) we sketch a projected structure whose atoms cluster rather closely about the lines corresponding to $x = \frac{1}{4}$ and $\frac{3}{4}$. In (*ii*) we concentrate attention on the reflexion from the 200 set of planes. As symbolised by the cosinusoidal curve drawn at the right-hand side, atoms in these positions will force $F(200)$ to be negative (see section 9.6.) and large. A

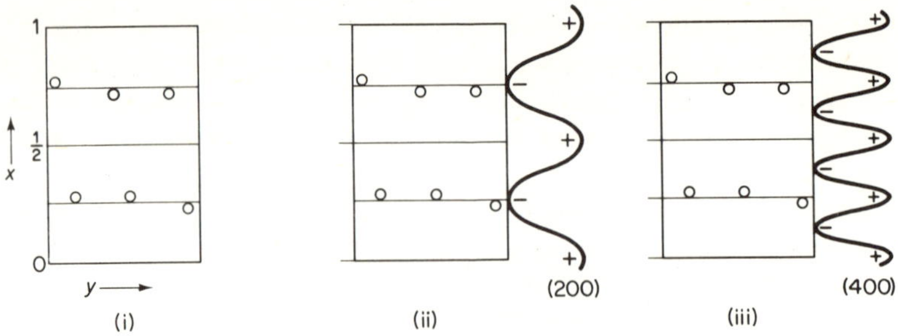

Fig. 9.9.3. An example of the physical principles upon which 'direct methods' of phase (or sign) determination are based; *(ii)* and *(iii)* correspond to 200 and 400 reflexions.

similar treatment of $F(400)$ in *(iii)* shows that its sign must be positive. Thus $s(200).s(200) = s(400)$, which is a special case of equation (1).

It is illuminating to notice that we are at liberty to move our origin by $\frac{1}{4}$; and, if we do this, $F(200)$ will become positive. But the equation still holds (+.+ = +). The essential point is that, always provided that $F(200)$ and $F(400)$ are strong enough, the sign of the latter must be positive.

As it stands, this equation is not very fruitful. It will apply only when nearly all the atoms lie close to a particular set of planes, as in Fig. 9.9.3. In an actual crystal such a favourable situation can apply only for a limited number of reflexions. However, for *any* triplet of reflexions with indices in the required relationship, it turns out that equation (1) is more likely to be true than not. Its probability can be statistically related to the amplitudes. The probability will be unity, and the equation certainly obeyed, only when all three $|F_o|$ are large; but it will be better than $\frac{1}{2}$ for weaker reflexions. For (1) we can substitute the equation $s(h).s(h') \approx s(h+h')$, where $\approx$ means 'is probably equal to', and the probability can be quantified. We then search our whole set of $|F_o|$ and seek all possible triplets. Besides 324 and $21\bar{2}$ for 532, for instance, we might also find that 004 and $53\bar{2}$, $\bar{3}11$ and $82\bar{3}$, etc. were also moderately strong. And thus combined probabilities can be found relating, and cross-relating, the signs of the stronger reflexions.

Of course, we cannot profit from these unless we can start with some 'given' signs. We do have some given, because we are free to allocate signs to a small number of terms (typically, three); to do this is merely to select one of several possible choices of origin. The detail in carrying out this scheme is technically complicated. But it can be programmed for a computer. With some luck perhaps, and certainly with some skill in the early stages, we may arrive in the end at perhaps 4 or 8 sets of signs for the 200 strongest reflexions. We then have merely to calculate Fourier syntheses with these sets of signs, starting with the one whose probability is indicated as the highest; and, before the end of the sets, we may come up with a density map which makes sense chemically and enables us to allocate provisional coordinates to a sufficient number of atoms for us to enter the refinement symbolised in Fig. 9.9.1.

We have entered at point B, whereas the older methods enter at A.

Though direct methods were developed during the 1950's, they did not begin to make much impact until the availability of computers made it practicable to apply them to three-dimensional data, rather than merely to projections. At the time of writing (1972) it is still true that the majority of crystal structures are solved by the heavy atom method; but a growing number are being solved directly. A recent example, without the benefit of a heavy atom, is a photo-product of thymine and uracil, $C_{24}H_{28}N_8O_5$, of unknown molecular structure (I. L. Karle and co-workers). This could not have been solved by any of the other methods described.

With non-centrosymmetric structures the problem is much harder. The phase-angles may have any value between 0 and 2π. Even when we simplify things by trying merely to place the phases of N reflexions in their correct 45°-sectors (*i.e.*, to restrict the angles to 0, $\pi/4$, $\pi/2$ . . . $7\pi/4$), we have 8^N phase-angle possibilities compared with 2^N sign possibilities for the centric case. Nevertheless progress has been made. A recent example, solved by I. L. Karle, is a diterpenoid of formula $C_{20}H_{30}O$ and of unknown structure.

9.10 Refinement of the Analysis by Fourier Synthesis

When once the signs (or phases) of a sufficient number of the stronger reflexions have been determined, or—what is equivalent—when approximate positions have been found for a sufficient number of the atoms, refinement of the analysis can proceed by the straightforward, if laborious, routine described earlier. After each cycle of refinement the agreement between observed, $|F_o|$, and calculated, $|F_c|$, structure amplitudes should improve, and it should prove possible to include more terms in the next synthesis, until all observations can be used. The increase of definition with the lengthening of the Fourier series has been illustrated in Fig. 9.6.3.

In practice one-dimensional work is hardly ever helpful, because, when the contents of the cell are all projected on to a single axis, far too many atoms overlap. One of the few examples of a useful employment of such a projection occurred in the classical work of Beevers and Lipson (1935) on the alums: an important stage in arriving at a proper understanding of their structures involved a comparison of projections on the cube diagonal, derived from *hhh*-reflexions, for sulphate and selenate alums.

Two-dimensional refinement had the advantage, in pre-computer days, of not being prohibitively laborious, and at the same time yielding a resolved picture of the structure provided the axis of projection was short enough. It still has its uses. The electron density, expressed as electrons per square Å and represented by contour-lines, may be plotted to give maps of the kind shown in Fig. 9.10.1, which cover the molecule of 4,5-diamino-2-chloro-pyrimidine (II). The various parts of this diagram strikingly illustrate the improvement of the resolution as more Fourier terms are included in the summation, running to shorter spacings (d) which is equivalent to higher values of the Bragg angle, θ. A density representation of this sort is

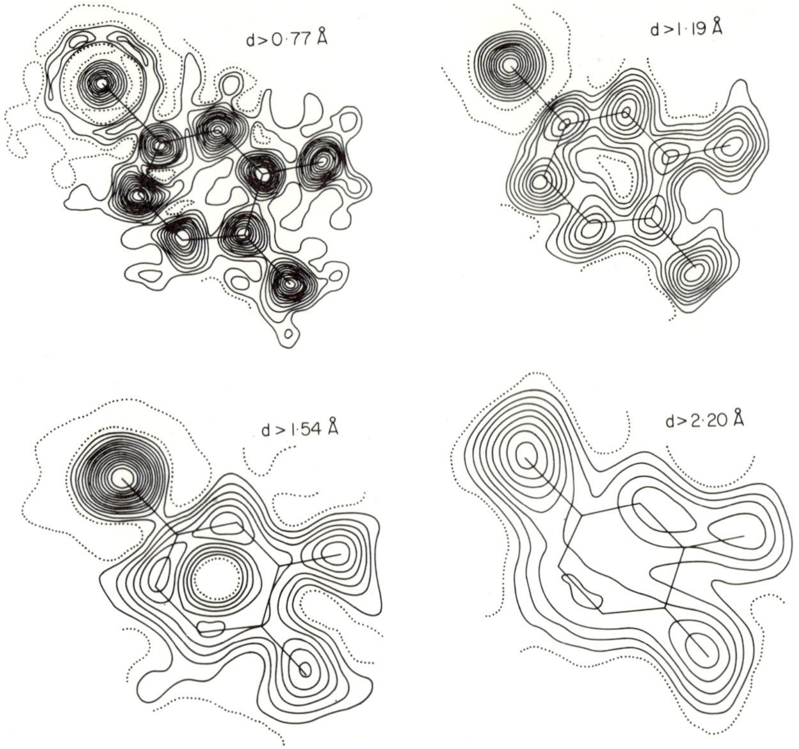

Fig. 9.10.1. Electron-density projections of the molecule of 4,5-diamino-2-chloropyrimidine. Successive maps show how the resolution and sharpness of the atomic peaks fall off as the Fourier series is curtailed: in each case there is shown the lower limit of spacing at which the series was terminated—corresponding to the upper limit in the value of sin θ. (These maps have been artificially 'sharpened', so as to represent atoms at rest; and this accounts for the negative regions in the diffraction rings round the heavy chlorine atom). (Reproduced from a photograph kindly supplied by Dr. D. C. Hodgkin).

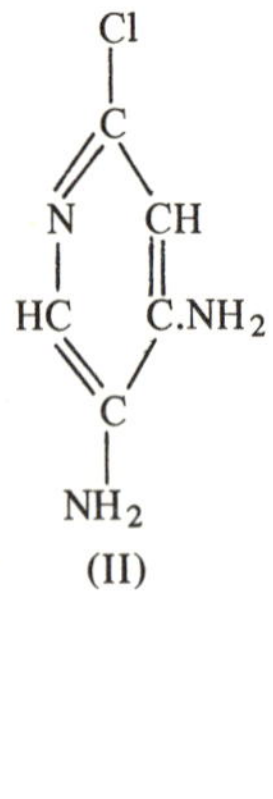

(II)

often referred to informally as a 'Fourier map'.

Nearly all structure analyses before 1955 were based on projections. However, two-dimensional analysis has severe limitations. The difficulty of obtaining resolution, unless the axis is very short, we have already stressed. Even more seriously, it suffers from the fundamental objection that it must be based on only a fraction of the accessible observational data. General reflexions of the type *hkl* are necessarily much more numerous than those confined to special zones, such as *hk*0; whereas a two-dimensional study might be based on 200 terms, three-dimensional studies are often based on 1000–5000 terms.

It is harder to reproduce the results of a triple Fourier synthesis for visual inspection on paper. The custom is to show a series of sections from the three-dimensional electron-density continuum, chosen at various levels to as to pass through the centre of every atom. Such a composite picture is reproduced in Fig. 9.10.2. The contour-lines now correspond to stated increments of electrons per cubic Å.

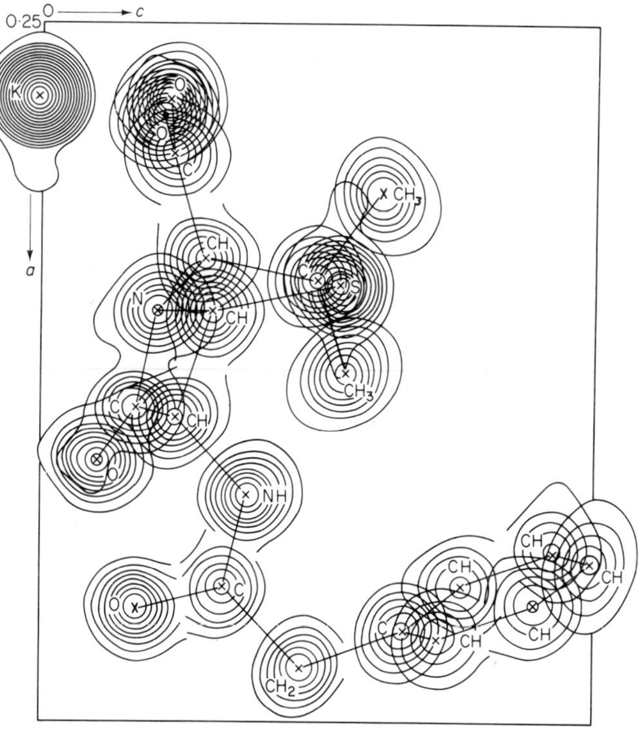

Fig. 9.10.2. Representation of the three-dimensional electron-density distribution in potassium benzyl-penicillin. (Reproduced, with permission, from Pitt, *Acta Cryst.*, 1952, **5**, 772).

9.11 Further Refinement–Least-squares Analysis

When the refinement has been taken to the stage where phases (or signs) can be allocated to all the observed reflexions, thus enabling them all to be included in the final Fourier synthesis, it by no means follows that the electron-density maxima indicate the best atomic coordinates attainable. For this, the Fourier series should be infinite. As it never is, termination-of-series errors will be present. The importance of correcting for these was not realised at first; and, for this reason, much of the early X-ray work may suffer from greater errors than its authors supposed.

Preliminary refinement by Fourier-series methods is now always followed by *least-squares* analysis, which, amongst other advantages, corrects for termination-of-series errors. As we have a large excess of observational data (F_o) over parameters (atomic coordinates in the first instance), we have the typical situation where the least-squares method can be applied. The computation can be set up so as to yield the small shifts in coordinates which will minimise the sum, $\Sigma(F_o - F_c)^2$, or some related quantity, for all observed reflexions. Since the relationship between F and the parameters is complicated, and non-linear, the parameters do not move right to their 'final' values after a single least-squares operation. Several (perhaps, many) cycles of refinement are necessary before there is convergence.

Adjustment of atomic coordinates is not the only way of improving the agreement between F_o and F_c. Atoms in crystals are always vibrating, and allowance must be made for this in work of even moderate accuracy. The atomic scattering functions, f, which are needed in deriving F_c-values, have been calculated for stationary atoms, of spherical shape. Vibration has the effect of spreading the electron density over a wider region of space, since it is necessarily the 'time-average' atom that is relevant during the long exposure needed in making an X-ray photograph. This spreading can be taken into account by diminishing f with a *Debye factor*, $\exp(-B \sin^2 \theta/\lambda^2)$, θ being the Bragg angle for the particular reflexion and B a parameter depending on the amplitude of the vibration, and hence on the temperature. (For this reason B is sometimes called the temperature factor–an inaccurate usage). For an harmonic vibration, $B = 8\pi^2\overline{u^2}$, where $\overline{u^2}$ is the mean square amplitude. For carbon or oxygen atoms in typical organic crystals at room temperatures, B may be 2–4 Å^2, which corresponds to vibrations with a root mean square (r.m.s.) amplitude of 0·15–0·23 Å.

When allowance is thus made for vibrations, a better agreement with experiment can be attained, but, since B-values cannot generally be determined independently, at the cost of introducing extra parameters into the analysis. Various degrees of finesse are possible: (*a*) a single isotropic B-value may be used for all the atoms; (*b*) different individual B-values may be used for each atom, and in particular larger values may be attributed to the lighter atoms, which will be vibrating more vigorously; (*c*) anisotropic factors may be used to allow for the fact that an atom may vibrate with different amplitudes in different directions; the atom is then represented by an ellipsoid of electron density rather than by a sphere, as–to be sure–is often observed in electron-density maps. In the fullest form of (*c*) nine parameters are applied for each atom: three coordinates of position, three constants to define the dimensions of the ellipsoid of vibration, and three more to define its orientation in the unit cell. Such a multiplication of parameters is justified only if the number of observed structure amplitudes still remains in adequate excess.

The study of thermal vibrations in crystals is an important field in its own right; it also has an indirect importance in accurate structure analysis. In so far as the disagreement between F_o and F_c is due to the neglect of vibrations, attempts to refine the structure further will merely force errors into the coordinates. Any method of refinement minimises $\Sigma\,|(F_o - F_c)|$, or some closely related quantity, in terms of the model structure used in calculating F_c-values. It does the best it can with the model provided; it cannot change the model if that is qualitatively at fault. At one stage during the precise analysis of the crystal structure of benzene it appeared that C—C was 1·378 ± 0·003 Å, which was significantly less than values given by other methods of precision. This anomaly disappeared after it had been realised that the benzene molecule is executing torsional oscillations (of r.m.s. amplitude $\sim 8°$) about its six-fold axis. Each atom therefore vibrates along an arc of a circle. Now the centre of gravity of such an arc lies within the circumference of the circle. Hence the original refinement programme was yielding carbon positions too close to the centre of the ring. When this new feature was included in the model, C—C rose to 1·392 Å.

As a measure of the refinement obtained in an analysis, crystallographers are accustomed to cite a value of the 'agreement index', R, defined as $\Sigma\,|(|F_o| - |F_c|)|/\Sigma\,|F_o|$; *i.e.*, the sum of the differences, irrespective of sign, between observed and calculated structure amplitudes expressed as a fraction or percentage of the sum of all observed amplitudes. Since R increases as the agreement gets worse, it would more properly be called the 'disagreement index'. Though a falling value of R satisfactorily indicates progress in refinement, the final value of R achieved is not a good measure of the reliability of the coordinates found. In a structure consisting of a dozen or more carbon and oxygen atoms, all in general positions, an R-value of 15% may be better evidence of a sound analysis than one of 8% in a structure whose X-ray diffraction is dominated by the effect of a heavy atom in a special position, the coordinates of which are necessarily correct as a symmetry requirement. Some better method for the assessment of accuracy is desirable; and we discuss this in the next section.

As the temperature is lowered thermal vibrations tend to become less vigorous, and therefore the electron-density peaks derived from X-ray analysis tend to become sharper and better resolved. The effect is similar to that of using a longer Fourier series (see sections 9.6 and 9.7, especially Fig. 9.10.1). Zero-point vibration, of course, persists even down to absolute zero; so that the benefit to be derived from going below the temperature of liquid nitrogen ($\sim$77 K) tends to meet a law of diminishing returns; the gain in resolution may not justify the experimental inconvenience. For some purposes X-ray diffraction studies at liquid hydrogen temperature (20 K)—or even lower—are profitable.

A further reason for developing low-temperature techniques has been that many substances with very simple, and normally gaseous, molecules can then be studied by crystallographic methods. Lipscomb's important structural work on numerous boron hydrides, with the discovery of their unusual type of three-centred bonds, has been based on low-temperature X-ray data.

9.12 Accuracy and the Assessment of Precision

The measurement of a physical property is of meagre value unless it can be accompanied by some indication of its reliability. In this book we are concerned with molecular dimensions, and particularly with interatomic distances. When we discover that a certain C—C bond has the length 1·537 Å, this piece of information is of substantial interest only when we can compare it with some other bond-length, or with the length predicted by some theoretical model; and then decide whether it is longer, or shorter, or the same. Before we can do this, we need some measure of the accuracy of our experimental result.

The true *accuracy* of a measurement is hard to come by; for it depends on every sort of error that may have affected our work, and we can never be sure that we know all sources of error. It is therefore customary to settle for a less satisfactory criterion–a measure of *precision*. In effect this means that we carry out the measurement a number of times, assume that the slightly different results we get each time are due only to random errors, and apply standard statistical methods to our results accordingly. The statistical quantity now always used is the *standard deviation*, σ. For example, we might qualify our C—C distance by writing 1·537 ± 0·015 Å, where the 0·015 is σ. This has a well understood meaning: the spread of our repeated measurements was such that, if we repeated the measurement again, there should be a 68% probability that the new result would lie between the limits $\pm\sigma$; *i.e.*, between 1·522 and 1·552 Å; a 99% probability that it would lie between $\pm 2\sigma$; and a 99·8% probability within $\pm 3\sigma$.

We have neglected unknown systematic errors. Our standard deviation is a less stringent criterion. The limits of true accuracy will be wider than those of precision. Despite its limitations, the standard deviation is of value. The experienced reader knows the limitations, and is duly wary. We must emphasise that the statement C—C = 1·537 ± 0·015 Å does not mean that the distance is certainly between 1·522 and 1·552 Å.

An example will stress this point. Suppose two C—C distances in a particular molecule were found to be 1·537 and 1·518 Å, each ± 0·015 Å. Since the difference between the lengths is only 0·019 Å, which is only a little greater than σ, we must not regard the difference as significant. But, if we could improve our technique and reduce σ to ± 0·005 Å, with the same distances, then, the difference being more than 3σ, the probability that the difference is genuine is high; the difference would now be regarded as 'highly significant'.

The statement above–that we carry out the experiment repeatedly–should not be taken literally in the context of crystal structure analysis. As we have already explained, any X-ray study uses a large excess of F_o over parameters. The problem is 'over-determined'. It is to this excess that we apply the techniques of statistics, and hence arrive at standard deviations for our parameters. Refinement is now always done by least-squares; and there are standard ways of estimating σ from the analysis. Application of these methods to crystallography was developed by Cruickshank (1959) in particular.

It is a common convention to put standard deviations in parentheses: 1·537(5) is a concise way of implying 1·537 ± 0·005.

9.13 Diffraction Analysis using Electrons and Neutrons

Hitherto we have discussed the crystal diffraction method wholly with reference to X-rays. But any form of wave motion can be used in principle, provided only that it is of appropriate wavelength. (The case has been described of a crystal, composed of very large protein molecules, which gave Laue-type diffraction effects with visible light . . .) Normally wavelengths have to be of the order of 1 Å. According to the de Broglie equation a wavelength, $\lambda = h/mv$, can be associated with radiations that are ordinarily regarded as made up of particles of mass, m, moving with velocity, v. Two such radiations that are now being used with crystals are beams of electrons and of neutrons.

Electrons accelerated to a uniform velocity by a potential of V volts are associated with a wavelength of $\sqrt{(150/V)}$ Å. The diffraction effects produced by such electron beams have been an important source of information about the structures of simple gaseous molecules since the nineteen-thirties, and this topic is discussed in Chapter 10. Electrons have also been used for a long time to study the surfaces of metals. Their use to examine the internal structures of crystals is a more recent development. There are experimental difficulties: the crystal must be extremely thin—not more than a few hundred Å thick; and it has to be exposed to a high vacuum. When an adequate diffraction pattern can be recorded, structure analysis can proceed in the same way as with X-ray data. Electrons are scattered by the potential field in the crystal due to the combined effects of the positive atomic nuclei and the negative electron density. The most important advantage of electron diffraction for the chemical crystallographer is that the positions of hydrogen atoms are rather more easily found than with X-rays. But the method has not been much used, since neutron diffraction proves to be much better.

'Thermal' neutrons have wavelengths in the region of 1 Å, and are very suitable for crystal diffraction. A powerful source of low-energy neutrons is needed because most of the flux has to be sacrificed in order to produce a monochromatic* beam, and because the detection of the reflected rays from the crystal is not easy. Hence neutron diffraction depended on the development of high-flux reactors. And for the same reason, much larger single crystals are needed than for X-ray diffraction.

* 'Monochromatic' means that all neutrons have the same velocity, and hence the same wavelength.

The technical difficulties and the general theory of neutron diffraction are described in a monograph by Bacon (1962).

Neutrons are scattered by the atomic nuclei in the crystal. Except in special circumstances, they are not affected by the electron density. This causes important differences as against X-rays. First, as the nucleus is of negligible size compared with the wavelength used, its scattering power is independent of $\sin\theta$ (see section 9.6). The vibration of the atom causes the effective scattering power to fall away (see section 9.11); but still the fall-off with $\sin\theta$ is much more gradual than with X-rays. Therefore the peaks in a Fourier map based on neutron structure factors are sharper.

Secondly, the neutron-scattering factor does not increase in any simple way with atomic number, and it depends on the nuclide. Deuterium (^{2}H), for example, is rather more effective than ^{16}O or ^{12}C, and nearly as effective as ^{239}U. Ordinary hydrogen (^{1}H) has a lower scattering amplitude than deuterium, and with a negative sign. This is because neutron scattering from ^{1}H, and from a few other nuclides, is accompanied by an unusual change of phase; and its consequence is that hydrogen atoms appear as negative peaks in a Fourier map based on neutron structure factors. An example is shown in Fig. 9.13.1, where the broken contour-lines signify levels of negative scattering density. This map is the result of a 'difference' synthesis the Fourier coefficients were $(F_o - F_c')$, rather than the usual F_o, where F_o means the observed neutron structure factor and F_c' the structure factor calculated for all the atoms *except hydrogen*. This tactic makes the protons positions stand out more prominently. 'When the main mountain chain is taken out, the foothills stand out

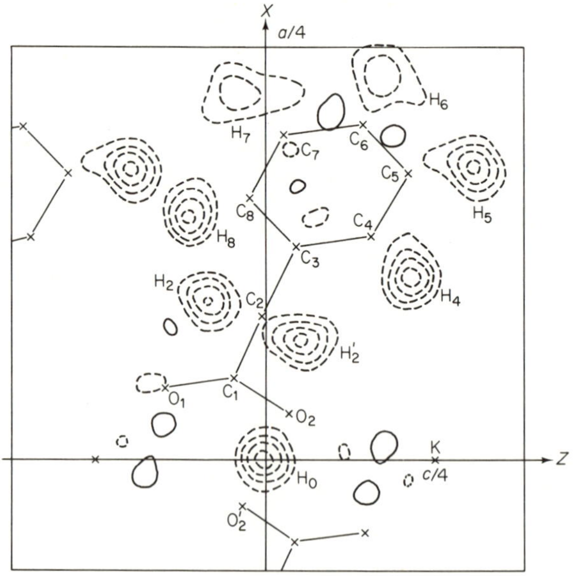

Fig. 9.13.1. Neutron-diffraction study of potassium hydrogen *bis*-phenylacetate: a difference map, with the effects due to K, O, and C-atoms subtracted, leaving the H-atoms as negative areas. (Reproduced, with permission, from Bacon and Curry, *Acta Cryst.*, 1957, **10**, 525).

in splendid isolation'.

Indeed the location of hydrogen atoms is one of the most important applications of neutron diffraction to chemical crystallography. It often happens that an ordinary X-ray analysis has located all the heavier atoms satisfactorily, but allowed the hydrogen atoms to be only roughly located. When special importance attaches to knowing their positions, recourse to neutron diffraction is valuable.

A still better way for finding hydrogen atoms is to deuteriate the material—to replace H by D. Deuterium has a larger scattering amplitude than has hydrogen, and a positive one. So D shows up about as well as C in a neutron Fourier map. An example is given in Fig. 9.15.1(*iii*), which shows the neutron-scattering density in the mean plane of the per-deutero-anthracene molecule, $C_{14}D_{10}$. It should be contrasted with the corresponding X-ray electron-density map of ordinary anthracene, $C_{14}H_{10}$, in (*i*). Another example may be taken from the many polymorphic forms of ice. It was important to study the varying schemes of hydrogen bonding between the water molecules in these polymorphs. To do this, the hydrogen atoms must be located. Recent neutron-diffraction studies of heavy ice, D_2O, in its various forms have achieved this.*

9.14 Some More Sophisticated Uses of X-Ray Diffraction

Though hydrogen atoms can be crucially important in deciding how a given type of molecule crystallises, they do not have much effect on the X-ray diffraction. There are two reasons for this: first, in a $C_6H_{12}O_6$ molecule for instance, only 12 of the 96 electrons belong to the hydrogen atoms; and secondly, and more important, the 84 electrons of the carbon and oxygen atoms are much more strongly condensed near their respective nuclei because of the higher nuclear charges. Hence the contributions of hydrogen to $|F_o|$ are small; and in consequence their positions much more difficult to determine. Indeed in early work hydrogen atoms were ignored; and an excellent modern textbook of physical chemistry (1971) states: 'it is not possible to deduce anything about the positions of hydrogen atoms from X-ray diffraction'. Such a statement ceased to be true in the early 1950's. Any good modern study of a moderate-sized organic molecule would be expected to locate all the hydrogen atoms, though with a lower precision† than the heavier atoms.

Hydrogen atoms are best made evident from X-ray measurements by the 'difference' Fourier synthesis explained in the previous section. Near the end of the refinement, F_c' are calculated for all the heavier atoms, and the differences $(F_o - F_c')$ are used as coefficients in the Fourier series. A typical (if now rather primitive) result is shown in Fig. 9.15.1(*ii*). The hydrogen atoms of the anthracene molecule appear as residual density-peaks about 1 Å out from each hydrogenated carbon atom.

* Kamb, Hamilton, LaPlaca and Prakash, *J. Chem. Phys.*, 1971, **55**, 1934.

† With lower accuracy too. For reasons that are now understood, X-ray analysis nearly always finds C—H, N—H or O—H distances that are too short by about 0·1 Å.

As X-rays are scattered by the electron density in a crystal, it should be possible to map out the density of a molecule by Fourier methods so accurately that details of electronic bonding might be revealed. In principle, we should be able to measure the electron density of a molecule experimentally, just as—given lavish computing facilities and a molecule that is not too large—we can calculate a complete molecular wavefunction from theory.

We face difficulties in this project. First, we need a very accurate set of $|F_o|$, near the limit of what is experimentally obtainable at this time. Least-squares refinement then depends on our choice of a model, which can be described by parameters that are to be adjusted to minimise some function of $\Sigma (F_o - F_c)^2$. Normally, and for strong practical reasons, we assume the atoms to be spherical masses of electron density of a form calculated for the free atom. The situation of each such atom within the unit cell is then defined by three positional parameters, and up to six vibrational parameters. With some reservations, this model works very well for ordinary structure analysis; for most of the electron density is concentrated into spherical blobs around the atomic nuclei. We can feel some confidence in the atomic positions found, when the least-squares analysis has converged. But any shape deduced for the residual electron density in the outer regions of the atoms is much more doubtful. The actual density depends on the original shape of the isolated atom, modified by smearing due to every form of vibration the atom may be undergoing, and also modified by changes resulting from whatever electron redistribution occurred when the molecule was formed. It is very difficult to separate these two effects. In Fig. 9.15.1(*ii*), for example, the small electron-density peaks near the centres of many of the covalent bonds may genuinely represent bonding electrons, but they may alternatively be an artefact of our over-simplified treatment of atomic vibrations.

A possible way round this difficulty has been developed in particular by Coppens (1968) by combining precise X-ray and neutron studies of the same crystal. So far as neutron diffraction is concerned, the model of a spherical (in fact, point) atom is an almost faultless approximation. Any smearing of the atomic nuclei must be due to vibration only.* Therefore vibrational parameters derived from neutrons are more reliable: they mean what they say. One way of taking advantage of this is to use the neutron parameters to calculate X-ray structure factors, F_c^N; and then to sum the Fourier series with $(F_o^X - F_c^N)$ as coefficients, F_o^X being the observed X-ray structure factors. The scattering factors used in obtaining F_c^N were based, as usual, on the spherical-atom model. Hence any features of the difference map indicate where the model is failing. Fig. 9.14.1 is an example of the results obtained by this device (Coppens and Vos, 1971). It covers half the cyclic molecule of cyanuric acid, $C_3H_3N_3O_3$. Electron density appears near the centre of each covalent bond. There is also evidence for lone-pair electrons on the oxygen atoms. Attempts are being made to render this method more quantitative—to measure, rather than merely demonstrate, as the figure does, the deviations from sphericity of the electron

* Or, possibly, disorder.

Fig. 9.14.1. An $(F_o^X - F_c^N)$ synthesis, based on X-ray and neutron-diffraction studies of the cyanuric acid molecule. The synthesis has been calculated in the mean plane of the molecule; positive density contours are represented by continuous lines, negative by dotted lines. Some evidence of bonding and lone-pair electrons is to be seen. (Reproduced, with permission, from Coppens and Vos, *Acta Cryst.*, 1971, **B27**, 146.)

density round the atoms. This may be done, for example, by deriving numerical values for the electron populations of the various atomic orbitals.

9.15 Structure Analysis of Naphthalene and Anthracene

The study of naphthalene and anthracene will be described in some detail because it illustrates the progressive development of X-ray crystal structure analysis. In 1921 W. H. Bragg published his measurements of the unit cell dimensions:

	a	b	c	β	Z
Naphthalene $C_{10}H_8$	8·24	6·00	8·66 Å	122·9°	2
Anthracene $C_{14}H_{10}$	8·56	6·04	11·16	124·7°	2

The only notable difference is an increase of 2·5 Å in c. This is about what would be expected to make room for the extra benzenoid ring. On this basis Bragg was able to suggest positions, which were essentially correct, for the two molecules in the unit cells. (The actual positions for naphthalene are shown in Fig. 2.13.2). In 1933 Robertson measured the intensities of about 70 reflexions for naphthalene and of

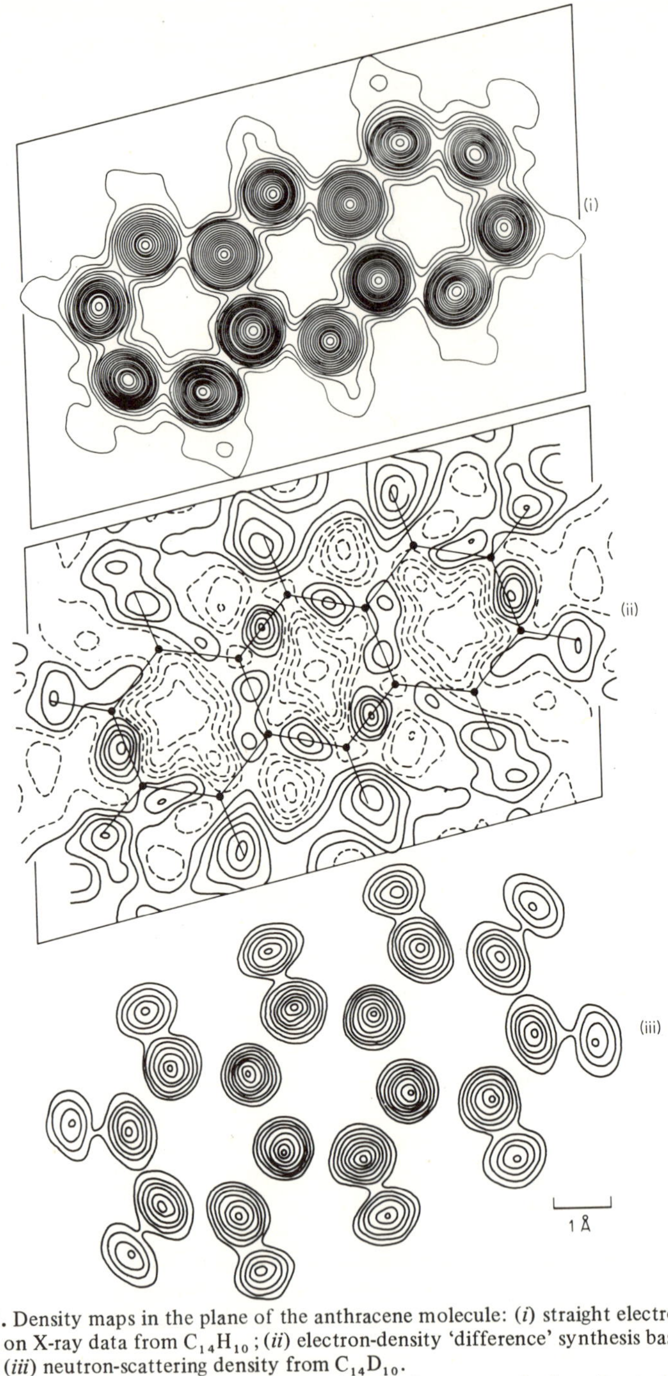

Fig. 9.15.1. Density maps in the plane of the anthracene molecule: (*i*) straight electron-density map based on X-ray data from $C_{14}H_{10}$; (*ii*) electron-density 'difference' synthesis based on the same data; (*iii*) neutron-scattering density from $C_{14}D_{10}$.

(*i*) and (*ii*) are reproduced, with permission, from Robertson *et al.*, *Acta Cryst.*, 1950, **3**, 254, and from Cruickshank *et al.*, *Acta Cryst.*, 1956, **9**, 917; (*iii*) is from a print kindly supplied by G. S. Pawley.

about 80 for anthracene, and used them to derive electron-density projections along the principal axes for each compound. The atoms were not all resolved, but, since the molecule was planar–or nearly so–their positions could be assessed within perhaps 0·05 Å. This accuracy was not sufficient to render the small differences between different C—C bond lengths significant, or to justify comparing them with the differences predicted by various theoretical treatments. In 1949–50 Robertson and his collaborators measured nearly all the reflexions accessible to copper X-rays–over 600 for naphthalene and about 700 for anthracene–and used them in three-dimensional syntheses. Fig. 9.15.1(*i*) is derived from this work and shows a section of the anthracene molecule in the mean molecular plane (not a projection). (All ten hydrogen atoms can be discerned in this map, by the way). Bond lengths could be determined now with an accuracy of about ± 0·01 Å; and the differences amongst them were significant, and broadly in the sense predicted by theory. This analysis stopped at the stage of the straight electron-density syntheses. Cruickshank (1956–7) used these same observations in the most thorough refinement hitherto applied in an analysis of this type. The final bond lengths had an e.s.d. of ± 0·004 Å. In the 1949 work small deviations of the carbon atoms from their mean plane had been noticed, but their significance was then doubtful. Similar deviations (up to 0·012 Å) persist in the later refinement, and they were found to be highly significant. They can be intelligibly accounted for as due to close contacts with neighbouring molecules; in other words, it is demonstrated that the molecule reacts like a flexible object to the irregular stresses exerted by its environment.

The parameters relating to the anisotropic vibrations of the different atoms were used to elucidate the molecular vibrations. As might be expected from the relative magnitudes of the restoring forces, intramolecular vibrational movements account for only a small percentage of the total mean square amplitudes. To a good approximation, the vibrations are those of a rigid body. They were resolved into translations and librations: for the former, the r.m.s. amplitudes varied from 0·22 to 0·19 Å in the case of naphthalene, and from 0·20 to 0·16 Å in that of anthracene, and in each case the maximum amplitude was along the greatest length–*i.e.*, in the direction in which the molecule might be expected to experience the least resistance to movement; for the latter, the r.m.s. amplitudes of oscillation about the various molecular axes averaged at about 4° and $2\frac{1}{2}$° respectively.

A difference map of anthracene is reproduced in Fig. 9.15.1(*ii*). In it the hydrogen atoms show up well. The small peaks near the middles of the C—C bonds arise* because, at this stage, the anisotropic motions of the carbon atoms had not been allowed for. The negative areas (contour-lines broken) at the centres of the rings and elsewhere are not to be taken literally; they imply that the electrons are more concentrated along the bonds then were those in the model used to calculate F_c, this model having been based on spherical atoms. Fig. 9.15.1(*iii*) shows the neutron-scattering density in the mean plane of the $C_{14}D_{10}$ molecule, which was discussed in section 9.13.

* Or mainly for this reason.

9.16 The Determination of Absolute Configuration

The important problem of determining the absolute configuration of an optically active molecule has exercised chemists for many years. It was solved by an X-ray method due to Bijvoet (1949). As was stated earlier, X-ray diffraction normally supplies a centre of symmetry to a crystal whether one is really present or not. This arises immediately because the intensities of reflexions from opposite faces of a crystal (hkl and $\bar{h}\bar{k}\bar{l}$) are identical; so that the diffraction pattern could be equally well explained by a structure in which the direction of x-, y- and z-axes has been reversed, which would have the effect of a centre of symmetry at the origin and would change a d-molecule into its l-enantiomorph. The circumstances of this identity of intensities are examined in Fig. 9.16.1(*i*), which represents reflexion from the hkl plane of a crystal supposed to consist of layers of atoms of two kinds, H and L. The atoms H are heavier, and the waves reflected from these layers are therefore more intense than those from the L-layers; the latter waves lag somewhat behind those from H, as is shown in Fig. 9.16.1(*ii*), and both combine to yield the resultant wave drawn with a thicker line. Reflexion from the opposite, $\bar{h}\bar{k}\bar{l}$, face is similarly represented in (*iv*); the wave from L is now ahead of that from H, as is shown in (*v*), where the combined wave differs in phase from that in (*ii*) but has the

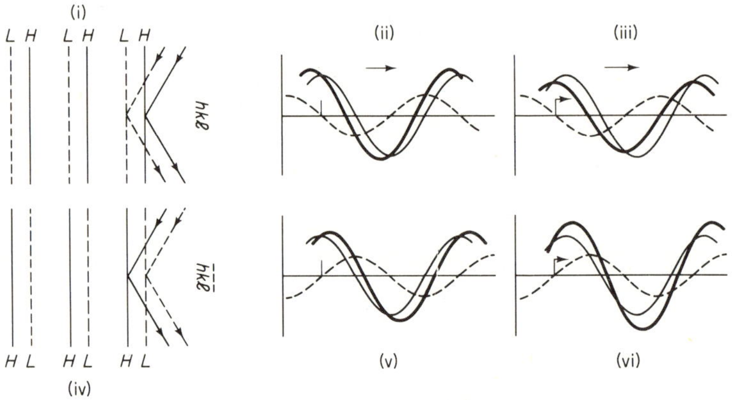

Fig. 9.16.1. (*i*) and (*iv*) Reflexion of X-rays from opposite faces, (hkl) and ($\bar{h}\bar{k}\bar{l}$), of a crystal; (*ii*) and (*v*) normal behaviour when Friedel's law is obeyed—$|F(hkl)| = |F(\bar{h}\bar{k}\bar{l})|$; (*iii*) and (*vi*) failure of Friedel's law when there is anomalous scattering from the H-atoms. (The anomaly is here betokened by a 30° advance in phase-angle, which has the consequence $|F(\bar{h}\bar{k}\bar{l})| > |F(hkl)|$).

same amplitude. Since only amplitudes can be recorded, the reflexions from hkl and $\bar{h}\bar{k}\bar{l}$ are indistinguishable. This is expressed in *Friedel's Law*, which states that

$$|F(hkl)| = |F(\bar{h}\bar{k}\bar{l})|,$$

or—more rigorously—that the structure factors are complex conjugates; $F(hkl) = A + iB$ and $F(\bar{h}\bar{k}\bar{l}) = A - iB$. (See section 9.6).

It has been found that Friedel's Law breaks down when the X-rays are of such a wavelength that they excite one of the inner-shell electrons of an atom. The scattering is then attended by an anomalous change of phase, the amount of which can be estimated from our knowledge of the electron distribution in the atom. Suppose now that this anomaly occurs at the *H*-layers, and that it consists of an advance of the wave amounting to 30°. Consideration of diagrams (*iii*) and (*vi*) will show that the resultant waves now differ in amplitude, as well as in phase; $|F(hkl)| < |F(\bar{h}\bar{k}\bar{l})|$.

X-rays from a tungsten target are anomalously scattered by rubidium. The principle was therefore applied to the crystal of rubidium sodium *d*-tartrate, the structure of which was already known apart from its absolute configuration. This configuration was provisionally supposed to be in accordance with Emil Fischer's arbitrary convention; and the intensities were calculated for various pairs of reflexions, *hkl* and $\bar{h}\bar{k}\bar{l}$, which should differ because of the effect at the rubidium atom. When these were compared with the corresponding observed intensities, the differences were unambiguously in the sense calculated. Hence, by a one-in-two chance, the Fischer convention corresponds to reality. This has since been confirmed for some other tartrates. Absolute configurations have been assigned to a number of key molecules, from which the configurations of a wide range of compounds can be deduced by chemical arguments.

BIBLIOGRAPHY

Robertson, J.M., *Organic Crystals and Molecules*, Cornell University Press, 1953. (History and coverage of some classical studies).

Lipson, H. and Cochran, W., *The Determination of Crystal Structures* (Vol. III of *The Crystalline State* by W. H. & W. I. Bragg), Bell, London, 1966.

Woolfson, M.M., *An Introduction to X-Ray Crystallography*, Cambridge University Press, 1971. (Elementary theory).

Stout, G.H., and Jensen, L.H., *X-Ray Structure Determination*, Macmillan, London, 1965. (Emphasis on experimental methods and their theoretical background).

Bacon, G.E., *Neutron Diffraction*, Oxford University Press, 1962.

Chapter 10

Electron diffraction by gases and vapours

10.1 Introduction

In a crystalline solid the atoms or molecules are arrayed in an ordered manner. A crystal therefore behaves as a diffracting medium for rays of appropriate wavelength. Provided the crystal is of finite size, very sharp diffraction patterns can be recorded, and we have seen in the preceding chapter how the arrangement of the atoms in the crystal may be deduced from them. In gases and liquids the molecules occupy relative positions that are wholly, or largely, randomised. Nevertheless, the waves scattered from neighbouring atoms will in general be out-of-phase with one another, so that diffraction effects may still be observed, though they are of a more rudimentary kind. The X-ray patterns given by liquids carry information about the packing of the molecules and about their structures; but, unless the molecules are very simple ones, it is difficult to disentangle the effects separately due to inter- and intra-molecular interferences. In gases and vapours, on the other hand, the molecules are so far apart that intermolecular effects can be neglected. The diffraction pattern is wholly due to rays scattered by different parts of the same molecule. It may therefore be possible to deduce the molecular structure from the X-ray- or electron-diffraction pattern. Both radiations have been used successfully, but with gases electrons have several advantages. In particular, exposure times are much less. A fruitful technique was developed by Wierl about 1931. A great many simple molecules have been studied, and, as long ago as 1950, Allen and Sutton* tabulated results for 500 molecules.

Since the material to be analysed must be in the vaporised condition, only small molecules can be studied readily; and indeed it is only for such molecules that the diffraction pattern can be interpreted. However, these are the molecules that cannot be obtained in the crystalline state at ordinary temperatures. In fact, the electron-diffraction method fills a gap between X-ray crystallography and molecular spectroscopy; it is applicable to many molecules which, though simple, are too complex for their spectra to be interpreted in detail, or too symmetrical to give pure rotational spectra. The accuracy is also intermediate; thirty years ago bond lengths could be

* *Acta Cryst.*, 1950, **3**, 46.

measured to ±0·01 Å, which was better than that attained in normal X-ray practice at that time, though less good than in spectroscopy.

Recently the electron-diffraction method has gained a new lease of life. This has been due to improvements in its precision, both on the experimental side and in the interpretation of the patterns (see sections 10.5 and 10.6). In the best work an accuracy of two or three units in the third place of decimals is now accessible.

10.2 Experimental Method

According to the de Broglie equation an electron of mass, m, moving with velocity, v, and with momentum, p, is associated with a wavelength $\lambda = h/p$; when a small relativity correction is ignored, this becomes $\lambda = h/mv$. After introducing appropriate values for the constants, we find that acceleration through a potential of V volts gives the electron-beam a wavelength of $12{\cdot}225/\sqrt{V}$ Å. Voltages of about 40 000 are found most useful, and accordingly they correspond to wavelengths in the region of 0·06 Å. During the short exposure time this voltage must be measured carefully and kept constant within close limits, so as to ensure constancy of wavelength.

A fine beam of such electrons passes perpendicularly through a fine jet of the gas issuing from a nozzle, and thence goes on to record the diffraction pattern on a photographic plate. Apart from the actual jet of vapour, the whole experiment must be carried out in a high vacuum; otherwise unwanted scattering by the residual gas blurs the pattern. Arrangements are therefore made for removing the material embodied in the jet immediately after it has passed out of the electron beam by condensation on a cold surface. Exposure times of a few seconds suffice.

When the plate is developed the diffraction pattern consists of a dark central region with a radial diminution of intensity away from the centre. The eye gains the impression of a succession of diffuse rings, alternately dark and light, and perhaps a dozen or more in number. Objective measurements with a densitometer prove this impression to be a remarkable optical illusion. Except close in near the centre sometimes, there are in fact no maxima or minima, but merely inflexions in a monotonously decreasing intensity. Nevertheless, until about 1946 the experimental raw material consisted simply of the apparent angular positions of these illusory diffraction rings, together with some semi-quantitative assessment of their relative intensities. With modern equipment it is possible to produce better data, as will be explained later. But the amount of information accessible in any diffraction pattern is limited, and this imposes a corresponding limitation on the complexity of the molecules that can be studied.

10.3 Elementary Theory of Electron Diffraction

When a beam of electrons or X-rays is scattered by matter, two kinds of interaction occur: coherent and incoherent, scattering. The latter is accompanied by a change of wavelength, and it is called the Compton effect in the case of X-rays. It is important only at low angles of deflexion. In the former there is no change of wavelength, and it is this type of scattering that is important in structural analysis.

The theory of scattering of X-rays was worked out in 1915 by Debye and Ehrenfest independently. It applies to electron scattering also when allowance is made for the difference that, whilst X-rays are scattered only by the orbital electrons, electron scattering is controlled by the total electrostatic potential field throughout the molecule, which field depends on the atomic nuclei as well as on the orbital electrons, and is indeed principally due to the former. Because they carry charges of opposite sign, nuclei and electrons exert opposite effects on the electrons being scattered; and accordingly the scattering factor for an atom is approximately $(Z-f)$, where Z is the atomic number and f the atomic scattering function for X-rays (see sections 9.6 and 9.7). . The two terms in $(Z-f)$ thus cover the effects of nucleus and electrons respectively.

The atoms in a gas are in motion by virtue of the translation, rotation, and vibrations of the moledules; but all such movements are negligible during the time-interval ($\sim 10^{-18}$ sec) a 40 kV electron needs to traverse a molecule. The atoms in a particular molecule can thus be regarded as in fixed positions during the passage of a particular electron-wave. The wavelets scattered from a given pair of atoms, and in a given direction, will follow paths of different lengths, and therefore they will in general interfere with each other at some distant point. This applies to each pair of atoms in the molecule. By combining the effects due to each pair, the diffraction pattern for the single molecule in its fixed position could be worked out. The total pattern is due to an immense number of different molecules, in all possible orientations with respect to the incident beam. The averaged effect can be derived by a suitable integration. Plainly the scattered intensity will depend on the angle of scattering, ϕ, but not on the azimuthal angle about the beam direction: the pattern must have circular symmetry. For the coherent scattering the result is approximately:

$$I(\phi) = I_0(8\pi me^2/h)^2(1/s^4)\sum_i\sum_j(Z_i-f_i)(Z_j-f_j)(\sin sr_{ij}/sr_{ij}). \quad (1)$$

I_0 is the intensity of the undeflected beam and $I(\phi)$ of that scattered through ϕ, m and e are the electronic mass and charge, and r_{ij} is the distance between atoms i and j, the summation being taken over all atoms in the molecule; s, which is defined as $(4\pi\sin(\phi/2))/\lambda$ and has the dimensions of reciprocal length, is a more convenient alternative measure of the scattering angle. It will be noted that ϕ is here the total angle through which the ray is deflected, and so corresponds to twice the Bragg angle, θ, used in X-ray crystallography.

The coherent scattering represented in (1) can be divided into two parts: the atomic scattering when i and j are the same (*i.e.*, each atom is 'paired' with itself), and the interatomic scattering when they are different. As r_{ii} is obviously zero for the former, $(\sin sr_{ii})/sr_{ii}$ assumes its limiting value of unity irrespective of s. The atomic terms merely give rise to a steadily falling background. In the latter, where r_{ij} is not zero, the sinusoidal factor gives a series of maxima and minima, of decreasing amplitude, for each pair of atoms, the maxima being more closely spaced in s the greater the value of r_{ij}. It is these series of superposed intensity waves due to the interatomic scattering that are of interest in structural analysis.

The factor $(1/s^4)$ in eqn. (1) is the chief cause of the severe decline in intensity with increase in ϕ (or s), and hence for the absence of genuine maxima and minima in the total diffraction function. A less important cause is the thermal vibration of the atoms. Though relative motion of the atoms during the passage of an electron is negligible, vibration is of significance because it implies that r_{ij} is distributed over a range of values for the different molecules that contribute to the averaged result. In other words, the atoms are effectively 'smeared' so that their scattering factors are diminished, and more so as s increases. As in X-ray analysis, this effect must be taken into consideration in accurate work, and this can be done by introducing a Debye temperature factor (see section 9.11). As it stands, the quantity $(Z - f)$ will actually increase with s, because Z is unchanged whilst f decreases. This means that the intense field near the atomic nuclei dominates the diffraction phenomena except at low angles. As we shall see later, in work of high accuracy attention is concentrated on that part of the diffraction that is due to the nuclei. In any case, $(Z - f)$ in eqn. (1) can be replaced by Z without serious error in very approximate work.

Because of the steep decline of intensity with angle, the range is very great—far beyond the latitude of a photographic plate, for instance. This was the root cause of the difficulty in measuring intensities by straightforward instrumental means. The eye, which seems subconsciously to 'subtract the background and divide by it', gave superior results, especially with the faint, outer rings. All the early work was based on visually estimated data; yet it often led to results of surprising accuracy.

10.4 Structural Analysis

In describing the interpretation of electron-diffraction patterns it is convenient to follow an historical line in so far as to deal with the traditional methods first. They usefully reveal the basic principles and the difficulties. Subsequently we will explain those modifications, both experimental and computational, that have enhanced the power and accuracy of the electron-diffraction method.

There are two ways of determining the molecular structure from the pattern, and they correspond to the trial-and-error and Fourier series methods in X-ray crystallography. Both require approximations.

In the *trial-and-error*, or *correlation*, method, a structure is postulated, giving a set of interatomic distances, r_{ij}. These distances are then used to compute some form of theoretical curve of intensity against s. The s-values at which maxima or minima occur (s_c) are then compared with those estimated visually (s_o) from the experimental record. Various structures may then be tried, and that one is taken to be most nearly correct which yields the closest agreement between the values of s_c and s_o. A simplified version of eqn. (10.3.1) is generally used, to bring out the interatomic diffraction and indeed to produce actual maxima in the calculated curve: the term $(1/s^4)$ is omitted, $(Z - f)$ may be replaced by Z, and the atomic scattering may be eliminated by summing the series only when $i \neq j$. The equation may then be reduced to

$$I(s) = K \sum_i \sum_j Z_i Z_j (\sin sr_{ij})/sr_{ij}, \quad (1)$$

in which K is a scale-factor. A graph calculated in this way is shown in Fig. 10.4.1, in which the positions of the observed features are indicated by the short vertical lines.

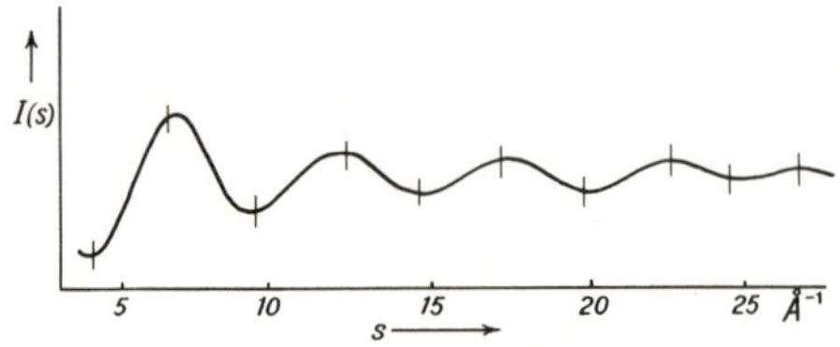

Fig. 10.4.1. Calculated intensity curve (interatomic) for formaldehyde. (Based on that of Stevenson, LuValle, and Schomaker, 1939.)

Though the comparison is chiefly between corresponding values of s_c and s_o, regard would also be had to the relative intensities of the various maxima. As a device for discovering a structure, this procedure was practicable only when the number of parameters needed to define the structure is very small.

What may be called the standard exercise throughout the development of the electron-diffraction method was carbon tetrachloride, CCl_4. In the most general way, nine parameters are needed to define the structure of this molecule. (At first sight it might seem that 15 would be necessary, since there are 5 atoms and since 3 co-ordinates would define the position of each atom in space. But 6 parameters go to define the orientation of the molecule and the position of its centre, both of which are irrelevant for our purpose. Hence 9 parameters suffice. These 9 can of course be chosen in various ways: for example, 6 Cl . . . Cl distances and 3 C—Cl; or 4 C—Cl distances and any 5 Cl—C—Cl angles. In general, a molecule with N atoms requires $(3N-6)$ parameters.)

Since in the earlier work only about ten diffraction rings were observed, a general solution of this structure was in practice impossible. However, if we are prepared to admit qualitative chemical knowledge, the molecule may be assumed to be regularly tetrahedral. Then a single parameter defines the structure; this could be the C—Cl distance, which might be the more obvious choice, though Cl . . . Cl($=\sqrt{8/3}$ C—Cl) would be equivalent. A value of 1·76 Å was chosen for C—Cl, and with this and the corresponding 2·876 Å for Cl . . . Cl, the interatomic scattering was worked out by equation (1) over the appropriate range of s, from 2·8 to 23 $Å^{-1}$. The curve would consist of two periodic components with $r = 1{\cdot}76$ and 2·876 Å and relative weights respectively $8 \times (17 \times 6)$ and $12 \times (17 \times 17)$. There was good agreement between observed and calculated maxima and minima, confirming the choice of the tetrahedral model, and–quantitatively–of the parameter 1·76 Å. This result was thought to be reliable to ± 0·01 Å.

When there are two or more parameters the problem becomes more difficult, though it can sometimes be simplified by concentrating upon the most important parameter only. This may be done when the molecule contains some atoms of relatively high atomic number, for it is these atoms that will be chiefly responsible for the diffraction pattern. An early analysis of benzene took advantage of this principle. The molecule C_6H_6 is a thirty-parameter structure, which reduces to a two-parameter problem when $6/mmm(D_{6h})$ symmetry is assumed: C—C and C—H would then suffice to define the molecule. The terms in eqn. (1) involving pairs of carbon atoms will be more important than those for carbon-hydrogen pairs, and still more important than those for pairs of hydrogens: $(6\times6) \gg (6\times1) \gg (1\times1)$. A substantially correct scattering curve can therefore be calculated, using only three terms of the first group, all based on a single C—C parameter. These terms have (1) $r =$ C—C for six pairs of neighbouring atoms, (2) $r = \sqrt{3}$(C—C) for six pairs of alternate atoms, and (3) $r = 2$(C—C) for three pairs of opposite atoms. A value of 1·39 Å for C—C was thus found by Pauling and Brockway in 1934. An accuracy of ± 0·005 Å was claimed, and certainly the result is in very close agreement with the present spectroscopic value.

This correlation method is open to the criticism that it is not objective, and that its findings may depend upon the particular trial structure chosen. The second method of dealing with the experimental data seeks to avoid this criticism, though–in its original form–it was no less open to objection on other grounds. We saw in Chapter 9 that a crystal structure can be directly deduced from the intensity data by means of a Fourier synthesis–always provided that the phases are known. A somewhat similar way of treating the electron-diffraction data was introduced by Pauling and Brockway in 1935. They made use of a *radial distribution function*, $p(r)$, which they described as 'the product of the scattering powers in volume elements the distance r apart as a function of r'.

A notion of what the radial distribution function represents may be gathered from the following approach. For the purposes of electron diffraction, the region of a molecule must be regarded as a continuum of varying electrostatic potential, which rises to maxima at the centres of the respective atoms. Suppose we place a straight line of length r anywhere in this region and note the values of the potential at either end of the line. These values are multiplied together. We then sum all such products for every possible position and orientation of the line in the volume occupied by the molecule. (For situations outside the molecule, the products will be zero.) This sum represents $P(r)$ for that particular length of line, r. We now repeat the summations for a range of lengths, and thus arrive at an ability to plot $P(r)$ against r. The resulting curve represents the radial distribution function. Evidently it will have peaks at all values of r which correspond to internuclear distances within the molecule. And, the higher the atomic numbers of the pair of atoms i and j, the higher will be the peak at r_{ij}.

This method is closely analogous to the Patterson synthesis used in crystallography (see section 9.9), rather than to the electron-density synthesis. It has the severe limitation that it is one-dimensional. The interatomic vectors in the actual molecule lie in three-dimensional space, but in the radial distribution function they are all superposed along a single direction. This is the inevitable consequence of the random orientation of the molecules in a gas. Just as the three-dimensional Patterson function for a crystal embodies all the information contained in its X-ray diffraction pattern, so does the one-dimensional radial distribution function for a molecule embody all the information implicit in the electron-diffraction pattern.

The radial distribution function and the function which shows how the intensity fluctuates in the diffraction pattern are Fourier transforms of each other (see section 9.8). A quantity $I'(s)$, related to the intensity, can be expressed as a function of the radial distribution of scattering power by an equation of the type,

$$I'(s) = K \int_{r=0}^{r=\infty} P(r)\,((\sin sr)/sr)\,\mathrm{d}r.$$

The Fourier integral inversion then leads to a similar equation for the distribution of scattering power as a function of the intensity:

$$P(r) = K' \int_{s=0}^{s=\infty} I'(s)\,((\sin sr)/sr)\,\mathrm{d}s. \tag{2}$$

The variables r and s are in fact distances in real and in reciprocal space. Therefore, if $I'(s)$ were known explicitly over a sufficient range of s, it should be possible to use eqn. (2) to derive an accurate distribution function, from which, provided its peaks were sufficiently resolved, the structure of any fairly simple molecule might be determined directly. Unfortunately, as we have seen, only a limited number of rings can be observed, and there are difficulties over their intensities; in any case, we cannot write an explicit, continuous function of s for the intensity variation, $I'(s)$. Therefore Pauling and Brockway made a drastic simplification in replacing the Fourier integral in (2) by the following summation:

$$P(r) = K \sum_k I_k (\sin s_k r)/s_k r, \tag{3}$$

where K is simply another scale-constant, and I_k is the visually estimated intensity of the k-the maximum at s_k Å^{-1}.

Despite its theoretical shortcomings, when used in this primitive form, the radial distribution method proved surprisingly useful. Fig. 10.4.2 (A) shows the distribution function for carbon tetrachloride, as it was first worked out in 1935, with only ten terms in the summation. The first peak lies at $r = 1{\cdot}74$ Å and corresponds to C—Cl, whilst the second and larger is at 2·85 Å and corresponds to Cl . . . Cl (which implies C—Cl = 1·75 Å), both in fair agreement with the value of 1·76 Å obtained by correlation. The other peaks represent spurious effects due to the limited number of terms available; they are in fact due to the 'termination-of-series' errors discussed

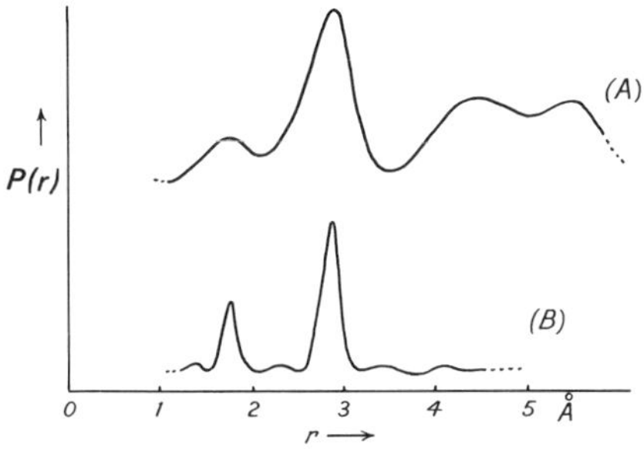

Fig. 10.4.2. Radial distribution curves for carbon tetrachloride. (Based (A) on that of Pauling and Brockway (1935), and (B) on that of Bartell, Brockway, and Schwendeman, 1955.)

in section 9.7. The presence of these false peaks, and the fact that the significant peaks tend to overlap one another with all but the very simplest molecules, imposed a severe limitation on the method. Recent developments, to be described in the next two sections, enable these difficulties to be surmounted in a considerable degree.

10.5 The Rotating-Sector Method.

Progress since 1950 in the scope and accuracy of the electron-diffraction method depends first on an improved technique for measuring intensities. The availability of better data has then justified and stimulated the development of more refined procedures for applying eqns. (10.3.1) and (10.4.2).

We have seen that the difficulty over measuring the scattering curve quantitatively, and instrumentally, arises chiefly from the factor $(1/s^4)$ in eqn. (10.3.1). Independently, and at various dates between 1935 and 1939, Trendelberg, Finbak, and Debye suggested that the difficulty might be overcome by introducing a suitably-shaped screen into the electron-diffraction apparatus between the jet of gas and the photographic plate. The screen must take the form of a sector, rapidly rotating about an axis coincident with the direction of the undeflected electron beam, and so shaped that diffracted rays are intercepted for a fraction of the rotation period which increases as s diminishes. The design of sector usually adopted allows rays to reach the plate for a fraction of the period proportional to s^3. To mount such a device close to the plate, and to cause it to rotate in the high vacuum that must be maintained, is an engineering problem of some severity, which has now been solved.

With a sector in operation, the steep fall in general intensity is largely offset, and the scattering curve shows genuine maxima. Furthermore, it can be measured quantitatively (and objectively) by scanning the plate with the minute spot of light of an optical densitometer. The pattern can also be traced to higher values of s; in a modern instrument the record can be taken to $s = 60$ Å^{-1}. The sector method was originally developed by a group of workers in Oslo, and subsequently at certain laboratories in the United States. At the time of writing (1973), British work on gas electron diffraction is based on the instrument at the University of Manchester Institute of Science and Technology.

10.6 Refinement of the Radial Distribution Function

With reliable scattering intensity curves available, it has become possible to subtract the steadily-falling background, mainly due to atomic scattering and to the effects of the orbital electrons, and thus to arrive at curves due to the interatomic scattering only. Instead of using merely the peak intensities in the approximate eqn. (10.4.3), one is now in a position to apply the integral eqn. (10.4.2). A numerical integration can be achieved by taking values of $I'(s)$ at regular small intervals δs, and summing the terms of the form, $I'(s)((\sin sr)/sr)\delta s$. Provided the intervals are small enough, the result approaches as closely to the true integral as may be necessary. However, the labour of making such a summation, for a large number of values of s and over a suitably wide range of r, is formidable; some type of electronic computer is essential.

The integration in (10.4.2) is from $s = 0$ to $s = \infty$. In fact, the intensity has been measured only over a limited range of s, say from s_1 to s_2. This will give rise to errors, which must therefore be minimised. The lack of coverage beyond s_2 causes termination-of-series errors; there is a background of small ripples, and a consequent liability to error in the apparent positions of the true interatomic peaks. Errors from this source can be largely eliminated by introducing an exponential damping-factor, $\exp(-bs^2)$, where b is an adjustable parameter. This factor, which is akin to the Debye temperature-factor used in X-ray analysis (see section 9.11), artificially reduces the influence of the outer reaches of the diffraction pattern, which are responsible for the ripples. The absence of experimental intensities below s_1 has little or no effect on the peak positions, but it introduces gradual fluctuations in the background to the distribution function, so that the function does not necessarily fall to zero at values of r not corresponding to interatomic distances, and therefore the heights of the peaks are not quantitatively reliable. This trouble can be countered by a process of successive approximations: from a preliminary knowledge of the structure, the intensity-curve is *calculated* for the region $s = 0$ to $s = s_1$, and this 'synthetic' information is then added to that obtained experimentally between s_1 and s_2 in order to calculate a revised distribution curve. The cycle of operations is repeated as may be necessary, but the insensitiveness of the peak positions to the intensity curve in the region of low s causes the process to converge rapidly. The quality of the distribution curves now attainable is illustrated in Fig. 10.4.2 (B), which is based on that derived for carbon tetrachloride by Bartell, Brockway, and Schwendeman in 1955.* It should be contrasted with curve A, from the original calculation of 1935.

From precise distribution curves of this type, when individual peaks are resolved, accurate interatomic distances can be measured. Indeed, the accuracy is such that the effects of molecular vibrations are significant: the distance derived from the peak will be an average value, and it may be necessary to consider just which kind of average is involved; and by examining the spread of a given peak it may be possible

* *J. Chem. Phys.*, 1955, **23**, 1854.

to estimate the mean amplitude of the molecular vibration. In their refinement of the carbon tetrachloride analysis, Bartell and Brockway obtained

$$\text{C—Cl} = 1{\cdot}766 \pm 0{\cdot}003\ \text{Å}$$

from the (internuclear) distribution function, and estimated 0·060 Å for the root mean-square amplitude of the vibrational displacement; the former value represents an *average* bond length, and by correcting for the vibrational anharmonicity, with the aid of spectroscopic data for similar C—Cl bonds, it was shown to correspond to 1·760 Å for the *equilibrium* length (see section 11.5).

An alternative form of the radial distribution function is now used: $P(r)/r$. It has the advantage that its peaks, for a given pair of atoms, correspond more closely to a Gaussian shape. The improvement may perhaps be seen by comparing Fig. 10.4.2(B) with Fig. 10.7.1. The width of such a peak can be better characterised, and so used to estimate the root mean square amplitude of the two atoms relative to one another.

Once a sound, but approximate, structure has been established for the molecule, the parameters defining the structure can be refined by the least-squares method, much as in crystal structure analysis (see section 9.11). The molecular intensity curve, $I(s)$, which is based on the experimentally observed diffraction pattern, can be sampled at several hundred separate values of s. These constitute the observational data, $I(s)_o$. From our approximate structure we may then calculate corresponding intensities, $I(s)_c$, for every value of s. The art is to calculate small changes in the parameters which will improve the agreement with observation–that is, to minimise $\Sigma w\Delta^2$, where $\Delta = I(s)_o - I(s)_c$, and w is a suitable weighting factor for each term. The parameters will primarily be the interatomic distances, r_{ij}, which describe the geometry of the molecule; but it will also include quantities, u_{ij}, representing the r.m.s. amplitudes of vibration, and may include parameters allowing for anharmonic vibration.

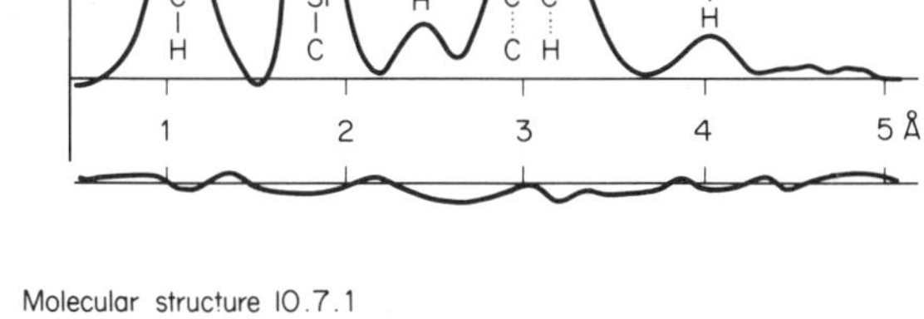

Molecular structure 10.7.1

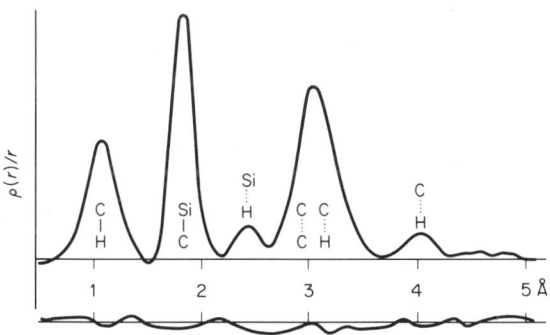

Fig. 10.7.1. Upper curve: the experimental radial-distribution function, $P(r)/r$, for tetramethyl-silane; lower curve: the difference function between experimental and calculated $P(r)/r$. (From a drawing kindly supplied by Dr. B. Beagley.)

As final evidence of the validity of the structure after refinement, observed and calculated curves are generally compared—either the curve showing the radial distribution $P(r)/r$, as a function of r, or that showing the molecular intensity, $I(s)$, as a function of s. Fig. 10.7.1 gives an observed radial distribution curve at the top, and underneath the corresponding 'difference' curve, which should show no features of significance.

10.7 Some Applications and Some Anomalies

Recent work* on tetramethyl-silane, $Si(CH_3)_4$, yielded results typical of the method as now applied. Chemical knowledge that the SiC_4 skeleton would probably have tetrahedral symmetry was assumed, and was confirmed by the successful outcome. This limited the number of parameters to nine: three interatomic distances, C—H, Si—C and Si . . . H, together with six vibrational amplitudes. The observed radial-distribution curve is shown in Fig. 10.7.1, with the 'difference' curve below. The least-squares values for the distances, with their standard deviations, were 1·115 (7), 1·875 (2) and 2·480 (11) Å, respectively. Examples of the r.m.s. amplitudes found were 0·082 (10) Å for C—H, and 0·053 (3) Å for Si—C, stretchings.

The conformations of small molecules can be studied. We have already mentioned that diphenyl ($H_5C_6 \cdot C_6H_5$), whose molecule is planar in the crystal, has an angle of 42° between the planes of its benzenoid rings in the gas. More complex situations can be elucidated. For instance, the diffraction pattern of *n*-butane ($H_3C \cdot CH_2 \cdot CH_2 \cdot CH_3$) shows that two conformational isomers are present in the vapour at about 290° C: 60 % of the molecules have the *trans*-conformation, with their carbon atoms in a flat zig-zag, whilst 40 % have the *gauche*-conformation. Furthermore, the energies of the two forms differ by about 0·6 kcal mole^{-1}.

In a number of cases precise structure determinations are available, for the same molecule, by both electron-diffraction and spectroscopic methods. The symmetrical, linear molecule of CS_2 was found to have r_e(C—S) = 1·5532(5) Å by Raman spectroscopy. Electron diffraction led to r(C—S) = 1·5592(11) Å. Most of this difference, however, arises because the molecule is vibrating and because the two methods yield average distances of different kinds. (See section 11.5.) When allowance is made for this, the spectroscopic value has to be raised to 1·5580 (5) Å, which does not differ significantly from 1·5592 (11).

These results illustrate another difference. For the equilibrium structure of CS_2 the S . . . S distance must of course be exactly twice C—S. But the electron-diffraction value for S . . . S is 3·1126(11) Å, which is less than 2 × 1·5592 Å by 0·0058 Å. This 'shrinkage' effect is due to the bending vibration of the molecule, in consequence of which the *averaged* molecule is non-linear; and hence, since any two sides of a triangle are together greater than the third, the S . . . S distance, as found

* Beagley, Monaghan and Hewitt, *J. Mol. Struct.*, 1971, 8, 403.

by electron diffraction, is slightly diminished.

There may be a fruitful interaction between these two methods. Spectroscopic knowledge of vibrational force-constants may be used to assess the r.m.s. amplitudes of certain vibrational modes in the early stages of an analysis. Conversely the final values found for the amplitudes may be used to estimate corresponding frequencies, and these may sometimes be frequencies that are not active in the infrared, and so cannot be measured directly.

Equation (10.3.1) is approximate in that it is based on the assumption that the scattering of electrons from each pair of atoms involved in r_{ij} occurs without relative change of phase. This assumption fails notably when the atoms i and j have very different atomic numbers. There is *anomalous scattering* (*cf.* section 9.16). If this is not allowed for—which may be done by using a more sophisticated version of the equation—erroneous structural conclusions may be drawn. A notorious example is UF_6. This molecule has O_h symmetry ($m3m$), with all six U—F distances equal. Electron diffractionists were therefore dismayed when it was found that the straight radial-distribution function had two prominent peaks at $r = 1{\cdot}9$ and $2{\cdot}2$ Å. This was at first interpreted as implying some unexpected distortion of the molecule. When the importance of anomalous scattering came to be recognised, the splitting of the U—F peak was shown to be due to the phase change between electrons scattered by U and F. After application of the necessary correction, the radial-distribution curve was dominated by a single peak corresponding to r(U—F) = 1·999(3) Å.

BIBLIOGRAPHY

Bastiansen and Skancke, article on 'Electron Diffraction in Gases and Molecular Structure' in *Advances in Chemical Physics*, Vol. III, Interscience, New York, 1961.

Hilderbrandt and Bonham, *Ann. Rev. Phys. Chem.*, 1971, **22**, 279–305.

Articles by H.M. Seif, B. Beagley and A.G. Robiette constituting Part I (Electron Diffraction) of *Molecular Structure by Diffraction Methods,* Volume I (Chemical Society Specialist Periodical Reports), 1973.

Chapter 11

Some dielectric and optical properties

11.1 Dielectric Constant

Dielectric properties illuminate a number of important aspects of molecular structure. In the first part of this chapter we shall examine the relation of the dielectric constant of a substance, and its variation with temperature and frequency, etc., to the consitution of its molecules.

Consider first how the orientation of molecules that possess a dipole moment contributes to the dielectric constant. Suppose that a substance composed of molecules having a permanent electric dipole moment is exposed somehow to a static (i.e., non-fluctuating) electric field. The field will try to orient the dipoles, whereas thermal motion will oppose any oriented arrangement and will tend to restore a random distribution.

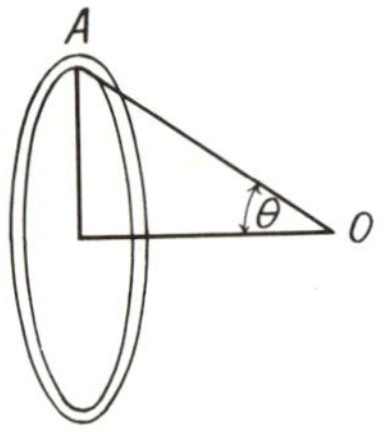

Fig. 11.1.1 Let OA be of unit length, then the circumference of the annulus is $2\pi \sin\theta$ and its area is $d\Omega = 2\pi \sin\theta \, d\theta$.

A dipole μ inclined at an angle θ to the direction of an electric field of intensity $\mathcal{E}$ has the potential energy $V = -\mu\mathcal{E} \cos\theta$. Interaction with the field alleviates the total disorder that otherwise prevails, for more molecules will point in the direction of the field ($\theta < 90°$) than against it ($\theta > 90°$). The relative number oriented with their dipole axes within an element of solid angle $d\Omega$ (Fig. 11.1.1) is, according to Boltzmann's law, given by,

$$Ae^{-V/kT}\,\mathrm{d}\Omega \tag{1}$$

in which A is a function of the total number of molecules but not of their inclinations. The total dipole moment in the direction of the field is calculated by multiplying the contribution of each dipole inclined at θ to the field (which is $\mu \cos \theta$) by the number of dipoles having that orientation, and integrating over all possible orientations. The result is

$$\mu A \int e^{-V/kT} \cos \theta \, \mathrm{d}\Omega. \tag{2}$$

Since the total number of molecules is $A\int e^{-V/kT}\,\mathrm{d}\Omega$, the average dipole moment $\bar{\mu}_0$ per molecule *in the direction of the field* is

$$\bar{\mu}_0 = \mu \int e^{-V/kT} \cos \theta \, \mathrm{d}\Omega / \int e^{-V/kT}\,\mathrm{d}\Omega. \tag{3}$$

For $\mathrm{d}\Omega$ we have the relation $\mathrm{d}\Omega = 2\pi \sin\theta \, \mathrm{d}\theta$ (Fig. 11.1.1). With the further substitution of $-\mu\mathcal{E} \cos\theta$ for V, eqn. (3) becomes

$$\bar{\mu}_0 = \mu \frac{\int_0^\pi e^{\mu\mathcal{E} \cos\theta/kT} \sin\theta \cos\theta \, \mathrm{d}\theta}{\int_0^\pi e^{\mu\mathcal{E} \cos\theta/kT} \sin\theta \, \mathrm{d}\theta}. \tag{4}$$

Let $\mu\mathcal{E}/kT = a$ and $\cos\theta = x$, then we have

$$\bar{\mu}_0 = \mu \frac{\int_{-1}^{1} e^{ax} x \, \mathrm{d}x}{\int_{-1}^{1} e^{ax} \, \mathrm{d}x}. \tag{5}$$

Under laboratory conditions $\mu\mathcal{E}/kT$ is very small. For instance, with $\mathcal{E} = 300$ volts cm^{-1} (1 esu) and $\mu = 5 \times 10^{-18}$ esu (5 Debye units), both quite high values, $\mu\mathcal{E}/kT \approx 10^{-4}$ at 300 K. Therefore we may write

$$e^{ax} = 1 + ax \tag{6}$$

and hence

$$\bar{\mu}_0 = \mu \frac{\int_{-1}^{1} x(1+ax)\,\mathrm{d}x}{\int_{-1}^{1} (1+ax)\,\mathrm{d}x} = \mu \frac{a}{3} = \frac{\mu^2 \mathcal{E}}{3kT}. \tag{7}$$

A similar calculation shows that the average dipole moment perpendicular to the axis of the electric field is zero, so that the average moment per molecule due to the orientation of dipoles is parallel to the direction of the external field. The expression (7) for its magnitude can be tested intuitively, for $\bar{\mu}_0$ must be proportional to the permanent moment μ, to the energy $\mathcal{E}\mu$ of one parallel dipole, and inversely dependent on the termperature T tending to restore random orientation.

To this point we have considered only the average moment due to orientation of molecules in an electric field. But the field also influences the molecules by displacement of the electrons relative to the nuclei and, to a smaller extent, by the displacement of the nuclei relative to one another. These effects give rise in each molecule to an *induced dipole moment* $\bar{\mu}_i$ parallel to the field, which is quite distinct from the average moment resulting from the orientation of dipoles. In a first approximation $\bar{\mu}_i$ can be assumed to be proportional to the field strength $\mathcal{E}$, and so we write

$$\bar{\mu}_i = \bar{\alpha}\mathcal{E}, \tag{8}$$

where $\bar{\alpha}$, the constant of proportionality, is known as the *polarisability* of the molecule. As the *polarisability* has, in general, different values for different directions in a molecule (section 7.2) the value used in (8) should be thought of as a mean, averaged over all possible orientations of the molecule relative to the field. The total average dipole moment per molecule is then the sum of the moments due to orientation ($\bar{\mu}_0$) and polarisability ($\bar{\mu}_i$), hence

$$\bar{\mu} = \bar{\mu}_0 + \bar{\mu}_i = \left(\bar{\alpha} + \frac{\mu^2}{3kT}\right)\mathcal{E}. \tag{9}$$

Evidently, when the molecules are non-polar ($\mu = 0$), only the contribution of the induced moments remains.

Next we seek a relation between $\bar{\mu}$ and the *dielectric constant* of a substance. Any electrical conductor charged with a quantity of electricity q at a potential Φ is said to have a capacity $C = q/\Phi$. A simple condenser consisting of two parallel plates of area A cm^2 set at a small distance r apart has the capacity

$$C = \epsilon_0 A/4\pi r \tag{10}$$

Here ϵ_0 is the dielectric constant of the material between the plates; and the suffix zero indicates that at present we are considering static fields, whose fluctuation with time is zero. In such a condenser the electric field between the plates is Φ/r. We denote this field by E, for the reason that the interaction of molecular dipoles may cause the field $\mathcal{E}$ acting on a representative molecule to be different from the external field E between the condenser plates. (E should not in any case be confused with the same symbol used for an energy eigenvalue.) Using the definition of capacity we find

$$E = \Phi/r = q/Cr \tag{11}$$

and hence, introducing the expression (10) for C,

$$E = \frac{4\pi}{\epsilon_0}\frac{q}{A}. \tag{12}$$

q/A is the charge per unit area, or surface density of charge, on the condenser plates. Writing $q/A = \sigma$ (12) becomes

$$E = 4\pi\sigma/\epsilon_0. \tag{13}$$

When the condenser is in vacuum ($\epsilon_0 = 1$) the field strength between the plates is simply

$$E^0 = 4\pi\sigma \tag{14}$$

and so the decrease in field strength when a material medium is present is

$$E^0 - E = 4\pi\sigma(\epsilon_0 - 1)/\epsilon_0. \tag{15}$$

Formally we attribute this decrease to charges induced on the surface of the dielectric opposite each plate of the condenser, as shown in Fig. 11.1.2. We recall that the average dipole moment per molecule, $\bar{\mu}$, is in the direction of the field (perpendicular to the plates); then, if the number of molecules per cm^3 of dielectric is n_0, the number of molecules enclosed by the plates is n_0 times the volume, Ar, and the total dipole moment of the material between the plates is $\bar{\mu}n_0Ar$. This represents a total charge of $+\bar{\mu}n_0A$ on one surface and $-\bar{\mu}n_0A$ on the other, corresponding to a surface density of charge on the faces of the dielectric of $\pm\bar{\mu}n_0$. The field, $4\pi\bar{\mu}n_0$, due to this surface charge equals $E^\circ - E$, and so we have

$$4\pi\bar{\mu}n_0 = 4\pi\sigma(\epsilon_0 - 1)/\epsilon_0. \tag{16}$$

Introduce eqn. (13), and we find,

$$\epsilon_0 - 1 = 4\pi\bar{\mu}n_0/E, \tag{17}$$

which is the equation we are seeking.

The molecules of a gas at low pressure are sufficiently far apart to neglect the electrostatic interaction of their dipoles. Therefore for a gas we may take

$$\mathcal{E} = E. \tag{18}$$

Eliminate the field strength $\mathcal{E} = E$ between eqns. (9) and (17), and we obtain

$$\epsilon_0 - 1 = 4\pi n_0(\bar{\alpha} + \mu^2/3kT). \tag{19}$$

Since n_0, the number of molecules per cm^3, equals $N_A d/M$ (where d is the density

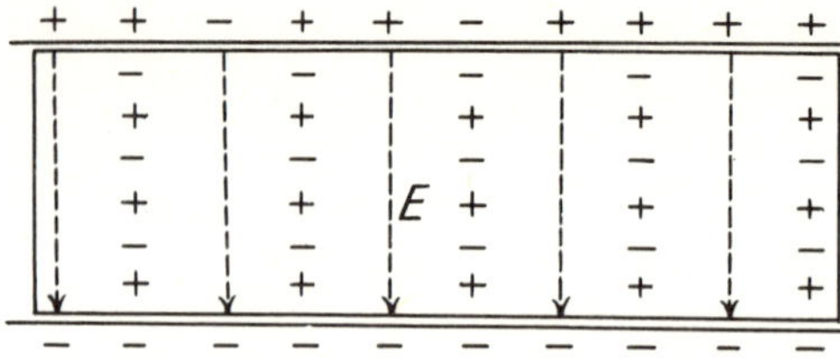

Fig. 11.1.2 Dielectric material with $\mathcal{E}_0 = 2$ between condenser plates. The dipoles in the material lead to surface charges that compensate (in this case) one-half the charges in the plates.

and M the molecular weight of the gas, and N_A is Avogadro's number), an alternative form of (19) is

$$(\epsilon_0 - 1)\frac{M}{d} = 4\pi N_A\left(\bar{\alpha} + \frac{\mu^2}{3kT}\right). \tag{20}$$

This is the Debye–Langevin equation for the dielectric constant of a gas in terms of the mean polarisability and permanent electric moment of its molecules.

For liquids and solids the situation is less simple. A rough evaluation of the field at a central molecule due to the dipoles of its neighbours gives

$$\mathcal{E} = \frac{\epsilon_0 + 2}{3}E. \tag{21}$$

Since $\epsilon_0 > 1$, it follows that the local field $\mathcal{E}$ at a given molecule exceeds the external applied field E, as might have been foreseen. However, the derivation of (21) neglects specific interaction (for instance, by hydrogen-bonding) between neighbouring molecules, and assumes that the molecules occupy spherical cavities: thus it cannot be expected that the equation will be valid for associated liquids and solids such as water or ice. Estimates of the local field are available for substances of this kind, but we shall not discuss them here.*

Combining eqns. (21), (17), and (9), we obtain

$$\frac{\epsilon_0 - 1}{\epsilon_0 + 2} = \frac{4\pi n_0}{3}\left(\bar{\alpha} + \frac{\mu^2}{3kT}\right). \tag{22}$$

Conversion to a molar basis by multiplication by M/d yields

$$\frac{\epsilon_0 - 1}{\epsilon_0 + 2}\frac{M}{d} = \frac{4\pi N_A}{3}\left(\bar{\alpha} + \frac{\mu^2}{3kT}\right). \tag{23}$$

* Kirkwood, *J. Chem. Phys.*, 1939, 7, 1911

(23) is the Debye–Langevin equation in a more general form. The quantity given by either side of the equation is known as the *molar total polarisability* and is denoted by the letter P. (It would have been more logical to take the molar polarisability as this quantity times $3/4\pi$, but this is not the convention.) In the gaseous state at low pressure $\epsilon_0 \approx 1$, and hence (21), (22), and (23) reduce to the simpler, but approximate, equations (18), (19), and (20), respectively.

The dielectric constant is a quantity that can be measured by experiment. The principle of the measurement is that, according to (10), the capacity of a condenser is proportional to the dielectric constant of the material between the plates; a filled condenser has the capacity $C = \epsilon_0 C^\circ$, where C° is the capacity in vacuum.

11.2 Dipole Moments

In eqn. (11.1.23) we have an expression for the molar total polarisability of a molecule. This quantity can be thought of as having two parts. We write

$$P = P_I + P_O. \tag{1}$$

Then $P_I = (4/3)\pi N_A \bar{\alpha}$ is the molar polarisability due to the elastic displacement of charges; and $P_O = \dfrac{4\pi N_A}{9k}\dfrac{\mu^2}{T}$ is the molar polarisability arising from the partial orientation of dipoles in the applied field. (We shall see later that P_I may be divided further into two components.) Since ϵ_0 is a pure number and M/d is the molar volume, $P = \dfrac{\epsilon_0 - 1}{\epsilon_0 + 2}\dfrac{M}{d}$ and its components have the dimensions of volume (cm^3).

The Debye–Langevin equation can now be written

$$P = P_I + (P_O \times T)/T. \tag{2}$$

In which the constants P_I and $P_O \times T = 4\pi N_A \mu^2/9k$ are independent of temperature. Evidently a plot of P against $1/T$ should be linear with slope $P_O \times T$ and intercept P_I. Measurements of ϵ_0 over a range of temperature therefore yield both the induced molar polarisability P_I and the permanent dipole moment μ. Some applications are shown in Fig. 11.2.1. Curve (*a*) is representative of a non-polar molecule, for which the polarisability is independent of temperature. In curves (*b*) and (*c*) we have results typical of a polar molecule, where the polarisability increases inversely with temperature. (The points (*d*) are for a molecule which does not conform to the Debye–Langevin equation, for the reason that its dipole moment changes with temperature. Behaviour of this sort is discussed in section 11.3. The method has been applied to the determination of the permanent moments of a large number of gases and vapours, but it is not generally satisfactory for liquids owing to the approximate nature of the expression (11.1.21) for the local field. For gases, whereas the Debye–Langevin formula may give accurate results for large dipole moments, it is more difficult to get a good determination of a small moment or even to distinguish a small moment from zero. For example, if we compare the moments 4·0 and 4·1 D, the μ^2 values differ (in the same units) by 0·81: on the other hand, for moments of 0·0 and 0·1 D, the difference in μ^2 is only 0·01.

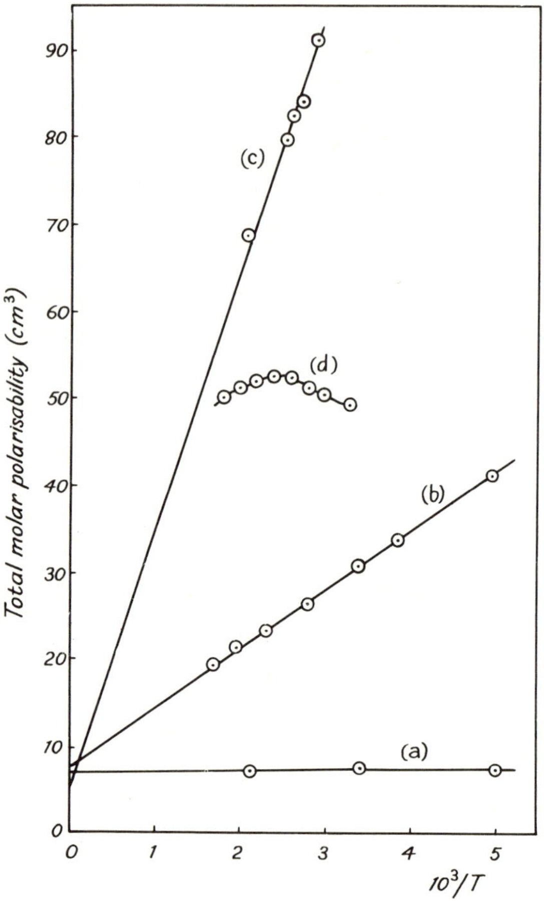

Fig. 11.2.1 Molar total polarisability of gases as a function of $1/T$. Curve (*a*) CO_2; (*b*) HC1; (*c*) HCN; (*d*) $C1CH_2.CH_2C1$.

Measurement on gases are, of course, limited to molecules with a sufficiently high vapour pressure. A feasible method for involatile compounds involves the measurement of the dielectric constant of a dilute solution in a non-polar solvent. Complications due to dipole-dipole interaction are then at a minimum and the approximation (11.1.21) for the local field is reasonably accurate. For a solution of polar compound 2 in a non-polar solvent 1, eqn. (11.1.22) takes the form

$$\frac{\epsilon_0 - 1}{\epsilon_0 + 2} = \frac{4\pi}{3}\left[n_1\bar{\alpha}_1 + n_2\left(\bar{\alpha}_2 + \frac{\mu_2^2}{3kT}\right)\right]. \tag{3}$$

Here $\bar{\alpha}_1$ is the mean polarisability of the solvent molecules 1, $\bar{\alpha}_2$ that of the solute 2, and n_1 and n_2 are the respective number of molecules per cm^3 of 1 and 2. μ_2 can be evaluated from a determination of ϵ_0 at a single temperature, provided that $\bar{\alpha}_1$ and $\bar{\alpha}_2$ are known independently (section 11.4). Moments deduced from solution measurements often differ appreciably from gas phase moments, and, to a smaller extent, from one solvent to another. This is not surprising when one considers that the expression (11.1.21) seeks to encompass all forms of interaction and aggregation between molecules by the factor $(\epsilon_0 + 2)/3$. The following data for chlorobenzene are typical.

	Gas	*C_6H_6*	*Dioxane*	*C_6H_{14}*	*CCl_4*
PhCl: μ(D)	1·70	1·60	1·62	1·59	1·65

It is useful to have an idea of the magnitude of quantities in the Debye-Langevin equation. $\bar{\alpha}$ for many molecules is of the order 10^{-23} cm^3: therefore according to (11.1.8) the induced moment in a field of unit intensity (1 esu or 300 volts cm^{-1}) equals 10^{-23} esu, or 10^{-5} D. The mean moment due to the tendency of the field to orient the dipoles is $\mu^2\mathscr{E}/3kT$ (eqn. 11.1.7): hence for a fairly polar molecule, say with $\mu = 1$ D, in the same field at $300\,K$, we find $\bar{\mu}_0 \approx 0{\cdot}8 \times 10^{-5}$ D. The average moments contributed by charge displacements and by dipole orientation are about equal in this example. As a field of 1 esu is of the order of magnitude used in dielectric constant measurements this illustration is not unrepresentative. Temperature is so effective in maintaining random orientation that for every 10^5 molecules the collective moment, $10^5\,\bar{\mu}_0$, due to orientation is less than the moment of one permanent dipole held parallel to the field.

11.3 Dipole Moment and Molecular Structure

Knowledge of the dipole moment is valuable in many problems of molecular structure. It was found in Chapter 3 that the dipole moment is a vehicle for the interaction of molecules with radiation: but dipole moments are in themselves a large field of study, and a few examples will be quoted to show its place in the scheme of things.

Atoms, centrosymmetrical molecules, and spherical tops have zero moments on account of symmetry. This idea can sometimes be used to determine the general shape of a molecule. The absence of a dipole moment in CO_2 and CS_2, for instance, shows that these molecules are centrosymmetrical and therefore linear; while the possession of a dipole moment by H_2O and SO_2 proves they are either non-symmetrical or non-linear. Likewise the moments of ammonia and like molecules exclude a symmetrical planar configuration. The magnitude of the dipole moment may answer qualitatively another fundamental question. Comparing moments of the hydrogen halides with those of alkali halide molecules (in the vapour state), the latter are seen to be much more strongly polar (Table 11.3.1). One may infer that the interatomic bond in, for instance, a gaseous molecule of KCl has higher *ionic character* than the bond in HCl. Since the interatomic distance in the gaseous KCl molecule is 2·67 Å, and the moment of two unit charges set this distance apart is 12·8 D, a primitive view of the observed moment might be that the bond has about 82% ionic character. This, however, neglects the polarisation of one ion in the field of the other, and other important considerations. The true ionic character is probably much closer to 100% (section 8.7).

The dipole moments of complex molecules are often analysed in terms of *bond moments.* Thus the substitution of hydrogen in benzene by chlorine leads to a dipole moment (p. 317) of 1·70D: this we equate with the C—Cl bond moment in aromatic compounds. Vector addition then gives 2·94, 1·70, and 0 D for the moments of *o*-, *m*-, and *p*-dichlorobenzene, compared with the measured values of 2·53, 1·67, and 0, respectively. Mutual influences are not negligible in the *o*-derivative and must be allowed for in a less approximate treatment. Also, the additivity of bond moments in aromatic compounds is often disturbed by mesomeric effects: for example, the moment of *p*-chloronitrobenzene (2·78 D) exceeds the difference of the moments of nitrobenzene (4·21) and chlorobenzene (1·70) by 0·27, the increment being due to the mesomeric charge-transfer,

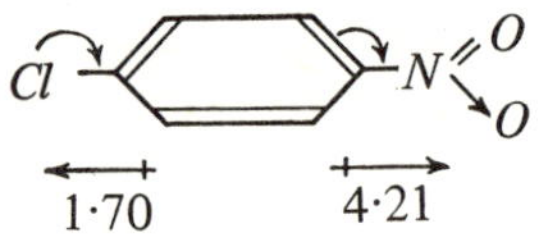

Equilibrium between different conformations can lead to an apparent variation of the dipole moment with temperature. 1:2-Dichloroethane is a popular example. The anomalous dependence of its molar polarisability upon temperature has been illustrated in Fig. 11.2.1. 1:2-Dichloroethane exists in three conformations, two of which are enantiomorphic. In the stable conformation, (*a*) in Fig. 11.3.1, the C—Cl bonds are *trans* to one another and the nuclei Cl—C—C—Cl forming the spine of the molecule are coplanar; the point group is then C_{2h} (see section 2.7). The *trans*-form is centrosymmetrical and so has zero dipole moment when the cuclei are in their equilibrium positions; but certain normal vibrations destroy the centre of symmetry, and their thermal excitation may lead to small moment of the order 0·1–0·2 D. (The major contribution is from the normal vibration which twists the central C—C bond. This vibration has a low frequency, and hence the nuclear displacements involve a large amplitude.) The two *gauche* conformations (*b*) and (*c*) are produced when one —CH_2Cl group is rotated through ±120° relative to the other: the point group is now C_2, and the *gauche* forms are dissymmetric. The simple bond moment hypothesis gives about 3·0 D for the moment of the *gauche* conformations, consequently from the results in Table 11.3.2 it follows that the

Table 11.3.1. Dipole Moments of Gases*

Diatomic Molecules	μ(Debye)	μ(Stark effect)
CO	0·14	0·112
NO	0·16	0·153
HF	1·91	1·82
HCl	1·08	1·12
HBr	0·80	0·83
HI	0·42	0·44
KCl		10·27
NaCl		9·00
KBr		10·41
Linear Polyatomic Molecules		
HCN	3·00	2·99
OCS	0·72	0·712
HCCCl	0·45	0·44
N_2O	0·17	0·167
Symmetric Tops		
CH_3F	1·85	1·85
CH_3Cl	1·87	1·88
CH_3Br	1·83	1·81
CH_3I	1·64	1·65
CH_3CN	3·96	3·92
NH_3	1·47	1·47
PH_3	0·55	0·58
Asymmetric Tops		
SO_2	1·63	1·62
CH_2Cl_2	1·57	1·63
H_2O	1·85	1·85
H_2S	0·98	0·974

*Selected Values of Electric Dipole Moments for Molecules in the Gas Phase, Nelson, Lide and Maryott, NSRDS–NBS 10, 1967.

proportion of the *gauche* form rises from about one-third to one-half as temperature is increased from 32° to 270°C. The shape of the barrier restricting internal free rotation is shown in Fig. 11.3.2. Evidently the energy of activation, *i.e.*, the barrier height, for the interconversion *trans* $\rightleftharpoons$ *gauche* is of the order of a few kcal only: therefore the unimolecular rate of change of one conformation into another is rapid even below room temperature. The equilibrium can also be studied through the vibrational spectrum which is characteristically different for the *trans* and *gauche* forms. In the crystalline state it appears that the molecules are present entirely in the *trans* form.

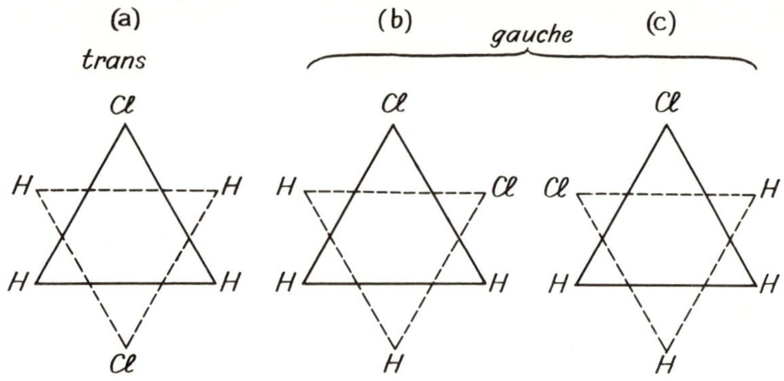

Fig. 11.3.1 Conformations of 1:2-dichloroethane.

Table 11.3.2. Dipole Moment of Gaseous 1:2-dichloroethane

T,K	$\mu(D)$	T,K	$\mu(D)$
304·9	1·12	419·0	1·40
341·0	1·24	479·8	1·48
376·2	1·32	543·7	1·54

The *trans* ($\phi = 0°$) and *gauche* ($\phi = \pm 120°$) conformations occur at minima of potential energy: they represent discrete species, since a small but definite energy of activation is needed to convert one form into another. The eclipsed structures ($\phi = \pm 60°, 180°$) correspond to energy maxima and thus have no stability: they are merely transition states in the interconversion of conformations. The potential energy function shown in Fig. 11.3.2 can be thought of as having two parts. The first, representing the electrostatic repulsion of the C—Cl bond dipoles, has its minimum and maximum 180° apart, corresponding to the *trans* ($\phi = 0°$) and (unstable) *cis* ($\phi = 180°$) forms, respectively. The contribution to the potential may be written $\frac{1}{2}A(1 - \cos\phi)$. The second part originates in the mutual repulsion of the C—H and C—Cl bonding electron pairs, and in consequence favours all three staggered conformations impartially in comparison with the eclipsed conformations. As this part of the potential has maxima and minima 60° apart, it can be represented by a $\cos 3\phi$ expression, $\frac{1}{2}B(1 - \cos 3\phi)$. The sum of these two terms,

$$V = \tfrac{1}{2}A(1 - \cos\phi) + \tfrac{1}{2}B(1 - \cos 3\phi),$$

with $A = 1830$ and $B = 2300$ cal mole^{-1},* correctly reproduces the qualitative shape of the true potential for $\phi < 130°$. A more detailed treatment is required to estimate the barrier height opposing internal rotation into the *cis* configuration.

* Gwinn and Pitzer, *J. Chem. Phys.*, 1948, **16**, 303.

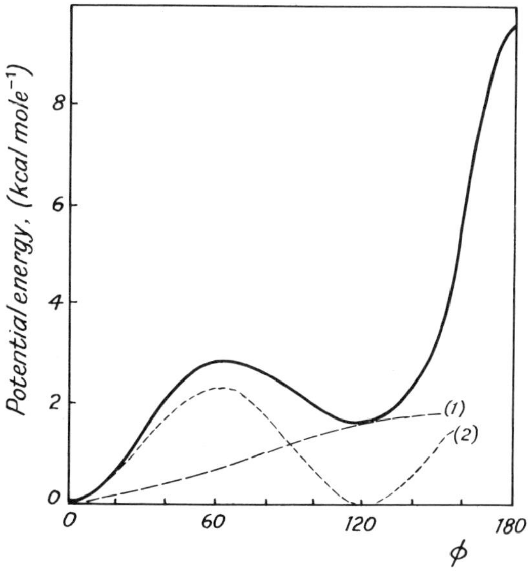

Fig. 11.3.2. Barrier to internal rotation in 1:2-dichloroethane. ϕ is the dihedral angle between the symmetry planes of the —CH_2Cl groups. The full curve is the best available approximation to the true potential: the broken curves show how the potential can be constructed from a superposition of potentials attributed to dipolar repulsion (1) and non-bonded interaction (2). All curves are symmetric about $\phi = 0°$.

11.4 Molar Refractivities

Up to this point we have considered only the dielectric constant ϵ_0, namely the value appropriate to a static electric field. Actually, for experimental reasons, the dielectric constant is always measured in an oscillating electric field, when ϵ_0 is interpreted as the limiting low-frequency value. The value is normally independent of the frequency of the electric field over a wide range, so that no difficulty arises in the extrapolation to zero frequency; but if measurements are made with the oscillating electric field associated with infrared or visible radiation, this high-frequency dielectric constant can be very different from the limiting low-frequency value.

Classical theory gives for the refractive index n of electromagnetic radiation the relationship $n^2 = \epsilon$, where ϵ is the *dielectric constant at optical frequencies*. ϵ is, in general, different from ϵ_0, as we shall see presently. If we substitute n^2 for ϵ in (11.1.23) we obtain the *molar polarisability at optical frequencies*, denoted by R,

$$R = \left(\frac{n^2 - 1}{n^2 + 2}\right)\frac{M}{d}. \tag{1}$$

R is known alternatively as the *molar refractivity.* In the visible region of the spectrum n changes somewhat with frequency, but for colourless substances the value is nearly constant for red or yellow light (H_α or D light). Molar refractivities, having the dimension of volume, are expressed in cm^3.

The refractivity of gases and liquids originates in the displacement of electrons relative to the nuclei under the influence of the electric field of the radiation. It can be analysed as a constitutive property of molecules, by regarding the molar refractivity as the sum of the refractivities of the atoms of which the molecule is built up. The effects associated with multiple bonds are allowed for by extra constants or by assigning special constants to one or other of the multiply bound atoms. Some atom and multiple bond refractivity constants are given in Table 11.4.1. An illustration will serve to show how these parameters are obtained. The molar refractivities of the saturated normal hydrocarbons can be represented by a linear equation

$$R = a + bN, \tag{2}$$

where N is the number of CH_2 groups in the molecule. The data for a number of such molecules (Fig. 11.4.1) then give $a = 2{\cdot}22$ and $b = 4{\cdot}64$ cm^3; hence

$$R = 2{\cdot}22 + 4{\cdot}64N. \tag{3}$$

Two atoms of hydrogen monovalently bound to carbon thus contribute 2·22 cm^3 to the molar refractivity. This checks with the facts that R for the H_2 molecule is 2·04, and that the difference of the refractivities of *n*-hexane and *cyclo*hexane is 2·01 cm^3.

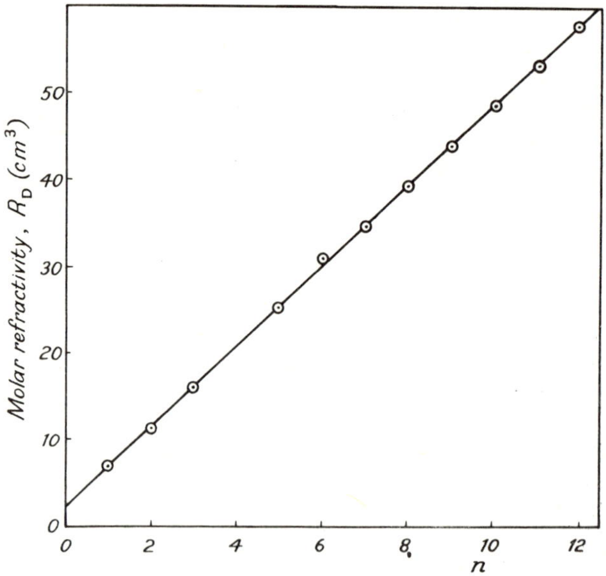

Fig. 11.4.1. Molar refractivity (in cm^3) of normal paraffins.

A value of $R[\mathrm{H}] = 1{\cdot}11\ \mathrm{cm}^3$ can then be assigned to the atom refractivity of hydrogen. Further, the refractivity of a saturated carbon atom will be that of a CH_2 group ($4{\cdot}64\ \mathrm{cm}^3$) less twice the atom refractivity of hydrogen: hence $R[\mathrm{C}] = 2{\cdot}42\ \mathrm{cm}^3$. The molar refractivity of a large molecule can be estimated with considerable accuracy as the sum of the atomic refractivities listed in the table.

Table 11.4.1. Atom and Multiple Bond Refractivities (Na D line)*

	R_D, cm³		R_D, cm³
H	1·11	N (primary amine)	2·32
C	2·42	N (sec. amine)	2·50
C:C	1·73	N (tert. amine)	2·84
C⋮C	2·40	Cl	5·97
O (ethers)	1·53	Br	8·86
O (carbonyl)	2·22	I	13·90

* Eisenlohr, *Spektrochmie organischer Verbindungen* (Enke, Stuttgart, 1912)

An alternative approach is based on the idea that, as the valence shell electronss are much more polarisable than the inner core of electrons, it is reasonable to interpret the molar refractivity as the sum of bond refractivities and the refractivity of unshared pairs. The difficulty is to separate the contribution of the bonding electrons from that of the unshared pairs, and in practice what is done is to allot different constants to a bond depending of whether or not the terminal atoms carry unshared pairs. The assumption of additivity holds rather better than in the older system based on atom refractivities.

Bond refractivities are useful to the theoretical chemist because they can be calculated from the wavefunctions of the valence shell electrons and so are available as a test of the wavefunctions. Otherwise atom (and bond) refractivities are now largely of historical interest. One such application should be mentioned. In conjugated, unsaturated hydrocarbons the molar refractivity always exceeds that calculated from the atom refractivities. Thus for C_6H_6,

$$R(\text{calc.}) = 6R[\mathrm{H}] + 6R[\mathrm{C}] + 3R[\mathrm{C{:}C}] = 26{\cdot}4,$$

the observed value being $27{\cdot}0\ \mathrm{cm}^3$. This 'exaltation' of the molar refractivity was recognised early in the development of the concept of mesomerism as due to delocalisation of the electrons in the multiple bonds.

11.5 Variation of the Dielectric Constant with Frequency

The Atomic Polarisability

The molar refractivity and the molar total polarisability of molecules do not normally have the same numerical value. The reasons have already been remarked upon, but in the present section we shall reassemble them for closer examination.

A relatively trivial reason is that, owing to the dispersion of the refractive index, the molar refractivity varies slightly with frequency. Since the molar polarisability is calculated from the limiting low-frequency dielectric constant, the action required to place refractivity on the same footing is to eliminate the frequency dependence by extrapolation of the optical refractive index to zero frequency. From this limiting refractive index, n_0, one may then calculate the molar refractivity at zero frequency, R_0, from an expression like (11.4.1), namely

$$R_0 = \left(\frac{n_0^2 - 1}{n_0^2 + 2}\right)\frac{M}{d}. \qquad (1)$$

For a majority of molecules the correction is very small. Representative data for H_2 are given in Table 11.5.1. At the lowest frequency quoted (in the infrared region) $R\,[H_2]$ is hardly distinguishable from the limiting, low-frequency value.

Table 11.5.1. Molar Refractivity of H_2 as a Function of Frequency

Frequency, cm^{-1}	24 700	22 390	18 250	16 960	14 900	1490	0
Wavelength, Å	4048	4360	5462	5895	6710	67 095	∞
R, cm^3	2·129	2·120	2·089	2·086	2·072	2·036	2·035

Except for a special class of molecules the limiting molar refractivity is numerically smaller than the molar polarisability. This brings us to a second, much more important, reason. At the frequency of visible radiation (*ca.* 5×10^{14} Hz for yellow light) the electric field fluctuates too rapidly for the nuclear positions to change during the period of oscillation. Consequently the contribution from the *orientation of molecules, as well as from the displacement of the nuclei relative to one another,* is missing from the molar refractivity: what is left, then, is the polarisability due to elastic displacement of the electrons of a molecule relative to its nuclear framework. The molar polarisability, on the other hand, is calculated from the dielectric constant measured at frequencies so low that the nuclei move in response to the changing electric field. In these conditions the contributions from orientation and nuclear vibration are not lost. It may happen, of course, that the last two contributions are zero, and for such molecules the molar refractivity and polarisability are equal: this is the special class of molecules referred to at the beginning of the paragraph.

These ideas can be cast in a more quantitative form by writing

$$P = P_E + P_A + P_O. \tag{2}$$

Here P symbolises the limiting, low-frequency polarisability,

$$P = (\epsilon_0 - 1)M/(\epsilon_0 + 2)d.$$

P_E is the polarisability due to the electrons, and hence $P_E \equiv R_0$; P_A, the *atomic polarisability,* is the part due to the normal modes of vibration of a molecule; whilst P_O is the contribution to the polarisability from the partial orientation of molecules. By comparison with eqn. (11.2.1) we also find

$$P_I = P_E + P_A. \tag{3}$$

It is worthwhile to recall here that P_I and P_O can be determined from, respectively, the intercept and slope of a plot of P *vs.* $1/T$ (section 11.2).

The shape of the curve representing the dependence of molar polarisability upon frequency is therefore a stepped one. At frequencies substantially above the range of the normal vibrations P_O and P_A are both zero; the residual level is that of the electronic polarisability P_E. The vibration frequencies normally fall in the range 10^{13}–10^{14} Hz ($\lambda = 0{\cdot}3 - 3\mu$). At frequencies below that of the vibrations but higher than that of the molecular rotation (10^{10}–10^{12} Hz) an experimental determination of the polarisability would give $P_E + P_A$, since the electron and the normal vibrations then contribute to the polarisability. Below the rotational frequency the orientational polarisability enters; and thereafter the molar polarisability is constant, equal to the limiting static field value. Naturally it may happen that for a given molecule one or both of the increments to the curve may be zero. Thus if the dipole moment is zero there will be no jump corresponding to P_O.

The curve of molar polarisability against frequency for a strongly polar molecule is illustrated qualitatively in Fig. 11.5.1. Proceeding from low towards high frequency the contribution from orientational polarisability is the first to drop out. The atomic polarisability disappears in the relatively narrow range of frequencies containing the normal vibrations. Here the detailed appearance of the curve is complex, because in the immediate vicinity of a normal vibration the polarisability varies sharply. As the frequency of the electric field comes into coicidence with that of a normal vibration frequency, the polarisability of the molecules becomes very large. The resulting *anomalous dispersion* of the polarisability dies out on both sides of the vibration frequency.

Each infrared-active vibration contributes separately to the atomic polarisability. Fig. 11.5.1 has been drawn on the assumption that there are three such vibrations and so is appropriate to a linear molecule such as HCN (which actually has four normal vibrations, two of which coincide in frequency), or to non-linear triatomic

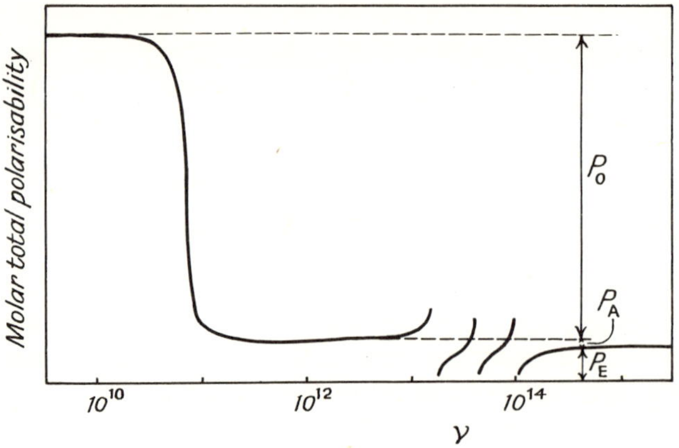

Fig. 11.5.1. Molar polarisability as a function of frequency (schematic).

molecules like H_2O or SO_2. The contribution of a normal vibration to the atomic polarisability can be calculated from the absorption intensity in the infrared spectrum (section 6.8). Data for a number of simple molecules are collected in Table 11.5.2.*

Table 11.5.2. Molar Atomic and Electronic Polarisabilities (in cm³) of Gases

	H_2	N_2	O_2	CO_2	C_2H_2	HCl	HCN
Point group	$D_{\infty h}$					$C_{\infty v}$	
P_A	0·00	0·00	0·00	0·78	1·05	0·013	0·40
$P_E(\equiv R_0)$	2·035	4·39	3·96	6·54	8·6	6·5	7·36
$P_I(=P_A+P_E)$*	2·051	4·39	3·96	7·35	9·8	7·8	7·77
	C_2H_4	NH_3	BF_3	C_2H_6	C_6H_6	CH_4	SF_6
Point group	V_h	C_{3v}	D_{3h}	D_{3d}	D_{6h}	T_d	O_h
P_A	0·33	0·54	1·98	0·12	0·73	0·91	4·4
$P_E(\equiv R_0)$	10·3	5·5	6·0	11·1	25·1	6·48	11·4
$P_I(=P_A+P_E)$*	10·7	6·2	8·4	11·2	26·2	6·53	15·7

* Independently, from the Debye – Langevin equation.

The atomic polarisabilities in the table are calculated from the absorption intensity of bands in the infrared spectrum; the electronic polarisabilities are obtained from the refractive index at optical frequencies; and their sum, $P_A + P_E$, is the induced molar polarisability found from the intercept of the Debye–Langevin plot,

* Whiffen, *Trans. Faraday Soc.*, 1958, **54**, 327.

effectively under conditions of a static electric field. The agreement between the induced molar polarisability and the sum of its parts is an impressive confirmation of dielectric theory.

In conclusion we should observe that our choice of a classical theory of dipole orientation has been merely a convenience. The quantum treatment* is based on the wavefunctions for the rotation of molecules: it leads to the same equation (eqn. 11.1.7) for the average dipole moment due to the orientation of molecules, but illuminates a different aspect of the process. In the wave-mechanical treatment the average moment $\bar{\mu}_0$ arises from molecules in the lowest ($J = 0$) rotational state. The mean moment from all higher states averages to zero. The dependence of the average moment upon reciprocal temperature is then explained by the depopulation of the rotational ground state with rise of temperature.

11. 6 The Kerr Effect

The relation between polarisability and refractive index, discussed in sections 11.4 and 11.5, implies that an anisotropic molecule will have a different refractive index along different directions in the molecule. This effect can be observed readily in the crystalline state where molecules are present in an ordered arrangement, for then the refractive index may differ along different axes in the crystal. It may also be observed in the liquid and gaseous states by the application of an external electric field, when the field introduces some degree of order to the otherwise completely random distribution of molecules. The resulting anisotropy in the refractive index of liquids and gases is known after Kerr, who discovered it.

Measurement of the *Kerr effect* involves the determination of the refractive index n of a substance for light polarised with its electric vector parallel ($n_{\parallel}$) and perpendicular ($n_{\perp}$) to the electric lines of force of the external field. The magnitude of the effect is measured by the *molar Kerr constant, K,*

$$K = A(n_{\parallel} - n_{\perp}) V_m / \mathcal{E}^2, \tag{1}$$

in which V_m is the molar volume of the material under investigation. The quantity A depends on the mean refractive index and dielectric constant of the material, and has the numerical value 2/27 for an ideal gas. Kerr first observed this effect in glass, which can be considered as an extremely viscous liquid; therefore there is an appreciable time lag as the anisotropy establishes itself. In normal liquids and gases the effect develops almost instantaneously. The Kerr constant of involatile solids can be measured in solution, in a manner analogous to the determination of the dipole moment (section 11.2). ($n_{\parallel} - n_{\perp}$) may be positive or negative, and the sign of the Kerr constant varies likewise. For isotropic molecules $n_{\parallel} = n_{\perp}$, and therefore $K = 0$.

Let us consider first a non-isotropic molecule with no permanent moment, such

* Fowler and Guggenheim, *Statistical Thermodynamics* (Cambridge, 1939), 1420.

as that of benzene. The polarisability ellipsoid is an ellipsoid of revolution, with the two axes of equal length lying in the plane of the molecule (perpendicular to the C_6 axis). Polarisability is greater in the molecular plane than perpendicular to it, because the π-electrons have freedom to move in the plane of the ring of carbon atoms. The moment induced by an electric field will be larger the more nearly the axes of greater polarisability lie parallel to the field, and therefore there will be a tendency for the molecules to lie with their planes parallel to the direction of the field. It follows that the refractive index $n_{\parallel}$, for light itself polarised in this direction will exceed that, $n_{\perp}$, for light polarised in the perpendicular direction. There will be a small *positive* Kerr constant. Thermal agitation tends to break up any oriented arrangement and thus the magnitude of the Kerr constant will decrease with temperature.

When, on the other hand, there is a permanent dipole moment its orienting effect in a strong electric field is much greater than that of the induced dipole.* The orientation of the polarisability ellipsoid in the electric field will be imposed by the permanent moment. Two limiting situations can be recognised. First, when the permanent moment lies in the direction of maximum polarisability, as in the molecule of chlorobenzene, the orientation will bring a positive value of K. Secondly, the permanent moment may coincide with the axis of least polarisability, as in the chloroform molecule. Here the dipole vector is in the C—H direction; while the polarisability is greatest in a plane parallel to the plane of the chlorine nuclei. The molecule will then tend to an alignment in which the axis of minimum polarisability is parallel to the field direction, and there will be a negative Kerr constant. With these remarks in mind the reader should have no difficulty in interpreting qualitatively the sign of the Kerr constant for the molecules listed in Table 11.6.1.†

Table 11.6.1. Molar Kerr Constants of some Simple Molecules in the Gaseous State

	Temp. °*C*	$10^{12}K$		*Temp.* °*C*	$10^{12}K$
Non-polar molecules			*Polar molecules*		
C_2H_6	18	1·06	CH_2Cl_2	83	−12·1
C_2H_4	18	1·93	$CHCl_3$	90	−17·0
CO_2	25	2·32	C_6H_5Cl	154	96·7
Cl_2	24	4·16	$(CH_3)_2CO$	83	69·5
C_6H_6	114	12·9	$(CH_3)_2O$	18	−8·85

The value of the Kerr constant can therefore be of significance in relation to

*Recall that an induced dipole moment is normally ~ 10^{-5} D (Section 11.2), *i.e.*, four or five orders of magnitude smaller than a permanent moment.

† Le Fevre and Le Fevre, *Rev. of Pure and Appl. Chem. (Australia)*, 1955, **5**, 261.

molecular structure. In many cases the information it supplies is already available from other sources; this would be true for chlorbenzene and chloroform, for example. Sometimes it may disclose something new. Thus for many years it was uncertain whether the free molecule of aniline was planar or not. A simple analogy with the known pyramidal structure of ammonia might lead one to expect a non-planar structure; whereas conjugation between the lone pair of electrons on the nitrogen atom and the π-electrons of the aromatic ring would tend to produce a planar structure. Liquid aniline has a negative Kerr constant, but since there is likely to be some association by hydrogen bonding this is not necessarily significant. In a non-polar solvent, aniline gives a positive Kerr constant which diminishes rapidly with concentration; and on extrapolation to infinite dilution yields a *small,* though still positive, value. It can be argued that the planar model should give a *large* positive Kerr constant; hence the deduction may be drawn that the aniline molecule is actually non-planar. This conclusion, though not definitive by itself, has been confirmed recently by microwave and ultraviolet spectroscopic studies of aniline vapour.

It is sometimes possible to apply the Kerr effect in a more quantitative way. The magnitude of the Kerr constant is related to the relative values of the polarisability along the principal axes of the ellipsoid. For linear molecules and symmetric tops we have the comparatively simple situation where $\alpha_{x'x'} = \alpha_{y'y'} \neq \alpha_{z'z'}$, so that only two principal component of the polarisability are independent. The mean polarisability gives one equation (eqn. 7.2.8) between these two unknowns, and the Kerr constant gives a second; hence both independent components of the polarisability can be determined from experimental observations. When all three components of α are independent a third measurable quantity dependent on the anisotropy, such as the depolarisation factor ρ of the Rayleigh scattering, is required to get a solution. The directional polarisabilities then obtained for a number of simple molecules are given in Table 11.6.2. The z' axis of the polarisability ellipsoid is taken to coincide with the principal symmetry axis of the molecule when such an axis is present.

Table 11.6.2. Principal Components (in Å^3 per Molecule) of the Polarisability α

	$\alpha_{z'z'}$	$\alpha_{y'y'}$	$\alpha_{x'x'}$		$\alpha_{z'z'}$	$\alpha_{y'y'}$	$\alpha_{x'x'}$
CH_4	2·6	2·6	2·6	C_2H_6	5·48	3·97	3·97
CCl_4	10·5	10·5	10·5	C_6H_6	6·35	12·31	12·31
Cl_2	6·60	3·62	3·62	C_6H_{12}	9·25	11·68	11·68
CO_2	4·49	2·14	2·14	C_6H_5Cl	15·93	13·24	7·58
CS_2	15·14	5·54	5·54	$C_6H_5NO_2$	17·76	13·25	7·75
C_2H_2	5·12	2·43	2·43	C_5H_5N	11·88	10·84	5·78

As we noticed in section 11.4, the mean polarisability of a molecule can be adequately represented as a sum of bond polarisabilities that are taken to be the same irrespective of the molecule in which the bond occurs. A similar analysis can be applied to the directional polarisabilities of a molecule. A chemical single bond has approximately axial symmetry; therefore its polarisability can be regarded as having a longitudinal component, α_L, parallel to the direction of the bond, and a transverse component, α_T, perpendicular to it. For want of information, double bonds are also thought of as having just the two components α_L and α_T, though in this case α_T must be considered merely as a mean. Values of α_L and α_T for some important bonds are given in Table 11.6.3.* A suitable vector summation of the

Table 11.6.3. Longitudinal and Transverse Polarisability of Bonds (in Å^3 per Molecule) for D Light

Bond	α_L	α_T	Bond	α_L	α_T
C—H	0·79	0·58	C═C	2·86	1·06
C—Cl	3·67	2·08	C≡C	3·54	1·27
C—C	1·88	0·02	C═O	1·99	0·75

bond contributions should give the directional polarisabilities of a molecule. This approach has been applied with success in elucidating the conformation of complex molecules with structures difficult to determine by more rigorous methods. The procedure is that the bond polarisabilities are first summed to give the directional polarisabilities of the molecule in its possible conformations; then the directional polarisabilities are used to calculate the Kerr constant, which can be compared with the experimental value. A few examples will suffice to show the scope of the method. Consider first the *cyclo*hexyl *mono*halides. If the *cyclo*hexane ring is assumed to have the *chair* conformation there are two structural possibilities: either the halogen substituent is bound *equatorially*, or it is bound *axially*. The calculated and observed Kerr constants, reproduced in Table 11.6.4,† show that the equatorial conformation

Table 11.6.4. Molar Kerr constant, $10^{12}K$, at 25°C

*cyclo*Hexyl halide	Calc. for equatorial halogen	Calc. for axial halogen	Observed (CCl_4 soln.)
Cl	130	42	122
Br	179	82	181
I	246	152	249

is the correct one. Only for gaseous *cyclo*hexyl chloride was this known in advance,

* Denbigh, *Trans. Faraday Soc.*, 1940, **36**, 936.
† Le Fevre and Le Fevre, *Rev. of Pure and Appl. Chem. (Australia)*, 1955, **5**, 261.

from electron-diffraction experiments.* Secondly, the molecule *cyclo*hexa-1:4 dione has a negative constant ($-41{\cdot}2 \times 10^{-12}$ in benzene soln. at 25°C) and so cannot be wholly in the non-polar chair conformation (*a*), Fig. 11.6.1, which would give rise to a positive Kerr constant; there must be present at least a proportion of molecules in the boat forms (*b*) and (*c*). The form (*b*) is an improbable one in view of the electrostatic repulsion of the C═O bond dipoles. The measured Kerr constant can be interpreted in terms of a mixture, in benzene solution, of 80% of the form (*a*) with 20% of (*c*), a conclusion which is also consistent with the fact that the dipole moment, 1·2 D, is obviously different from the zero moment anticipated for the centrosymmetrical form (*a*)

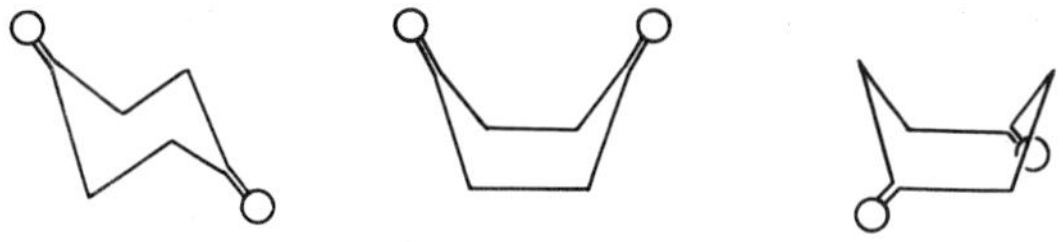

(a) (b) (c)

Fig. 11.6.1. Possible conformations of *cyclo*hexa-1:4-dione. The calculated molar Kerr constants for the forms (*a*), (*b*), and (*c*) are $+7{\cdot}6$, $+16{\cdot}8$, and -233 (all $\times 10^{-12}$), respectively, at 25°C.

Further illustrations of the application of the Kerr effect to chemistry are given by the Le Fevres in their review.

BIBLIOGRAPHY

Frolich, *Theory of Dielectrics,* Oxford, London, 1948.
Smyth, *Dielectric Behaviour and Structure,* McGraw-Hill, New York and London, 1955.

* Hassel, *Quart. Revs.,* London, 1953, 7, 221.

Chapter 12

Appraisement and conclusions

12.1 Compilations of Molecular Data

In this book we have attempted explanations of the more general physical methods for studying molecular structure. Except by way of an occasional illustration, we have not been explicitly concerned with the results obtained by these methods. It seems appropriate to offer the reader some guidance through the inumerable papers that are being published.

The methods are being actively developed and applied, and a phase of exponential growth brings its inevitable problems. Review articles and compilations of results are never comprehensive by the time they achieve publication, and they rapidly become out of date. Three compilations deserve special mention however. *Structure Reports* are issued under the auspices of the International Union of Crystallography. They contain a critical review of virtually all published work in structural crystallography, as well as some coverage of electron-diffraction and microwave studies. A volume is devoted to each year, though there is a time-lag of several years before it appears—and the time-lag itself shows signs of exponential growth. A wealth of information on molecules is contained in Parts 2, 3 and 4 of Volume I of the Sixth Edition of the *Landolt–Börnstein Tables* (1951–5). The Chemical Society's *Special Publications* Nos. 11 and 18 contain a critical collection of interatomic distances and angles from papers published up to 1959. Continuations of these volumes are being published by the International Union of Crystallography; and Volume A1, 'Inter-atomic Distances 1960–5, Organic and Organometallic Crystal Structures', is to appear in 1973. From time to time more specialised reviews are published giving detailed lists of parameters for compounds of particular types.

12.2 The Molecule and Van der Waals Radii

In Chapter 1 we made a number of general remarks on the molecular concept. It may now be pertinent to reconsider some of them in the light of the intervening text.

The concept of the molecule arose chiefly from the chemical study of organic compounds and a few simple inorganic ones which could readily be studied in the gaseous state. Had the early chemists been able to study the structure of matter in the solid state, the concept would have developed in a different way; the idea of a 'sodium chloride molecule', for instance, would have had a different history. We may now ask how well the notion of a molecule has survived the application of the physical methods which constitute the subject matter of this book. The answer must be that it survives convincingly—even in Science, 'seeing is believing'—and that it is confirmed and clarified. Over a wide range of materials we find assemblages of atoms, with distances of 1–2 Å between neighbouring atoms, and with distances that are nearly always greater than 3 Å between the atoms of neighbouring assemblages. In the vaporised condition, in which it is often most convenient to study molecules, the distances between molecules are so great that we can, for most purposes, ignore their interactions. But even in the solid state the discreteness of the molecular aggregate is often unmistakable.

The notion of the discrete molecule carries two implications: first that the atoms within a given molecule are held together by strong, short-range forces, which chemists term—and think of as—*bonds*, between one atom and its immediate neighbours; and secondly that some kind of repulsive force must come into play to maintain a certain minimal distance between the atoms of different molecules, or between different portions of the same molecule. The first we will discuss at some length in the next section. As to the second, we may point out that crystal structure analysis, and the study of the phenomena brought together under the general description of 'steric hindrance' are our chief sources of information about the minimum distances between non-bonded atoms.

Repulsive forces seem to obey an inverse-power law of high order, so that they rise from the negligible to the very great during a small decrease in separation of the atoms. For this reason it is reasonably satisfactory to imagine atoms as bodies with firm boundaries, which come sharply into 'contact' when their centres are a certain, critical distance apart. A fairly characteristic radius can thus be attributed to any species of atom. This is usually known as the *van der Waals radius* because of its connexion with the b term in the most celebrated equation of state. (This is the term which takes account of the appreciable volume taken up by the molecules of a gas, and hence reducing the effective volume in which the molecules can move to $(V-b)$.) Values for these radii were suggested by several early workers; those commonly in use now were given in a table by Pauling (1939), and some entries from this table are as follows:

H	1·1 Å	O	1·40 Å	N	1·5 Å
F	1·35	Cl	1·80	Br	1·95
S	1·85	I	2·15	$-CH_3$	2·0

Half-thickness of aromatic ring 1·7 Å.

Broadly speaking, a pair of non-bonded atoms cannot come closer together than the sum of their van der Waals radii without severe steric repulsion coming into play.

This will reveal itself in observable ways, such as a distortion of the molecule from its normal shape, or an abnormal reactivity. An important exception occurs with two atoms themselves not directly linked, but each linked to the same atom. For example, the two chlorine atoms in methylene dichloride, CH_2Cl_2, are only 2·9 Å apart, which is much less than twice 1·80 Å, though the molecule shows no signs of strain. We may explain this by supposing that the repulsion is not fully effective in directions lateral to the bond which connects an atom to its neighbour. Again, two methyl groups may come closer to one another than 4·0 Å, even when they are not attached to the same carbon atom, provided their hydrogen atoms can 'enmesh', with a consequent restriction on the freedom of the methyl group to rotate. In hexamethylbenzene the carbon atoms of neighbouring methyls are only about 2·9 Å apart.

A simple trust in van der Waals limits also fails where there is hydrogen bonding. A typical situation discovered in early X-ray work was of two oxygen atoms, belonging to contiguous molecules, separated by less than the sum of their van der Waals radii (2·8 Å). In very strong hydrogen bonds the O . . . O distance may be less than 2·5 Å, for instance. This was firm evidence of hydrogen bonding even at a time when hydrogen atoms were not normally located by X-rays. However an O . . . O distance up to 3·2 Å may also indicate hydrogen bonding when we take account of an intervening hydrogen atom attached covalently to one of the oxygen atoms, and to the O—H distance (1·0 Å) add the van der Waals radii of the hydrogen atom itself and of the other oxygen, all this totalling 3·5 Å.

It is noteworthy that the van der Waals radius of an atom is roughly the same as its anionic radius, when the latter is known. This is further evidence for the short range over which repulsive forces build up; the effective size of the chlorine atom, for instance, is about the same when the repulsion is balanced against a strong attraction as in an ionic compound, or against a much weaker one as in a molecular solid.

12.3 The Chemical Bond

Structural theory makes use of the hypothesis that neighbouring atoms are held together by bonds, of which each type of atom has the capacity to form a small, and usually fixed, number. The concept of the carbon atom with its four bonds directed tetrahedrally has been particularly successful. Most chemists have acquired an intuitive feeling for a bond as some physical entity which links two atoms firmly together, and which may be crudely represented—but usefully—by a model with a coiled-wire spring between two spherical balls. At a different level of convention there are familiar 'bond diagrams', in which the chemical symbols for the atoms are shown joined by lines representing covalency bonds. We may now enquire what is the status of this bond in the light of the physical methods for studying molecular structure.

In the valency-bond approach to the theory of molecular structure, the bond appears as a pair of electrons, with anti-parallel spins, shared between two atoms, and hence giving rise to an accumulation of electron density between their atomic kernels. Since X-rays are scattered by the electrons we might hope that a sufficiently careful X-ray analysis would reveal this accumulation. In fact, this has proved to be difficult; for the enhancement of the electron density is relatively small compared with the high density around the atomic kernel, and it is near the limit of experimental significance at present. We have discussed this problem in an elementary way (section 9.14). At the time of writing (1974) there are signs of progress by an approach based on a comparison of, on the one hand, a combination of the results of precise X-ray and neutron-diffraction analyses with, on the other, precise theoretical calculations of the electron distributions in simple molecules.

The study of nuclear quadrupole coupling gives information about the shape of the electric field, due to the orbital electrons, at the nucleus. Variations in the coupling constant in a series of compounds have been interpreted as an index of bond formation (section 8.10).

The answer to our question on the status of the chemical bond must therefore be that it finds rather little *direct* support in the physical methods we have covered in this book. (To be sure, it might be objected that our exclusion (apart from Chapter 4) of electronic spectra may have curtailed our chances of eliciting information about an essentially electronic phenomenon.) Compared with the totality of electronic energy in a molecule, that associated with its covalency bonds is small. The changes an atom undergoes when it enters a molecule are superficial. But that does not prevent them from being all-important to the chemist.

12.4 Bond Properties

Though the direct study of covalent bonding has had only limited success hitherto, the expression of experimental data in terms of *bond properties* has proved extraordinarily useful. A range of somewhat disparate quantities can be collected under this heading: we speak of the distance between neighbouring atoms as the *bond length*; the angle subtended by two atoms at a third, to which the first two are supposed to be bonded, is called the *bond angle*; the constant governing the vibration of one atom with respect to another is called the *bond-stretching force constant*; the energy change accompanying the dissociation of a molecule into two radicals is the *bond-dissociation energy*; the dipole moment of the water molecule is resolved into two O—H *bond moments*. We will consider these properties and any generalisations that have been discerned amongst them.

Bond properties can be considered in two aspects: *constancy*, and (in some cases) *additivity*. We deal with covalent bond lengths first. Constancy would require the internuclear distance to be always the same between bonded atoms of given types. There is a complication due to bonds of multiple order; but, if we concentrate on bonds of the same order–single ones, for instance–a general impression of constancy is gained. Thus a great many C—C bonds have been measured, in substances as different as diamond and straight-chain fatty acids, and their lengths nearly always lie in the range 1·52–1·55 Å, which does not exceed the limits of error of most determinations. When a formally single C—C bond is significantly shorter than this–as is $\equiv$C—CH_3 in methylacetylene (1·46 Å) or Ph—CO_2H in benzoic acid (1·48 Å)–it is usually reasonable to invoke some special explanation, such as conjugation, hyper-conjugation, of a change in the hybridisation at the carbon atom.

However, as more precise data accumulated, it became clear that a strict constancy does not obtain. This is especially well shown by the microwave results. For example, C—Cl has been found to be 1·782 Å in methyl chloride, CH_3Cl; 1·772 in methylene dichloride, CH_2Cl_2; and 1·767 in chloroform, $CHCl_3$; whilst an accurate electron-diffraction analysis of carbon tetrachloride indicates C—Cl = 1·766 Å. An even more marked drift occurs in C—F in the corresponding fluorides: according to microwave measurements, it ranges from 1·385 Å in CH_3F to 1·332 in CHF_3; according to electron diffraction, from 1·391 to 1·334, and then on to 1·323 in CF_4.

Additivity of bond lengths means that the distance between the nuclei of bonded atoms can be consistently expressed as the sum of contributions characteristic of each atom. The length r_{12} of the bond between atoms 1 and 2 can be put equal to the sum of *covalent-bond radii*, $r_1 + r_2$. Such a principle necessarily depends on the prior assumption of constancy. A test of it would be to determine whether r_{12} were the arithmetic mean of the bond lengths r_{11} and r_{22}; for if in general $r_{12} = r_1 + r_2$, then $r_{11} = 2r_1$ and $r_{22} = 2r_2$, so that $r_{12} = \frac{1}{2}(r_{11} + r_{22})$. In the comparatively small number of cases where this test is applicable, the agreement is satisfactory. That most often cited involves the bonds C—C, C—Cl, and Cl—Cl. The unit-cell parameter of diamond can be measured with extreme accuracy, and, as the structure is so simple, the parameter at once leads to C—C = 1·5445 Å at 20°C. Though no other C—C bond has been measured with such precision, and though small variations certainly occur, we may take $1{\cdot}54_4$ Å as a fair average value for the single bond. The electronic spectrum of molecular chlorine yields Cl—Cl = 1·988 Å. Data of the kind mentioned above indicate an average value of about 1·76 Å for C—Cl. This differs but insignificantly from 1·766, the mean of 1·544 and 1·988.

Tables of covalent-bond radii have accordingly been in use since the early nineteen-twenties. Some entries from Pauling's table (1939) are reproduced here. (Where a second or third value is given, it applies to double or triple bonding.)

Covalent-Bond Radii (Å)

						H	0·30
C	0·77	N	0·70	O	0·66	F	0·64
	0·67		0·61		0·57		
	0·60		0·55		0·51		
Si	1·17	P	1·10	S	1·04	Cl	0·99
					0·95		0·90
				Se	1·17	Br	1·14
				Te	1·37	I	1·33

There are also strange variations in the lengths of certain C—C and C═C bonds. For instance, electron-diffraction study of the 1 : 3 : 5-hexatrienes, CH_2═CH—CH═CH—CH═CH_2, finds that the central double bond is much longer than are those at the ends: 1·368(4) against 1·337(2) Å for the *trans*-isomer, with a similar difference for the *cis*. Such irregularities are of great interest, and their interpretation constitutes an important check on theory. Nevertheless, a fair measure of constancy amongst bonds of a particular sort exists, and the rule of constancy is useful. Along with a similar rule for bond angles, it enables us to predict the structures of molecules which have not yet been measured, or which may even be hypothetical; and to do so with some confidence that the results will not be too far from the truth.

Emendations of these radii and modifications of the system of summation have been suggested from time to time. For example, Schomaker and Stevenson proposed a different set of radii which were to be used along with a correction for any difference of electronegativity of the atoms; having an extra parameter, this scheme gave a somewhat better agreement with the observed bond lengths. However, the original scheme has its simplicity as a recommendation, and, though admittedly an approximation, it is still in common use.

The evidence for the constancy and additivity of bond lengths is heavily weighted by the results for carbon compounds. Possibly this circumstance imports a fallacious appearance of conformity.

A bond angle is more sensitive to environmental influences than is a bond length. However, the angle between two bonds of given multiplicity at a given atom is likely to be constant within 3° or 4° unless very bulky groups are present to open up the angle, or unless ring-closure operates to contract it. We may cite some values for the angle between two single bonds at an oxygen atom: in the water molecule, 104·5° from the spectrum (equilibrium positions of nuclei); in dimethyl peroxide, CH_3·O·O·CH_3, 105° from electron diffraction; in solid mercuric oxide,* 109·8° from X-ray and neutron diffraction; in dioxan, O<(CH_2·CH_2)(CH_2·CH_2)>O, 108° from electron diffraction. None of these values differs greatly from the tetrahedral value of 109·4°, though there is a distinct tendency for them to be less. Much wider deviations can be reasonably ascribed to special circumstances; such are values exceeding 120° in diphenyl ether, Ph·O·Ph, and some of its substitution products, or a value of 61·6° from a microwave study of ethylene oxide, O<(CH_2—CH_2). (In the latter case, the concept of 'bent bonds' would imply that the angle between the actual bonds is greater than that subtended by the nuclei.) X-ray analyses of a wide range of aromatic hydrocarbons give values close to 120° for the valency-angles at the carbon atoms, though small, significant variations often occur.

Even more liable to variation than bond angles is the *dihedral angle* between alternate bonds—*i.e.*, in a non-linear chain of atoms A—B—C—D, the angle

* Solid mercuric oxide contains infinite —Hg—O—Hg—O chains.

between bonds A—B and C—D when these are viewed in projection along the direction B—C. We have seen (section 6.8) that in saturated carbon compounds the bonds tend to adopt a staggered conformation; the dihedral angles in a chain of carbon atoms will thus tend to be near either to 180° or to 60°, corresponding to the *trans*- and *gauche*-forms, of which the former is usually the more stable. With elements other than carbon, molecules embodying chains of atoms are much less common, and of only a few such have the structures been determined. Chains of sulphur atoms occur, however, and in them the dihedral angles tend to be about 90°, but range through roughly 20° on either side of the right-angle.

Next we consider the forces needed to distort bonds. The foregoing discussion of bond lengths, bond angles, and dihedral angles suggests that distortions can be categorised as (1) bond stretching, (2) bond bending, or deformation, and (3) bond twisting, or torsion; and that the necessary force will diminish in the order, (1) > (2) > (3). The idealised shape of a molecule is most easily altered by twisting its single bonds, least easily by stretching them.

In a diatomic molecule a force-constant can be simply related to the bond-stretching frequency. This constant is a measure of the bond's strength for a small displacement. Generalising this, we would expect the trend, (1) > (2) > (3), to be reflected in the corresponding force-constants. This is in fact true, though the comparison is complicated by two considerations. First any distortion of a molecule will usually involve simultaneous strains of all three types. Fortunately, the normal modes of vibration each approximate to displacements of one type only; the flexing of the water molecule about the oxygen atom will include only slight changes of bond lengths, and to regard it as involving only a change of angle is a fair approximation. Secondly, it is not competent to make a direct comparison between a constant expressed as dynes cm^{-1} (or as ergs cm^{-2}, which is equivalent) and one expressed as dynes $radian^{-1}$ (or ergs $radian^{-2}$, which is not equivalent in this case). All must be converted into the same units, and this can be done roughly.

For the stretchings of single bonds, the constant is usually around 5×10^5 dynes cm^{-1} For C—C in ethane it is $4{\cdot}5 \times 10^5$. With multiple bonds the force increases in rough proportionality: for the double bond in ethylene the constant is $9{\cdot}57 \times 10^5$; for the triple bond in acetylene, $15{\cdot}72 \times 10^5$. In benzene, where the bond-order may be regarded as 1·5, the corresponding constant is $7{\cdot}6 \times 10^5$. The deformation of the benzene ring by displacements of the carbon atoms in the plane, so that alternate angles will become greater and less than 120°, can be interpreted with a constant of $13{\cdot}7 \times 10^{-12}$ erg $radian^{-2}$; a simple calculation will show that this corresponds to a force of about $0{\cdot}7 \times 10^5$ dynes per cm displacement at a carbon atom 1·4 Å from the point of flexure. For the torsional, out-of-plane vibration, the constant is 1·0 × $1{\cdot}0 \times 10^{-12}$ erg $radian^{-2}$, corresponding to roughly $0{\cdot}06 \times 10^5$ dynes cm^{-1}. Thus the relative values 7·6, 0·7 and 0·06 strikingly illustrate the trend we have been discussing.

Torsion is of course much stiffer about a double, than about a single, bond. The torsional constant in the ethylene molecule is about 5×10^{-12} erg $radian^{-2}$, which corresponds to about $0{\cdot}6 \times 10^5$ dyne cm^{-1} displacement at the hydrogen atoms.

An alternative index of bond-strength can be sought in the energy needed to disrupt the bond, though this topic is really outside the scope of this book. (It is the theme of Cottrell's *The Strengths of Chemical Bonds* (1958).) With a diatomic molecule the interpretation of the experimental result is unambiguous: when we can measure the heat change accompanying the dissociation of such a molecule into unexcited atoms, the value is clearly a possible indicator of the strength of the bond. We may call it the bond dissociation energy, or simply the bond energy. With a molecule of the type A—B—A, the situation is more complicated: the dissociation energies to first A—B + A and then 2A + B may each be measurable, and in general they will differ. Their sum is however a well-defined quantity, and it is permissible to divide it by two, and to allocate the resulting bond energy to each A—B bond. The bond energy will then differ from the bond dissociation energy. With a molecule such as that of *n*-propane, $CH_3 \cdot CH_2 \cdot CH_3$, the complexity is much worse. The energy needed to dissociate the molecule into eleven atoms may be measured, but the distribution of it amongst the ten bonds broken is an arbitrary procedure. For instance, it by no means follows that the quantity to be associated with each of the two methylenic C—H bonds should be the same as that for the six methylic bonds. Nevertheless, when we make it a convention that the same energy is to be attributed to a bond of the same multiplicity between atoms of given types, the outcome is roughly self-consistent. This convention is used in the usual method for estimating resonance energy in a conjugated compound: the sum of the appropriate bond energies for a molecule of hypothetical bond structure is subtracted from the dissociation energy of the actual molecule into its atoms. The bond-energy terms thus conventionalised are not generally additive; E_{12} is not usually equal to the arithmetic mean of E_{11} and E_{22}.

A similar difficulty occurs with *bond moments*. The ordinary experimental methods give the electric dipole moment of the whole molecule. Only with a diatomic molecule can we simply equate this with a bond moment; 1·03 D for the hydrogen chloride molecule can be taken as the moment appropriate to the H—Cl bond. When we find a moment of 1·85 D for the water molecule, it is an assumption, though no doubt a reasonable one, to suppose that this total moment (μ) can be attributed to two equal bond moments (m) along the O—H bonds. Adopting this convention, we resolve μ by the relation, $\mu = 2m \cos (105°/2)$, so that m is found to be about 1·5 D. The outcome is fairly successful. We can tabulate bond-moment values that are supposed to be constant for a given type of bond, irrespective of the molecular environment. This scheme is useful in that it enables us to calculate the overall dipole moment for a molecule of hypothetical structure, by vector summation of the bond moments. By comparison of observed and calculated dipole moments in this way, deductions as to the configuration of the molecules of a compound may be made. The rule of constancy has thus a certain pragmatic value.

However, there have been grave uncertainties about the absolute values of some bond moments, particularly that for C—H. This was originally supposed to have a value of 0·4 D with hydrogen positive. Subsequently reasons have been advanced for even reversing the direction of this moment when the carbon atom is in its usual sp^3-hybridised state with four tetrahedrally directed valencies, though it is conceded that the hydrogen atom may still be the seat of a positive pole in an acetylenic C—H bond, where the hybridisation is of *sp*-type. The C—H moment would thus be far from constant, and smaller variations might occur with other bonds. (The use of an erroneous value for the C—H moment, provided it is constant, is less serious than might be supposed at first sight. It would merely put a uniform error into other bond moments, and the effect of this would tend to cancel out in many problems.)

Granted some degree of constancy of bond moments, we might expect the moments of different bonds to be a *subtractive,* rather than an additive, function of the two atoms concerned. In an unsophisticated view, a moment arises in the H—Cl bond because the pair of shared electrons is drawn closer to the halogen atom than formally equitable sharing might seem to require. The bond moment is thus the result of the opposing electron-attracting powers of the two atoms. This power is often known as the *electronegativity* (x) of the element; and the A—B bond moment would be a function of $|x_A - x_B|$. Sets of electronegativity values have been deduced in various ways. The best known is that of Pauling, some part of whose table (1939) is shown here:

Electronegativity Values

C	2·5	N	3·0	O	3·5	F	4·0
		P	2·1	S	2·5	Cl	3·0
				Se	2·4	Br	2·8
H	2·1					I	2·5

The difference between the x-values is usually within 0·2 D of the conventional bond moment. Thus $x_O - x_H = 1{\cdot}4$ D.

This view of the bond dipole moment envisages it as a property of the bond itself, as something within the bond. In simple cases, the electron density over a molecule can now be calculated by quantum mechanics. It becomes clear that the overall dipole moment is not in fact simply due to charge separations along the bonds. In the water molecule for instance, the 1·5 D conventionally allocated to each O—H bond carries a substantial contribution from the lone pairs of electrons on the other side of the oxygen atom. Thus the notion of a constant moment resident in the bond itself, and derivable from the electronegativities of the two elements forming the bond, has no support from current views of the physical structure of the molecule. All the same, it has some pragmatic value. It enables us, by an easy calculation, to estimate the dipole moment of a given molecule, with a fair prospect that the result will not be too far wrong. The proper *ab initio* calculation of the charge-density distribution would be more formidable, and much more expensive.

12.5 A General Comparison of the Physical Methods

We end with a general comparison of the methods we have described in this book.

We might ask first whether the different methods give concordant results when applied to the same molecule. Since our table of covalent-bond radii is based on molecular dimensions obtained by a variety of techniques, it follows that their results must be consistent in a general way. In fact, only a few structures have been accurately measured by two or more methods; where the test has been made, the agreement is normally excellent. In benzene C—C is 1·397 Å according to a high-resolution Raman spectrum study of the vapour, $1{\cdot}39_3$ according to an electron-diffraction study, and 1·392 according to a well-refined X-ray analysis of the solid at − 3°C. A neutron-diffraction study of the ammonium halides led to 1·03(2) Å for N—H, whilst proton magnetic resonance led to 1·03(1) Å. Chlorine trifluoride, ClF_3, has a *T*-shaped molecule, in which Cl—F is 1·716(4) Å along the top of the *T* and 1·621(6) in the downstroke from an X-ray study of the solid at − 120°C, and 1·698(5) and 1·598(5) Å from a microwave study of the vapour. (See also section 10.7.)

The small differences observed in the latter example may be just significant. It would not be surprising if molecular dimensions were slightly affected by the transition from solid to vapour; the search for such differences is at any rate a matter of interest. In a few compounds profounder differences have been found. The most remarkable example is phosphorus pentachloride, which exists in the vapour as covalent PCl_5 molecules of D_{3h} symmetry, whilst the solid consists of the ions PCl_4^+ and PCl_6^-. Such cases are exceptional. (Antimony pentachloride has similar $SbCl_5$ molecules in both gas and solid.) In general, differences of structure between solid and gas are slight, because the van der Waals forces between molecules are of low energy compared with the covalent forces within the molecule. In a number of cases a compound has been studied by X-ray analysis in two polymorphic forms, and no significant variations of molecular structure have been detected.*

One other feature is relevant in comparing bond-lengths precisely determined by different methods. (We have already illustrated it in section 10.6.) Molecules are always vibrating. The distance found between two particular atoms therefore represents an averaged value. But different methods imply different sorts of averaging. There are various reasons for this, notably anharmonicity of the molecular vibration. Before we can compare precise bond-lengths, we must allow for this discrepancy between the quantities measured. One way is to correct each averaged distance to its *equilibrium* value: *i.e.* the distance between atomic nuclei when the molecule is in an hypothetical, non-vibrating state. To do this we need a thorough understanding of the vibrations, and there are technical difficulties in estimating the correction properly. Fortunately, it lies only in the third place of decimals (of Å), and need not concern us unduly in the present context.†

Although the various physical methods generally are in accord with each other, they should be regarded as complementary rather than alternatives; only one of them might be deployed to full advantage in a given structural problem. A complete

* Except, occasionally, for minor differences in the conformation of a terminal group.

† It is the equilibrium distances that are usually produced from theoretical calculations of molecular structure.

spectroscopic structure analysis would be attempted only for a very simple molecule that could be obtained in the vapour state. For a more complex molecule, X-ray analysis might have to be adopted, and a much lower standard of accuracy accepted. Were it a question of locating hydrogen atoms, neutron diffraction might be most appropriate.

The non-equivalence of the methods and of the ends to which they are adapted may be illustrated by a series of graded examples. In some cases the structure of the molecule in a qualitative sense is so simple as to be of merely trivial interest. That the hydrogen chloride molecule consists of a hydrogen atom bonded to a chlorine must have been—at least implicitly—realised a hundred years ago. The spectroscopist who studies this molecule is concerned with attaining precise, quantitative information about the structure: the internuclear distance, the stretching-force constant, the dipole moment. Again, once its non-linearity and C_{2v} symmetry have been recognised, the qualitative structure of the water molecule is solved; the object of spectroscopic examination of water vapour is to determine exact molecular parameters. The X-ray crystallographer may then take up the study of ice in order to find out how the molecules are linked together in the solid state, or how far the dimensions may be modified by the crystalline environment. For the more complex molecule of naphthalene, the constitutional formula was well known on chemical grounds. Since this system is too complicated for detailed study of the spectrum, the structure has been elucidated by crystallographic methods. The first structure analyses confirmed the accepted qualitative formulation, and added the quantitative information that the sides of the hexagonal rings of carbon atoms are about 1·4 Å in length. Subsequent work added first one, then two, significant figures to the bond lengths, and made it possible to discuss small differences between individual C—C lengths—differences which were to be expected on theoretical grounds. The general constitution of the cholesterol molecule was known with some certainty in 1946; but the X-ray analysis of cholesteryl iodide then completed not merely confirmed this constitution, but also answered some outstanding questions about the stereochemical configuration and conformation. At a much further stage of complexity stands the molecule of vitamin B_{12}: the formula is $C_{63}H_{88}O_{14}N_{14}PCo$, but even this was not exactly known when the analysis was undertaken (about 1950), whilst the structural formula was almost completely unknown. The X-ray analysis, completed in 1957, revealed the structure, apart from some minor details. Neutron diffraction has recently (1971) been applied to one of the B_{12}-acids; by locating some of the hydrogen atoms, it cleared up some of these details.

A different order of difficulty faces X-ray work on the structures of crystalline proteins. For proteins which, because of the relative smallness of their molecules and other favourable circumstances, are suitable for study, the amino-acid sequence is likely to be known beforehand. This is a great help in interpreting electron-density maps. The problem then consists in deciphering the secondary and tertiary structure of the protein—the conformational details which characterise the shapes of these large molecules and which are so essential to their biological functions. For this purpose, the electron density does not need to be resolved down to the atomic level (see

Fig. 9.10.1). Indeed for many protein crystals there is so much 'disorder' in the structure that resolution to 2 Å, at which atomic detail may start to be discernible, is unattainable. In 1972 it was reckoned that the structures of about 35 proteins had been determined with a resolution of 3·5 Å, or better. Neutron diffraction, too, is beginning to be applied to proteins.

When we consider the natures of the various physical methods, we find differences in their approach. Most of them employ some well-understood physical phenomenon to deduce certain parameters, with the aid of which a molecular model may be described. For example, the basic principles of X-ray diffraction by a crystal have been established for half a century. Admittedly, there has been spectacular progress in the field recently; but this has been due to better devices for solving the phase problem and for handling large masses of data. No new principles have been applied. The immediate product is a representation of the electron-density distribution over the molecule—a distribution averaged in time and in space over the vast number of molecules in the crystal, and not, unless very special precautions have been taken (see section 9.14), directly correspondent to the density distribution of the theoretical chemist. Despite the difference, we deduce atomic positions and other molecular parameters from the experimental distribution. The strategy of gas electron diffraction is similar.

The spectroscopic approach is more radical. Its immediate findings are transitions between pairs of energy levels. These levels depend upon the eigenvalues of the real wavefunction, and the transitions are strictly the only observable properties of the molecule. Were we able to write down the wave equation and to solve it explicitly for every energy state, we could then predict all properties of the molecule. This we cannot accomplish yet, even for simple molecules. Furthermore, the spectroscopist has to be content with an incomplete—usually a very incomplete—exploration of the total spectrum of his material. Nevertheless the spectroscopic method operates in a setting that seems to be closer to whatever reality may underlie our concept of the molecule. To this extent it differs in kind from the other methods, and this is why we have treated it in more detail in this book.

12.6 'Labs or Computers?'

The internuclear distance in the hydrogen molecule and its binding energy can now be calculated by quantum mechanics with an accuracy better then that with which these quantities can be determined experimentally. How far can such theoretical calculations be extended to less simple molecules than H_2? And how soon may the methods described in this book become obsolete?

The answers must certainly be: not far, yet; and not in the immediately forseeable future. Though useful and instructive calculations can be made on molecules of considerable size – *e.g.* the hydrogen-bonded base-pairs that play an essential role in *DNA* – increasingly severe approximations become more and more necessary as the complexity of the system increases. Only with simple molecules can theoretical calculations compete with experiment for determining molecular parameters.

The molecule of propylene, $CH_3—CH=CH_2$, illustrates what can now be achieved by rather rigorous calculation.* Certain technical simplifications have to be made, and the molecular symmetry is taken to be $m(C_s)$, as is indicated by the spectrum. When the total energy has been minimised, the mathematical precision of the derived bond lengths is about 0·001 Å. Some of the molecular dimensions thus found are listed in Table 12.6.1. Corresponding experimental values, derived from microwave spectroscopy,† are also listed for comparison. The agreement is impressive, though the calculated inter-carbon distances seem to be significantly in error. The Table also compares molecular dipole moments. Here the calculation does better than experiment: it can find the direction of the moment, whereas the spectroscopic determination leads, directly, only to its magnitude. (The second microwave study made a fuller use of deuteriated species, and so was able to assess an approximate direction for the moment. This agrees fairly well with the calculation.)

Table 12.6.1. A Comparison of Some Theoretical and Experimental Parameters for the Propylene Molecule

	Theoretical	Spectroscopic
C—C	1·520 Å	1·501 (4) Å
C=C	1·308	1·336 (4)
C—H (CH_2)	1·081	1·091 (3)
	1·085	1·081 (3)
C—(CH)	1·085	1·090 (3)
C—H (CH_3)	1·085	1·085 (4)
	1·088 (twice)	1·098 (14)
C—C=C	125·1°	124·3 (0·3)°
Dipole moment	0·34 D	0·36 D

To the question with which this section is introduced, we must surely agree with the answer given by C.A. Coulson: in studying molecular structure, we need both laboratories and computers; but neither is used properly unless it is used with due regard to what is important in Chemistry.

Shortly before his untimely death in 1973, the distinguished American chemist, Walter Hamilton, put the following question to his fellow-crystallographers: 'Do you feel that you are now using the theory to calibrate the experiment, but – as you go along to larger and larger molecules – (do you not also feel that) the experiment will play the dominant role in getting the really ... interesting information?'**

* Radom, Lathan, Hehre and Pople, *J. Amer. Chem. Soc.*, 1971, **95**, 5339.

† Lide and Christensen, *J. Chem. Phys.*, 1961, **35**, 1374; Hirota and Morino, *J. Chem. Phys.*, 1966, **45**, 2326.

** Quoted by P. Coppens, *Acta Cryst.*, 1974, **B30**, 255.

Appendix I

Character tables and direct product rules

Tables I. 1–17

Character Tables of Some Important Point Groups

1.

C_1	I
A	1

2.

C_2	I	C_2			
A	1	1	z	R_z	x^2, y^2, z^2, xy
B	1	−1	x, y	R_x, R_y	xz, yz

3.

$C_i \equiv S_2$	I	i			
A_g	1	1		R_x, R_y, R_z	x^2, y^2, z^2
A_u	1	−1	x, y, z		xy, yz, xz

4.

C_{2v}	I	C_2	$\sigma_v(xz)$	$\sigma_v'(yz)$			
A_1	1	1	1	1	z		x^2, y^2, z^2
A_2	1	1	−1	−1		R_z	xy
B_1	1	−1	1	−1	x	R_y	xz
B_2	1	−1	−1	1	y	R_x	yz

5.

C_{3v}	I	$2C_3$	$3\sigma_v$			
A_1	1	1	1	z		$x^2 + y^2, z^2$
A_2	1	1	−1		R_z	
E	2	−1	0	(x, y)	(R_x, R_y)	$(x^2 - y^2, xy)$ (xz, yz)

6.

$C_s \equiv C_{1h}$	I	σ_h			
A'	1	1	x, y	R_z	x^2, y^2, z^2, xy
A''	1	-1	z	R_x, R_y	xz, yz

7.

C_{2h}	I	C_2	σ_h	i			
A_g	1	1	1	1		R_z	x^2, y^2, z^2, xy
A_u	1	1	-1	-1	z		
B_g	1	-1	-1	1		R_x, R_y	xz, yz
B_u	1	-1	1	-1	x, y		

8.

$D_{2h} \equiv V_h$	I	$C_2(z)$	$C_2(y)$	$C_2(x)$	i	$\sigma(xy)$	$\sigma(zx)$	$\sigma(yz)$			
A_g	1	1	1	1	1	1	1	1			x^2, y^2, z^2
A_u	1	1	1	1	-1	-1	-1	-1			
B_{1g}	1	1	-1	-1	1	1	-1	-1		R_z	xy
B_{1u}	1	1	-1	-1	-1	-1	1	1	z		
B_{2g}	1	-1	1	-1	1	-1	1	-1		R_y	xz
B_{2u}	1	-1	1	-1	-1	1	-1	1	y		
B_{3g}	1	-1	-1	1	1	-1	-1	1		R_x	yz
B_{3u}	1	-1	-1	1	-1	1	1	-1	x		

9.

D_{3h}	I	$2C_3$	$3C_2$	σ_h	$2S_3$	$3\sigma_v$			
A_1'	1	1	1	1	1	1			$x^2 + y^2, z^2$
A_1''	1	1	1	-1	-1	-1			
A_2'	1	1	-1	1	1	-1		R_z	
A_2''	1	1	-1	-1	-1	1	z		
E'	2	-1	0	2	-1	0	(x, y)		$(x^2 - y^2, xy)$
E''	2	-1	0	-2	1	0		(R_x, R_y)	(yz, xz)

10.

D_{6h}	I	$2C_6$	$2C_3$	C_2	$3C_2'$	$3C_2''$	i	$2S_3$	$2S_6$	σ_h	$2\sigma_d$	$3\sigma_v$			
A_{1g}	1	1	1	1	1	1	1	1	1	1	1	1			$x^2 + y^2, z^2$
A_{1u}	1	1	1	1	1	1	-1	-1	-1	-1	-1	-1			
A_{2g}	1	1	1	1	-1	-1	1	1	1	1	-1	-1		R_z	
A_{2u}	1	1	1	1	-1	-1	-1	-1	-1	-1	1	1	z		
B_{1g}	1	-1	1	-1	1	-1	1	-1	1	-1	1	-1			
B_{1u}	1	-1	1	-1	1	-1	-1	1	-1	1	-1	1			
B_{2g}	1	-1	1	-1	-1	1	1	-1	1	-1	-1	1			
B_{2u}	1	-1	1	-1	-1	1	-1	1	-1	1	1	-1			
E_{1g}	2	1	-1	-2	0	0	2	1	-1	-2	0	0		(R_x, R_y)	(yz, xz)
E_{1u}	2	1	-1	-2	0	0	-2	-1	1	2	0	0	(x, y)		
E_{2g}	2	-1	-1	2	0	0	2	-1	-1	2	0	0			$(x^2 - y^2, xy)$
E_{2u}	2	-1	-1	2	0	0	-2	1	1	-2	0	0			

11.

$D_{2d} \equiv V_d$	I	$2S_4$	C_2	$2C_2'$	$2\sigma_d$			
A_1	1	1	1	1	1			x^2+y^2, z^2
A_2	1	1	1	−1	−1		R_z	
B_1	1	−1	1	1	−1			x^2-y^2
B_2	1	−1	1	−1	1	z		xy
E	2	0	−2	0	0	(x,y)	(R_x, R_y)	(yz, xz)

12.

D_{3d}	I	$2C_3$	$3C_2$	i	$2S_6$	$3\sigma_d$			
A_{1g}	1	1	1	1	1	1			x^2+y^2, z^2
A_{1u}	1	1	1	−1	−1	−1			
A_{2g}	1	1	−1	1	1	−1		R_z	
A_{2u}	1	1	−1	−1	−1	1	z		
E_g	2	−1	0	2	−1	0		(R_x, R_y)	$(x^2-y^2, xy), (yz, xz)$
E_u	2	−1	0	−2	1	0	(x, y)		

13.

D_{5d}	I	$2C_5$	$2C_5^2$	$5C_2$	i	$2S_{10}^3$	$2S_{10}$	$5\sigma_d$			
A_{1g}	1	1	1	1	1	1	1	1			x^2+y^2, z^2
A_{1u}	1	1	1	1	−1	−1	−1	−1			
A_{2g}	1	1	1	−1	1	1	1	−1		R_z	
A_{2u}	1	1	1	−1	−1	−1	−1	1	z		
E_{1g}	2	$2\cos\epsilon$	$2\cos 2\epsilon$	0	2	$2\cos\epsilon$	$2\cos 2\epsilon$	0		(R_x, R_y)	(yz, xz)
E_{1u}	2	$2\cos\epsilon$	$2\cos 2\epsilon$	0	−2	$-2\cos\epsilon$	$-2\cos 2\epsilon$	0	(x, y)		
E_{2g}	2	$2\cos 2\epsilon$	$2\cos\epsilon$	0	2	$2\cos 2\epsilon$	$2\cos\epsilon$	0			(x^2-y^2, xy)
E_{2u}	2	$2\cos 2\epsilon$	$2\cos\epsilon$	0	−2	$-2\cos 2\epsilon$	$-2\cos\epsilon$	0			

$\epsilon = 360/5 = 72°$

14.

T_d	I	$8C_3$	$3C_2$	$6\sigma_d$	$6S_4$			
A_1	1	1	1	1	1			$x^2+y^2+z^2$
A_2	1	1	1	−1	−1			
E	2	−1	2	0	0			$(x^2+y^2-2z^2, x^2-y^2)$
F_1	3	0	−1	−1	1		(R_x, R_y, R_z)	
F_2	3	0	−1	1	−1	(x, y, z)		(xy, yz, xz)

15.

O_h	I	$8C_3$	$6C_2$	$6C_4$	$3C_2''$	i	$6S_4$	$8S_6$	$3\sigma_h$	$6\sigma_d$			
A_{1g}	1	1	1	1	1	1	1	1	1	1			$x^2+y^2+z^2$
A_{1u}	1	1	1	1	1	−1	−1	−1	−1	−1			
A_{2g}	1	1	−1	−1	1	1	−1	1	1	−1			
A_{2u}	1	1	−1	−1	1	−1	1	−1	−1	1			
E_g	2	−1	0	0	2	2	0	−1	2	0			$(x^2+y^2-2z^2, x^2-y^2)$
E_u	2	−1	0	0	2	−2	0	1	−2	0			
F_{1g}	3	0	−1	1	−1	3	1	0	−1	−1		(R_x, R_y, R_z)	
F_{1u}	3	0	−1	1	−1	−3	−1	0	1	1	(x, y, z)		
F_{2g}	3	0	1	−1	−1	3	−1	0	−1	1			(xy, yz, xz)
F_{2u}	3	0	1	−1	−1	−3	1	0	1	−1			

16.

$C_{\infty v}$	I	$2C_\phi$	$2C_\phi^2$	...	$\infty\sigma_v$			
Σ^+	1	1	1	...	1	z		x^2+y^2, z^2
Σ^-	1	1	1	...	−1		R_z	
Π	2	$2\cos\phi$	$2\cos(2\phi)$	...	0	(x, y)	(R_x, R_y)	(xz, yz)
Δ	2	$2\cos(2\phi)$	$2\cos(4\phi)$	...	0			x^2-y^2, xy
Φ	2.	$2\cos(3\phi)$	$2\cos(6\phi)$	...	0			
	...	...	...	...	...			

17.

$D_{\infty h}$	I	$2C_\phi$	...	$\infty\sigma_v$	i	$2S_\phi^2$	...	∞C_2			
Σ_g^+	1	1	...	1	1	1	...	1			x^2+y^2, z^2
Σ_u^+	1	1	...	1	−1	−1	...	−1	z		
Σ_g^-	1	1	...	−1	1	1	...	−1		R_z	
Σ_u^-	1	1	...	−1	−1	−1	...	1			
Π_g	2	$2\cos\phi$	...	0	2	$-2\cos\phi$	...	0		(R_x, R_y)	(xz, yz)
Π_u	2	$2\cos\phi$	...	0	−2	$2\cos\phi$	...	0	(x, y)		
Δ_g	2	$2\cos 2\phi$	...	0	2	$2\cos 2\phi$	...	0			(x^2-y^2, xy)
Δ_u	2	$2\cos 2\phi$	...	0	−2	$-2\cos 2\phi$	...	0			
...	...	...	...	...	...	...	...	...			

Table I. 18. Rules for the Direct Product

(i) General (1-, 2-, 3- *or* 6-*fold principal axis*)

$A \times A = A$ $\quad B \times B = A$ $\quad A \times B = B$

A or $B \times E = E$ $\quad A$ or $B \times F = F$ $\quad A \times E_1 = E_1$

$B \times E_2 = E_1$ $\quad A \times E_2 = E_2$ $\quad B \times E_1 = E_2$

$g \times g = g$ $\quad u \times u = g$ $\quad g \times u = u$

$' \times ' = '$ $\quad '' \times '' = '$ $\quad ' \times '' = ''$

except that for D_{2h}, $B \times B = B$

(ii) Subscript numerals on A or B

$1 \times 1 = 1$ $\quad 2 \times 2 = 1$ $\quad 1 \times 2 = 2$

except that for D_{2h}, $1 \times 2 = 3$, $2 \times 3 = 1$, $1 \times 3 = 2$

(iii) Degenerate representations

$E_1 \times E_1 = E_2 \times E_2 = A_1 + A_2 + E_2$, $\quad E_1 \times E_2 = B_1 + B_2 + E_1$

except that for D_{2d}, $E \times E = A_1 + A_2 + B_1 + B_2$;

and for D_{5d}, $E_1 \times E_2 = E_1 + E_2$.

(When the symbols *A*, *B*, or *E* occur without subscript in the group table, read $E_1 = E_2 = E$, *etc*. When no symbol *B* appears, read *A* for *B*.)

For T_d: $E \times F_1 = E \times F_2 = F_1 + F_2$,

$F_1 \times F_1 = F_2 \times F_2 = A_1 + E + F_1 + F_2$,

$F_1 \times F_2 = A_2 + E + F_1 + F_2$.

(iv) Linear molecules

$\Sigma^+ \times \Sigma^+ = \Sigma^+$ $\quad \Sigma^- \times \Sigma^- = \Sigma^+$ $\quad \Sigma^+ \times \Sigma^- = \Sigma^-$

Σ^+ or $\Sigma^- \times \Pi = \Pi$ $\quad \Sigma^+$ or $\Sigma^- \times \Delta = \Delta$

$\Pi \times \Pi = \Sigma^+ + \Sigma^- + \Delta$ $\quad \Pi \times \Delta = \Pi + \Phi$ $\quad \Delta \times \Delta = \Sigma^+ + \Sigma^- + \Gamma$

Appendix II

II .1

We desire to show that the polynomial solution of the differential equation

$$\frac{d^2v}{dx^2} - 2x\frac{dv}{dx} + 2vv = 0 \tag{1}$$

can, except for an arbitrary constant, be written in the form

$$v = e^{x^2}\frac{d^v}{dx^v}e^{-x^2}. \tag{2}$$

Substitution of (2) in (1) yields

$$\frac{d^{v+2}}{dx^{v+2}}e^{-x^2} + 2x\frac{d^{v+1}}{dx^{v+1}}e^{-x^2} + 2(v+1)\frac{d^v}{dx^v}e^{-x^2} = 0. \tag{3}$$

To demonstrate that (2) is a solution of (1) we must show that (3) holds for $v = 0$, 1, 2, 3 . . . Suppose $v = 0$, then the left-hand side of (3) becomes

$$\frac{d^2}{dx^2}e^{-x^2} + 2x\frac{d}{dx}e^{-x^2} + 2e^{-x^2}$$

which equals 0, so that (3) is certainly valid when v is zero. But differentiation with respect to x yields the same equation as that obtained by writing $(v + 1)$ for v in (3): therefore if (3) holds for one integer v it must hold for the next higher integer, and so forth. Thus, as (3) holds for $v = 0$, it does also for $v = 1$, hence for $v = 2$, and so for all positive integral values of v.

The Hermite polynomials are the expressions (2) multiplied by $(--1)^v$,

$$H_v(x) = (-1)^v e^x \frac{d^v}{dx^v} e^{-x^2}. \tag{4}$$

The first three members of the set are

$$H_0(x) = 1; \quad H_1(x) = 2x; \quad H_2(x) = 4x^2 - 2.$$

In general,

$$H_v(x) = (2x)^v - \frac{v(v-1)(2x)^{v-2}}{1!} + \frac{v(v-1)(v-2)(v-3)(2x)^{v-4}}{2!} + \ldots \quad (5)$$

Differentiation yields

$$\frac{dH_v(x)}{dx} = 2v\ (2x)^{v-1} - \frac{(v-1)(v-2)(2x)^{v-3}}{1!} + \ldots$$

$$= 2vH_{v-1}(x) \quad (6)$$

Differentiating a second time we obtain

$$\frac{d^2H_v(x)}{dx^2} = 2v\frac{dH_{v-1}(x)}{dx} = 4v(v-1)H_{v-2}(x). \quad (7)$$

Substitute $v = H_v(x)$ into (1), then we find

$$4v(v-1)H_{v-2}(x) - 4xvH_{v-1}(x) + 2vH_v(x) = 0, \quad (8)$$

whence

$$(v-1)H_{v-2}(x) - xH_{v-1}(x) + \tfrac{1}{2}H_v(x) = 0. \quad (9)$$

Replacing $(v-1)$ in (9) by v, the equation becomes

$$xH_v = vH_{v-1} + \tfrac{1}{2}H_{v+1}. \quad (10)$$

(10) is known as the recursion formula for Hermite polynomials.

Next, we seek to evaluate the integral

$$\int_{-\infty}^{\infty} H_m(x)H_n(x)e^{-x^2}\,dx.$$

Recalling the definition (4) we have

$$\int_{-\infty}^{\infty} H_m(x)H_n(x)e^{-x^2}\,dx = (-1)^n\int_{-\infty}^{\infty} H_m(x)\frac{d^n}{dx^n}e^{-x^2}\,dx. \quad (12)$$

Integrating by parts, one obtains

$$\int_{-\infty}^{\infty} H_m(x)\frac{d^n}{dx^n}e^{-x^2}dx = \left[H_m(x)\frac{d^{n-1}}{dx^{n-1}}e^{-x^2}\right]_{-\infty}^{\infty} - \int_{-\infty}^{\infty}\frac{dH_m(x)}{dx}\frac{d^{n-1}}{dx^{n-1}}e^{-x^2}dx \tag{13}$$

The first term on the right equals zero, since e^{-x^2} and its derivatives vanish at $x = \pm\infty$. Substitution of $2mH_{m-1}(x)$ for $dH_m(x)/dx$ then yields

$$\int_{-\infty}^{\infty} H_m(x)H_n(x)e^{-x^2}dx = (-1)^{n+1}2m\int_{-\infty}^{\infty} H_{m-1}(x)\frac{d^{n-1}}{dx^{n-1}}e^{-x^2}dx. \tag{14}$$

Repeating the integration by parts, we obtain finally

$$\begin{aligned}\int_{-\infty}^{\infty} H_m(x)H_n(x)e^{-x^2}dx &= (-1)^{n+m}2^m m!\int_{-\infty}^{\infty} H_0(x)\frac{d^{n-m}}{dx^{n-m}}e^{-x^2}dx\\ &= (-1)^{n+m}2^m m!\left[\frac{d^{n-m-1}}{dx^{n-m-1}}e^{-x^2}\right]_{-\infty}^{\infty} = 0.\end{aligned} \tag{15}$$

When $n = m = v$, (15) becomes

$$\begin{aligned}\int_{-\infty}^{\infty} [H_v(x)]^2 e^{-x^2}dx &= (-1)^{2v}2^v!\int_{-\infty}^{\infty} e^{-x^2}dx\\ &= 2^v v!\sqrt{\pi}\end{aligned} \tag{16}$$

since $(-1)^{2v} = 1$ and $\int_{-\infty}^{\infty} e^{-x^2}dx = \sqrt{\pi}$.

II.2. Application to the Harmonic Oscillator

The normalising condition for the wavefunctions of the harmonic oscillator is

$$\int_{-\infty}^{\infty} \psi_v{}^2 dx = 1 \tag{1}$$

with

$$\psi_v = N_v H_v(\eta)e^{-\frac{1}{2}\eta^2}; \quad \eta = x/r. \tag{2}$$

As $r d\eta = dx$, eqn. (1) becomes

$$N_v^2 r\int_{-\infty}^{\infty} [H_v(\eta)]^2 e^{-\eta^2}d\eta = 1. \tag{3}$$

Eqn. (II.1.16) gives $2^v v!\sqrt{\pi}$ as the value of the integral in (3). Thus $N_v{}^2 r 2^v v!\sqrt{\pi} = 1$, and so

$$N_v = (r2^v v!\sqrt{\pi})^{-\frac{1}{2}} \tag{4}$$

It was shown in section 3.11 that a transition between the states m and n of an oscillator will occur if, and only if, the integral

$$M_x = \int_{-\infty}^{\infty} \psi_m{}^* x \psi_n \mathrm{d}x \tag{5}$$

differs from zero. From the recursion formula (II.1.10) we see that (using the relation $\eta = x/r$)

$$\begin{aligned} x\psi_n &= rN_n \eta H_n(\eta) e^{-\frac{1}{2}\eta^2} \\ &= rN_n n H_{n-1}(\eta) e^{-\frac{1}{2}\eta^2} + \tfrac{1}{2} r N_n H_{n+1}(\eta) e^{-\frac{1}{2}\eta^2}. \end{aligned} \tag{5}$$

Therefore

$$\psi_m{}^* x \psi_n = nrN_m N_n H_m(\eta) H_{n-1}(\eta) e^{-\eta^2} + \tfrac{1}{2} r N_m N_n H_m(\eta) H_{n+1}(\eta) e^{-\eta^2}. \tag{6}$$

Recalling that $\mathrm{d}x = r\mathrm{d}\eta$, we have

$$\begin{aligned} M_x &= \int_{-\infty}^{\infty} \psi_m{}^* x \psi_n \mathrm{d}x \\ &= nr^2 N_m N_n \int_{-\infty}^{\infty} H_m(\eta) H_{n-1}(\eta) e^{-\eta^2} \mathrm{d}\eta \\ &\quad + \tfrac{1}{2} r^2 N_m N_n \int_{-\infty}^{\infty} H_m(\eta) H_{n+1}(\eta) e^{-\eta^2} \mathrm{d}\eta. \end{aligned} \tag{7}$$

The first integral on the right is zero unless $m = n - 1$, the second unless $m = n + 1$. The condition for M_x to be different from zero is, therefore

$$m = n \pm 1, \tag{8}$$

i.e., the only transitions allowed to take place in absorption or induced emission are between adjacent states of the oscillator.

Appendix III

Demonstration of the Relationship Between Structure Factors and the Coefficients of the Fourier Series Representing the Electron Density (See section 9.7)

In a centrosymmetric, one-dimensional crystal—and provided the origin is taken at a centre—the electron density may be represented by the simple, cosinusoidal Fourier series,

$$\rho(x) = \sum_{n=-\infty}^{n=+\infty} C(n) \cos 2\pi(nx), \tag{1}$$

whilst the structure factor of order h is given by

$$F(h) = \int_0^1 \rho(x) \cos 2\pi(hx)\,\mathrm{d}x. \tag{2}$$

Substituting (1) into (2), we have

$$F(h) = \int_0^1 [\Sigma C(n) \cos 2\pi(nx)] \cos 2\pi(hx)\,\mathrm{d}x. \tag{3}$$

The right-hand side of (3) comprises a series of integral terms, each of the form,

$$I = \int_0^1 C(n) \cos 2\pi(nx).\cos 2\pi(hx)\,\mathrm{d}x,$$

which, since $\cos A \times \cos B = \frac{1}{2}\cos(A+B) + \frac{1}{2}\cos(A-B)$, can be expanded to

$$I = \tfrac{1}{2}\int_0^1 C(n) \cos 2\pi(n+h)(x)\,\mathrm{d}x + \tfrac{1}{2}\int_0^1 C(n) \cos 2\pi(n-h)(x)\,\mathrm{d}x.$$

Since, over a complete period, the positive and negative parts of such cosinusoidal functions must cancel one another, these terms are in general zero. The only exceptions are when $n = +h$ or $-h$, for then

$$I = \tfrac{1}{2}\int_0^1 C(n) \cos 2\pi(0)(x)\,\mathrm{d}x = \tfrac{1}{2}\int_0^1 C(n)\,\mathrm{d}x = \tfrac{1}{2}aC(n).$$

Therefore

$$F(h) = \tfrac{1}{2}a\{C(h) + C(\bar{h})\}.$$

In the centrosymmetric case $C(h)$ and $C(\bar{h})$ are not physically distinguishable, so that they may be combined to give

$$C(h) = (1/a)F(h).$$

In the most general case, and in three dimensions, we have correspondingly,

$$\rho(x, y, z) = \sum_n \sum_o \sum_p C(nop) \exp\{2\pi i(nx + oy + pz)\}$$

and

$$F(hkl) = \int_0^1 \int_0^1 \int_0^1 \rho(x, y, z) \exp\{2\pi i(hx + ky + lz)\} \, dx . dy . dz,$$

where $C(nop)$ and $F(hkl)$ are now complex numbers. In the combined equation corresponding to (3), all the integral terms are zero unless

$$n = -h, \quad o = -k, \quad \text{and} \quad p = -l, \qquad (4)$$

when

$$C(hkl) = F(hkl)/\text{volume of the cell}.$$

Because of the inverse signs in (4), equations (9.6.4) and (9.7.7) formally carry opposite signs in their exponentials.

Problems

These miscellaneous problems are mostly taken from examination papers set at the University of Glasgow. They roughly follow the sequence of topics discussed in the text. Where appropriate, answers are given in brackets at the end of each problem.

1 The formaldehyde molecule is planar and belongs to the point group C_{2v}. Classify the normal vibrations according to their symmetry types and comment on their infrared activities. [$3A_1 + B_1 + 2B_2$].

2 A study is being made of the microwave spectrum of a planar molecule with the general formulation

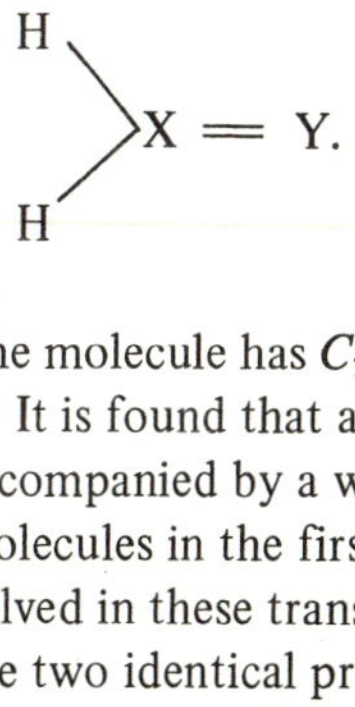

The molecule has C_{2v} symmetry.

It is found that a rotational transition arising from the ground vibration state is accompanied by a weaker line of exactly one tenth the intensity. This line is due to molecules in the first excited state of a certain vibration. The rotational levels involved in these transitions are known to be symmetric with respect to interchange of the two identical protons.

The observations are made at a temperature of 25°C. Calculate the frequency of the vibrational mode in question if

(*a*) it is the out-of-plane motion of symmetry B_1,

(*b*) It is essentially the X═Y stretching mode of symmetry A_1.

[(*a*) 726 cm^{-1} (*b*) 477 cm^{-1}]

3 OCS is a linear molecule with the following dimensions:

0—C = 1·161 Å, C—S = 1·561 Å.

Taking the approximate atomic masses C = 12, 0 = 16, S = 32 calculate the frequency of the $J = 1 \rightarrow J = 2$ rotational transition. [24 324 MHz].

4 When millimetre-wave spectroscopy was developed in the late 1950's the following rotational transitions were measured for carbon monoxide in its ground vibrational state:

$$J = 0 \rightarrow J = 1 \Rightarrow 115\,271{\cdot}20\,\text{MHz},$$

$$J = 2 \rightarrow J = 3 \Rightarrow 345\,795{\cdot}90\,\text{MHz}.$$

Knowing that the centrifugal distortion constant for a diatomic molecule D_J, is given by $4B^3/\omega^2$, estimate the vibrational frequency, ω, from the above data. [2149 cm^{-1}].

5 For the short lived diatomic species CS: $B_0 = 24\,495{\cdot}58$ MHz and $\alpha = 177{\cdot}54$ MHz. Calculate the equilibrium internuclear distance r_e. [1·535 Å].

6 The $J = 0 \rightarrow J = 1$ transition of $D^{35}Cl$ is split into three components by ^{35}Cl quadrupole coupling. The measured frequencies of the triplet are 323 282·28, 323 299·17 and 323 312·52 MHz. Calculate the ^{35}Cl quadrupole coupling constant eQq for this molecule. [−67·3 MHz].

7 The dipole moment of propyne, $CH_3C{\equiv}CH$, has been measured as 0·781 D. Estimate the component of the total dipole moment of propyne in the direction of an applied, uniform electric field for those molecules in the state specified by $J = 3$, $K = 2$, $M_J = 1$. [0·130 D].

8 An apparatus consists of two parallel plates mounted in a vacuum chamber. A low pressure of gas can be admitted into the apparatus and can be subjected to a uniform electric field by applying a steady voltage across the plates. Microwave radiation can be passed through the gas.

With carbonyl sulphide (OCS) in the apparatus and zero voltage across the plates no absorption of microwave energy is found until a frequency of 12 162·97 MHz is reached. This figure is found to vary with the applied voltage (V) as in the table. The experiment is repeated with a low pressure of fluoroacetylene (FCCH) in the apparatus with the result shown.

Knowing that the dipole moment of OCS is 0·71 D, calculate the dipole moment of FCCH.

V (volts)	0	200	400
OCS	12 162·97	12 169·09	12 187·45
FCCH	19 412·37	19 416·65	19 429·49

[0·75 D]

9 Derive an expression for the S-branch frequencies in the pure rotational Raman spectrum of a linear molecule.

The pure rotational Raman spectrum of acetylene is being studied. The S-branch

interval in the spectrum of C_2H_2 is 4·708 cm^{-1} while that in C_2D_2 is 3·392 cm^{-1}. Calculate the internuclear distances in the acetylene molecule. Neglect centrifugal distortion effects in this calculation and take $h/8\pi^2c = 16{\cdot}85$ cm^{-1} u Å^2.
[C—C = 1·208 Å; C—H = 1·062 Å.]

10 Estimate the internuclear distances in O_2 and N_2 from the pure rotational Raman spectrum of air shown in Fig. 7.5.1. [O_2 1·21 Å; N_2 1·09 Å]

11 The fundamental vibrational band of carbon monoxide is centred at 2143 cm^{-1} and the spacing between successive members of the rotational R-branch is 3·8 cm^{-1}.

If there are 10^{20} molecules in the $J = 0$ rotational level of the ground vibrational state at a temperature of 25°C, calculate:

(*a*) The population of the $J = 5$ level in the ground vibrational state at 25°C.

(*b*) The population of the $J = 6$ level in the first excited vibrational level of CO at the same temperature. [(*a*) 8·34 × 10^{20}; (*b*) 2·84 × 10^{16}]

12 Estimate the force constant and internuclear distance in CO from the data given in the previous question. [18·6 m dynes Å^{-1}; 1·13 Å]

13 Given that the fundamental vibrational transition of $H^{35}Cl$ occurs at 2886 cm^{-1} estimate the vibrational amplitudes for both hydrogen and chlorine nuclei in the $v = 0$ and $v = 1$ states. [$v = 0$: H = 0·106 Å, Cl = 0·003 Å; $v = 1$: H = 0·184 Å, Cl = 0·005 Å]

14 The fundamental I.R. band of CO is centred at 2143·3 cm^{-1} and the first overtone band at 4259·7 cm^{-1}.

If the vibrational energy, E_v, is given by:

$$E_v = \omega_e(v+\tfrac{1}{2}) - \omega_e x_e(v+\tfrac{1}{2})^2,$$

calculate:

(*a*) the harmonic force constant for CO,

(*b*) the dissociation energy D_e.

[(*a*) 19 m dynes Å^{-1}, (*b*) 10·9 eV]

15 The fundamental vibrational frequency of H_2 is 4159 cm^{-1}. How might this frequency be determined?

Calculate the harmonic force constant for hydrogen and hence predict the fundamental frequencies in HD and D_2. Comment of the infrared activities of the vibrations for the three species mentioned.

Estimate the differences between the bond dissociation energies measured from the ground vibrational states (i.e.) D_0's for H_2, HD and D_2.
[$k = 5{\cdot}1$ m dynes Å^{-1}; $\nu(HD) = 3602$ cm^{-1}; $\nu(D_2) = 2941$ cm^{-1}; $D_0(HD) - D_0(H_2) =$ 279 cm^{-1}; $D_0(D_2) - D_0(H_2) = 609$ cm^{-1}.]

16 When hydrogen gas is excited with radiation of wavelength 4358 Å the first S-branch rotational Raman line occurs at 4427 Å. Calculate the internuclear distance in H_2. [0·75 Å]

17 Sodium hydrogen di-acetate, $NaH(C_2H_3O_2)_2$, forms crystals whose density is 1·40 g cm^{-3} and which belong to the cubic system with a = 15·92 Å. How many molecules, of the formula given, are there in the cell? [24]

18 Durene, 1:2:4:5-tetramethylbenzene, crystallises in the monoclinic system with a = 10·90, b = 5·77, c = 7·03 Å and β = 103°. Show that the cell volume is 431 $Å^3$; and that – the density being 1·03 g cm^{-3} – there are two molecules per cell. If the space group is C_{2h}^5 (see Section 2.13), show that the molecule must be centrosymmetric in the crystal.

19 Di-*p*-chlorophenyl hydrogen phosphate, $(ClC_6H_4O)_2PO(OH)$, crystallises in the orthorhombic system with a = 12·43, b = 4·61, c = 23·78 Å, and with a density of 1·53 g cm^{-3}. Show that there are four molecules per cell.

The space group is *Pnaa* which is order 8 (i.e. any atom at a position x, y, z is multiplied by the symmetry elements to give seven others in the cell). These elements include twofold axes and centres of symmetry. What deductions may be made about the symmetry of the structure? [The P atom must lie on a twofold axis; the acidic H atom must also be in a special position, either on an axis or at a centre. This implies that there must be symmetrical OHO bonding between molecules, and this is borne out by structure analysis.]

20 An imaginary compound $(C_8H_8S)^{2+}2Cl^-$ crystallises in a cell with a = 8, b = 5, c = 10 Å and none of the angles between these axes differing significantly from 90°. The density of the material is 1·7 g cm^{-3}. Show that there are two formula units per cell.

X-ray photographs correspond to a minimum symmetry of 2 (or $\bar{2}$). The absent reflexions are $0k0$ when k is odd, and $h0l$ when l is odd (see Section 2.13). What deductions may be drawn regarding the symmetry of the cation? Discuss the prospects for a successful structure analysis.
[The cation must be centrosymmetric, with the S atom in a special position.]

21 The water molecule has O—H = 0·95 Å and the H—O—H angle 105°. For this molecule, sketch (*a*) the vector-map (cf, Fig. 9.9.2), and (*b*) the radial-distribution function, as it might be found by X-rays, roughly assessing the relative heights of the peaks.
[(*b*) Apart from a large peak at the origin, there will be other peaks at 0·95 and 1·51 Å, with relative weights 18:1.]

22 Taking the benzene molecule to have $6/mmm$ symmetry, with C—C = 1·4 and C—H = 1·0 Å, draw its two-dimensional vector-map. Allowing for multiplicity, this map must have 144 peaks: 36 of type C/C, 36 H/H, and 72 C/H. Account for all these in your diagram. Sketch the radial-distribution function in the manner suggested in the preceeding problem.

23 Use Plate I*A* (frontispiece) to determine the length of the c-axis in lead thiocyanate. The formula $n\lambda = c \sin \phi$ relates the number (n) of a particular layer-line and the angle (ϕ) subtended, at the crystal, by that layer-line with the equatorial line (see Fig. 9.3.3). The camera had a radius of 2·865 cm, and the wavelength of the X-rays used (λ) was 1·542 Å. In the original photograph the vertical distances between corresponding upper and lower layer-lines were 1·10, 2·33, 3·89 and 6·55 cm, for $n = 1, 2, 3$ and 4. (*E.g.* for the first pair of lines $\tan \phi = 1{\cdot}10/(2 \times 2{\cdot}865)$.) [8·22 Å]

24 The fourth minimum of the electron-diffraction pattern of hydrogen chloride gas occurs at $\phi = 10°\ 8'$, when the accelerating voltage is 40 000. What is the interatomic distance? Use equation (10.4.1); $(\sin sr)/sr$ is an oscillating function of sr, whose fourth minimum occurs at 23·5 radians. [1·29 Å]

25 Using covalent bond and van der Waals radii from Chapter 12, estimate the volume of a molecule of durene and compare it with the result of problem 18. [About 200 Å^3]

Index